Monographs on
Theoretical and Applied Genetics 2

Edited by
R. Frankel (Coordinating Editor), Bet Dagan
G. A. E. Gall, Davis · M. Grossman, Urbana
H. F. Linskens, Nijmegen · D. de Zeeuw, Wageningen

R. Frankel · E. Galun

Pollination Mechanisms, Reproduction and Plant Breeding

With 77 Figures

Springer-Verlag
Berlin Heidelberg New York 1977

Professor Dr. RAFAEL FRANKEL, Agricultural Research Organization,
The Volcani Center, P. O. Box 6, Bet Dagan/Israel

Professor Dr. ESRA GALUN, Weizman Institute of Science, Rehovot/
Israel

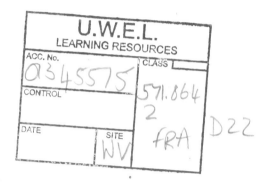

ISBN 3-540-07934-3 Springer-Verlag Berlin Heidelberg New York
ISBN 0-387-07934-3 Springer-Verlag New York Heidelberg Berlin

Library of Congress Cataloging in Publication Data. Frankel, Rafael, 1922. Pollination mechanisms, reproduction, and plant breeding. (Monographs on theoretical and applied genetics;
2) Bibliography: p. Includes index. 1. Plant—breeding. 2. Fertilization of plants. 3. Plants—Reproduction. I. Galun, E., 1927. joint author. II. Title. III. Series. SB123.F7.631.5′3.76-42450.
Printed in Germany.
The use of registered names, trademarks, etc. in this publication does not imply, even in the
absence of a specific statement, that such names are exempt from the relevant protectice laws and
regulations and therefore free for general use.
Typesetting, printing, and bookbinding: Zechnersche Buchdruckerei, Speyer.

Preface

"...Nature has something more in view than its own proper males should fecundate each blossom."
ANDREW KNIGHT
Philosophical Transactions, 1799

Pollination mechanisms and reproduction have a decisive bearing upon rational procedures in plant breeding and crop production. This book intends to furnish under one cover an integrated botanical, genetical and breeding-methodological treatment of the reproductive biology of spermatophytes— mainly angiosperms; it is based on an advanced topical course in plant breeding taught at the Hebrew University of Jerusalem.

We have tried to present a coverage which is concise, but as comprehensive as possible, of the pollination mechanism and modes of reproduction of higher plants, and to illustrate topics, whenever practicable, by examples from cultivated plants. Nevertheless, some relevant publications may have escaped our attention or may not be mentioned because of various limitations.

The book is organized into three parts. The first part starts with an evaluation of the significance of the different pollination mechanisms for plant breeding and crop production, describes modes of reproduction in higher plants and discusses ecology and dynamics of pollination. The second part is devoted to crops propagated by self pollination and describes specific breeding procedures for such crops. The third part details sexual reproduction in higher plants and handles three mechanisms involved in the prevention of self pollination and their utilization in plant breeding: sex expression, incompatibility, and male sterility.

We hope this book will prove useful to biology and agriculture students at the graduate level, to plant breeders and other people interested in acquiring a broader knowledge of the reproductive biology of higher plants, and serve for intra- and interdisciplinary communication between botanists, geneticists, plant breeders and agriculturists.

We owe thanks to Mrs. V. ASSCHER for her careful typing of the manuscript, to Mrs. V. PRIEL for English editing, and to Mrs. N. GESTETNER for preparing the line drawings. One of us (E. G.) is grateful to the Israel Academy of Sciences and Humanities and to the Royal Society for a grant which enabled him to perform a literature study at the Royal Botanic Gardens, Kew. Thanks are also due to many colleagues and students for valuable discussions and suggestions and to authors and publishers for permission to use or reproduce their illustrations.

Bet Dagan/Rehovot,   R. FRANKEL · E. GALUN
January 1977

Contents

Chapter 1. Introduction 1

1.1   Implications of Pollination Mechanisms in Plant
        Breeding and Crop Production 1
1.1.1   Pollination Mechanisms and Breeding of New Culti-
        vars 1
1.1.2   Pollination Mechanisms and Cultivar Maintenance . 2
1.1.3   Pollination Mechanisms and Agricultural Yield . . 3
1.2   Reproduction in Higher Plants 3
1.2.1   Historical and General Background 3
1.2.1.1 Morphological-Structural Flower Biology 4
1.2.1.2 Functional Flower Biology 4
1.2.1.3 Correlation between Structure and Function 5
1.2.2   Modes of Reproduction 5
1.2.2.1 Sexual Reproduction 6
1.2.2.1.1 Spatial Separation of Sex Organs as an Outbreeding
        Device 6
1.2.2.1.2 Basic Concepts of Structural Differentiation 6
1.2.2.1.3 Models of Floral Differentiation 7
1.2.2.1.4 Definitions of Sex Types in Flowering Plants . . . 10
1.2.2.1.5 Temporal Separation of Sex Organs as an Outbreeding
        Device 13
1.2.2.2 Asexual Forms of Reproduction 14
1.2.2.2.1 Asexual Propagules outside the Floral Region . . . 14
1.2.2.2.2 Asexual Propagules within the Floral Region . . . 15
1.2.2.3 Distribution of Modes of Reproduction among Culti-
        vated Plants. 16
1.3   Ecology and Dynamics of Pollination 29
1.3.1   Specificity of Flowers and Pollen 29
1.3.2   Pollen-Dispersal Agents 31
1.3.2.1 Biotic Vectors 31
1.3.2.2 Abiotic Vectors 34
1.3.3   Timing and Climatic Factors in Pollination Dynamics 36
1.3.4   Location and Mass Effects in Pollination Ecology . 41
1.3.5   Competition Effects in Pollination Dynamics . . . 43
1.3.6   Determination of the Natural Cross-Pollination Rate
        (NCP) 43
1.3.6.1 Progeny Testing of Dominants. 45
1.3.6.2 Progeny Testing of Recessives 46

1.3.6.3 Progeny Testing of Heterozygotes   46
1.3.6.4 Progeny Testing of Recessives and Heterozygotes. . 47
1.3.7   Artificial Control of Outcrossing   48

Chapter 2. Autogamy   51

2.1   Evolutionary Aspects of Autogamy.   51
2.1.1   Strategies for Adjustment of Recombination 51
2.1.2   Origin of Mating Systems in Higher Plants 52
2.1.3   Variation in Autogamous Populations   53
2.2   Mechanism of Autogamy   56
2.2.1   Cleistogamy   56
2.2.2   Chasmogamic Selfing   58
2.3   Management of Pollination in Autogamous Crops .   62
2.3.1   Emasculation   62
2.3.1.1 Mechanical Removal of Microsporophylls.   62
2.3.1.2 Male Gametocide   66
2.3.1.3 Circumvention of Emasculation Requirements . . .   69
2.3.2   Controlled Pollination   70
2.3.2.1 Isolation   70
2.3.2.2 Pollen Collection and Storage   70
2.3.2.3 Pollen Transfer Methods   73
2.3.2.3.1 Utilization of Natural Pollen Vectors   73
2.3.2.3.2 Forced Pollination   76

Chapter 3. Allogamy   79

3.1   Sexual Reproduction—Structures and Functions . .   79
3.1.1   The Anther and the Male Gametophyte   79
3.1.1.1 Differentiation of the Anther   79
3.1.1.2 The Tapetum   80
3.1.1.3 Development of the Sporogeneous Tissue   81
3.1.1.4 Microsporogenesis   82
3.1.1.5 From Microspore to Pollen Grain   83
3.1.2   Androgenesis: Production of Haploid Plants by
         Anther and Pollen Culture   85
3.1.2.1 Haploid Plants—Occurrence, Induction, and Identi-
         fication.   85
3.1.2.2 The Production of Haploid Callus and Embryoids by
         Anther and Pollen Culture   86
3.1.2.3 Pathways of Pollen Embryogenesis   89
3.1.2.4 Factors Affecting Androgenesis   92
3.1.2.4.1 Culture Conditions   92
3.1.2.4.2 Donor Plants   92
3.1.2.4.3 Pollen Age   92
3.1.2.4.4 Anther Stage and Ploidy   93
3.1.2.5 Application of Androgenesis for Breeding and Genetic
         Studies   93

3.1.3   The Pistil and the Female Gametophyte.   96
3.1.3.1 The Pistil.   97
3.1.3.2 The Female Gametophyte.   98
3.1.4   Fertilization   100
3.1.4.1 Contact between Pollen and Stigma   100
3.1.4.2 Pollen Germination   101
3.1.4.3 Pollen Tube Discharge and Double Fertilization . .   102
3.1.5   Sexual Reproduction in Conifers.   103
3.2   Control and Modification of Sex.   104
3.2.1   The Genetic Control of Sex Determination   104
3.2.1.1 General Considerations Concerning the Genetic
         Control of Dioecism   105
3.2.1.2 Artificial Dioecism in *Zea mays*   107
3.2.1.3 Artificial Dioecism Caused by Suppressive Genes—
         Linkage between Genes as a Prerequisite for Sex
         Dimorphism   109
3.2.1.4 Chromosomal Control of Sex Determination . . .   110
3.2.1.5 Main Methods for the Study of the Genetics and
         Cytology of Sex Determination   112
3.2.1.6 Genetic Regulation of Sex: Representative Examples   113
3.2.1.6.1 The Active Y Chromosome System.   113
3.2.1.6.2 The X-Autosomal Balance System   118
3.2.1.7 Genetics of Sex Determination in Some Economic
         Crops   119
3.2.2   Modifications of Sex Expression   120
3.2.2.1 Introduction   120
3.2.2.2 Mineral Nutrition and Edaphic Factors   127
3.2.2.3 Light   128
3.2.2.4 Temperature   132
3.2.2.5 Chemical Agents   134
3.2.2.5.1 Auxins and Related Compounds   135
3.2.2.5.2 Gibberellins.   137
3.2.2.5.3 Kinins   139
3.2.3   Sex Expression in Some Economic Crops and its
         Application to Breeding and Crop Improvement . .   141
3.2.3.1 The Cucumber and Other Cucurbit Crops   141
3.2.3.1.1 Patterns of Sex Expression   141
3.2.3.1.2 Effects of Day Length and Temperature on Flowering
         and Sex Expression   145
3.2.3.1.3 Effects of Growth Regulators   147
3.2.3.1.4 Inheritance of Sex Expression and Breeding Proce-
         dures   151
3.2.3.1.5 Scheme of Sex Expression in Cucumber   157
3.2.3.2 Hemp *(Cannabis sativa)*   157
3.2.3.3 Maize *(Zea mays)*   159
3.2.3.3.1 Reproductive Morphology   160
3.2.3.3.2 Environmentally Induced Sex Modification   160

3.2.3.3.3 Sex Modification by Chemical Agents 161
3.2.3.3.4 Inheritance of Sex Expression 163
3.3   Incompatibility 163
3.3.1   Genetics of Incompatibility 165
3.3.1.1   Gametophytic Incompatibility 165
3.3.1.1.1 One Multiallelic S Locus 165
3.3.1.1.2 Two Multiallelic S Loci 169
3.3.1.1.3 Three or More S Loci 171
3.3.1.2   Sporophytic Incompatibility 172
3.3.1.2.1 Heteromorphic Incompatibility 173
3.3.1.2.2 Homomorphic Incompatibility 176
3.3.2   Pollen-Pistil Interaction 179
3.3.2.1   Pollen Cytology and Pollen-Stigma Interaction . . . 179
3.3.2.2   Pollen Tube-Style Interaction 183
3.3.2.3   Pollen Tube-Ovule Interaction 184
3.3.3   Incompatibility, Crop Production, and Breeding . . 184
3.3.3.1   Transfer of Incompatibility into Cultivars 185
3.3.3.2   Permanent Elimination of Incompatibility 186
3.3.3.3   Surmounting the Incompatibility Barrier 188
3.3.3.3.1 Treatment of Pollen 188
3.3.3.3.2 Bud Pollination 188
3.3.3.3.3 Delayed Pollination 188
3.3.3.3.4 Heat Treatment 188
3.3.3.3.5 Surgical Techniques 189
3.3.3.3.6 Double Fertilization 189
3.3.3.3.7 Other Methods 189
3.3.3.4   Hybrid Seed Production 190
3.3.3.4.1 Characteristics of the Gametophytic and Sporophytic
          Incompatibility Systems which are Related to Hybrid
          Seed Production 190
3.3.3.4.2 Hybrid Seed Production in Brassica and Raphanus . 191
3.3.3.4.3 Problems in Breeding Aimed at Hybrid Seed Produc-
          tion 195
3.4   Male Sterility 196
3.4.1   Gynodioecy: Male Sterility in Natural and Artificial
          Populations 196
3.4.1.1   Male Sterility as an Outbreeding Mechanism in
          Gynodioecious Species 196
3.4.1.2   Nuclear Male Sterility Genes in Cultivars 199
3.4.1.3   Male-Sterility-Inducing Plasmatypes in Cultivated
          Plants 201
3.4.1.4   Utilization of Gynodioecy in Plant Breeding 203
3.4.2   Inheritance of Male Sterility 204
3.4.2.1   Genic Male Sterility (Mendelian Male Sterility) . . 204
3.4.2.2   Cytoplasmic and Gene-Cytoplasmic Male Sterility . 206
3.4.3   Structural, Developmental, and Biochemical Charac-
          terization of Male Sterility 209

3.4.3.1   Developmental Modifications Leading to Breakdown
           in Microgametogenesis 209
3.4.3.1.1 Modifications in the Structural Differentiation of the
           Stamen 210
3.4.3.1.2 Faulty Differentiation of the Anther 212
3.4.3.1.3 Breakdown in Microsporogenesis 214
3.4.3.1.4 Abortion of the Microgametophyte 216
3.4.3.1.5 Functional Male Sterility 219
3.4.3.2   Biochemistry of Male Sterility 219
3.4.3.3   Male Sterility Genes and Their Action 221
3.4.3.3.1 Site of Male Sterility Factors 221
3.4.3.3.2 Pleiotropic Effects 224
3.4.4   Utilization of Male Sterility in Plant Breeding . . . 225
3.4.4.1   Comparison of Hybrid Production Using Genic,
           Cytoplasmic and Gene-Cytoplasmic Male Sterility . 226
3.4.4.1.1 Genic Male Sterility 226
3.4.4.1.2 Cytoplasmic Male Sterility 227
3.4.4.1.3 Gene-Cytoplasmic Male Sterility 227
3.4.4.2   Application of Genic Male Sterility 228
3.4.4.2.1 Genetic and Field Management Programs to Provide
           a Homogeneous Stand of the Genic Male Sterile Seed
           Parent 228
3.4.4.2.2 Pollination Control 230
3.4.4.3   Use of Gene-Cytoplasmic Male Sterility 233

References 235

Subject Index 269

Chapter 1. Introduction

The pollen grain is the discrete and mobile stage of the male gametophyte of higher plants. We shall regard the pollination process in the broad sense, constituting conveyance of the pollen grain to the sessile female gametophyte as well as the germination and growth of the male gametophyte, up to fertilization. *Pollination mechanisms* are thus, in this sense, the means by which transference of functional pollen is achieved, or avoided, from the pollen-bearing structure to the compatible and receptive surface of the ovule, and by which ultimately fusion of the sperm nuclei of the pollen with the egg cell and the polar nuclei of the female gametophyte is effected. As such, pollination mechanisms are considered by us to comprise sexual differentiation, microsporogenesis, macrosporogenesis, structural modifications of flowers, pollen dispersal ecology, and structural or physiological barriers and facilities to fertilization. We shall deal with these aspects, emphasizing their implications for plant breeding and crop production.

1.1 Implications of Pollination Mechanisms in Plant Breeding and Crop Production

1.1.1 Pollination Mechanisms and Breeding of New Cultivars

Plant breeding activities may be characterized by a closed cycle (Fig. 1.1). Desired traits are drawn from the natural or artificially induced store of accessible variability. Traits are isolated, evaluated and recombined to enrich the store of variability. The creation of a new cultivar constitutes a break in the cycle and ingress into multiplication, maintenance and distribution of the new cultivar (Fig. 1.1).

Pollination mechanisms in plants have a decisive bearing upon rational procedures in all mentioned plant breeding activities. The nature of the assemblage of genotypes depends to a high degree on the extent of natural recombination of traits, on genetic uniformity within a population, on degrees of homozygosity—all being subject to the past and prevalent modes of pollination. Isolation or selection procedures must be based on the behavior of the individual or the population of phenotypes (and not of genotypes) under the pressures of the chosen separation method. Thus, homozygosity and dominance relations come into play in the isolation process. The modus operandi of the evaluation of a phenotype (which often has to be preceded by an initial multiplication)

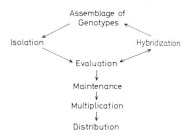

Fig. 1.1. Plant breeding activities

must obviously be adopted with the breeding behavior of the plant in mind.
Sexual hybridization, finally, constitutes controlled pollination, implementation
of which obviously depends on pollination mechanisms.

Various plant breeding methods have been designed to adjust to the pollination
pattern of the crop. However, pollination patterns may not remain consistent, but
vary in time and location. Thus, for instance, different degrees of homozygosity
may be encountered in the same crop, a situation which is important for pedigree
breeding and backcrossing. Among other factors, selection breeding depends
on the extent and range of variation, population breeding on stability of population
structure, and synthetic varieties on requirements for components maintenance
and renewed assembly of the cultivar. Thus, the importance of pollination
mechanisms and dynamics is indicated in every phase of breeding of higher
plants.

1.1.2 Pollination Mechanisms and Cultivar Maintenance

The identity of a cultivar must be. kept constant by proper maintenance of
its original genetic makeup. Gene frequency changes may cause loss of the
unique characteristics of cultivars such as adaptation to ranges of environments
or quality characteristics. Agriculturists term such development "running out"
or "loss of trueness to type" of cultivars.

Proper procedures for maintenance and multiplication depend largely on
modes of pollination within the variety population, on reproductive efficiency
of components of the population, and on isolation from foreign pollen. Avoidance
of transfer of foreign pollen may be achieved by spatial or temporal isolation
of one cultivar community of mass in the maintenance and multiplication of
an open-pollinated cultivar. In composite or synthetic cultivars, differential reproductive efficiency of genotypes may cause shifts in the genetic composition of
the variety population; such differential may be enhanced by certain environments
and minimized by others. Thus, both male and female outcrossing as well as
the reproductive efficiency of genotypes in the prevailing environment, must
be considered in stages of reassembly, maintenance and multiplication of composite cultivars.

Pollination mechanisms are of extreme importance in multiplication of seeds
and in the production of hybrids. Maintenance of parents, as well as the economic

production of hybrid seed, utilize genetic variation in flower biology. Some crops are propagated asexually. If vegetative parts of asexually propagated crops serve as agricultural yield, the pollination mechanism may be of importance in the breeding phase only (e.g. Irish potato), or in the breeding and maintenance phases only (e.g. sugar cane). If reproductive organs of asexually propagated crops serve as agricultural yield, pollination becomes important for both the breeding and the final production phases (e.g. strawberry and fruit trees). In some cases, where the parent genotypes are difficult to derive or to maintain, asexually reproduced clones may serve as parents for F_1 hybrid seed production (e.g. double-flowered petunia or gerbera). Thus, pollination mechanisms become important in the breeding and multiplication phases only.

1.1.3 Pollination Mechanisms and Agricultural Yield

Agricultural yield may be composed of vegetative or reproductive parts of the plant. In cabbage, onion or forage plants, vegetative organs serve as yield and pollination mechanisms are of no importance in the production. For that matter, it may be desirable that cultivars do not flower at all to insure quantity and quality of yield. "Bolting", as such unwanted production of flower stalks is termed, may in fact make the agricultural yield useless. Even in plants in which the agricultural yield is related to reproductive parts, seed set is sometimes unwanted. Thus, the "seedless" characteristic of parthenocarpic fruits such as oranges, pineapples and cucumbers, or sterility of ornamental flowers increasing vaselife, depend on the absence of pollen or pollen vectors, incompatibility, or sterility.

On the other hand, pollination mechanisms influencing proper pollination are of extreme importance for yield where seed set is required. Self-incompatibility and vegetative propagation in certain fruit trees may cause problems in production. The morphology of tomato flowers may become important in greenhouse production where air vibration is minimal. The presence of insect vectors for pollination may become critical in cucumber cultivation under plastic tunnels. Various climatic factors, such as rain, wind and temperature, directly affect efficiency of pollination in many crop plants.

1.2 Reproduction in Higher Plants

1.2.1 Historical and General Background

The causal relationship between pollen and the formation of fruits and seeds was recognized at the dawn of agricultural civilization. Artificial pollination of the dioecious date palm was practiced as early as 2,000 BC during the reign of the Assyrian king Hammurabi (ROBERTS, 1929). CAMERARIUS' statement (1694) that "pollen is necessary for fertilization and formation of seeds" may be regarded as the first published recognition of the universal role of pollination in the sexual reproduction of higher plants. AMICI (1824, 1830) traced the fate of the germinating pollen and its tube up to the ovule, and STRASBURGER (1884) confirmed the fusion of one of the male gametes with the egg nucleus. Finally, at the end of the 19th Century, the double fertilization process was discovered as a normal feature in the angiosperms (NAWASHIN, 1898; GAUGNARD, 1899).

1.2.1.1 Morphological-Structural Flower Biology

It appears to have been DARWIN who primed the vigorous quest for knowledge of pollination mechanisms. As a keen observer and integrator of the common knowledge of plant and animal breeders (among them notably CHRISTIAN KON-RAD SPRENGEL), he concluded that cross-fertilization is not only beneficial in long-term evolution, but also required for maintenance of vigor and fertility:

"I have collected a large body of facts, showing ... that close interbreeding diminishes vigour and fertility; that these facts alone incline me to believe that it is a general law of nature ... that no organic being selffertilizes itself for an eternity of generations; but that a cross with another individual is occasionally—perhaps at very long intervals—indispensable". (DARWIN, 1859)

Although DARWIN had vague or confused ideas about the origin of genetic variation, he elaborated on the variation in pollination mechanisms. The various adaptations and interactions in flower biology were recognized as serving the necessity that "in no organic being can selffertilization go on for eternity". DARWIN described almost all known outbreeding mechanisms in plants (separation of sexes, prepotency of distinct varieties of pollen over own pollen, protandry, protogyny, self-incompatability, style extrusion and other flower modifications, insect vector adaptations). It is astonishing how "modern" was DARWIN'S interpretation of the apparent contradiction between the hermaphrodite makeup of flowers and superimposed cross-pollination mechanisms:

"How strange that the pollen and stigmatic surface of the same flower, though placed so close together as if for the very purpose of selffertilization, should in so many cases be mutually useless to each other! How simply are these facts explained on the view of an occasional cross with a distinct individual being advantageous or indispensable". (DARWIN, 1859)

DARWIN'S observation of inbreeding depression in plants and the fact that "a single cross with a fresh stock restores their pristine vigour" (DARWIN, 1878) initiated extensive investigation of structural adaptations of flowers to cross-pollination. The morphological-structural approach to flower biology was prevalent during the second half of the 19th Century; the extensive reference work by KNUTH (1898–1905) dominates the literature of that period.

1.2.1.2 Functional Flower Biology

Only in the last thirty years has the descriptive, morphological-structural viewpoint been superseded by a dynamic functional approach to flower biology.

This was due mainly to the influence of contributions of workers in the field of the interrelations between pollen transference vectors and flower structure (see recent discussions on the subject by FAEGRI and VAN DER PIJL, 1971; PROCTER and YEO, 1973). Studies were initiated or extended in fields such as flower induction, photoperiodic and thermoperiodic influences, sex expression, environmental and hormonal influences, genetic sex determination, inheritance of sex types, natural cross-pollination rates, developmental aspects of apomixis, inheritance and physiology of self-incompatibility, inheritance of male sterility, and pollination ecology and dynamics.

1.2.1.3 Correlation between Structure and Function

The causal interrelation between flower structure and mode of pollination should be obvious from a consideration of evolution. Nevertheless, structural adaptations to cross-fertilization may often be only an evolutionary remnant upon which a new reproductive mechanism is superimposed. This can be illustrated by a large number of examples. Thus, we find commonly on one hand assignment of sexes to different plants (dioecy) serving obligate cross-fertilization (e.g. date palm, spinach), vs dioecy upon which asexual reproduction is superimposed by apomixis (e.g. *Simmondsia sinensis*). We find adaptation to insect pollination (entomophily) in legumes serving the goal of outcrossing (e.g. *Vicia villosa, Medicago sativa, Trifolium pratense*), against adaptation to entomophily achieving self-fertilization (e.g. *Vicia sativa, Medicago hispida, Trifolium fragiferum*). In the Gramineae adapted normally to wind-pollination we find both cross-pollinated species (e.g. *Setaria sphacelata, Hordeum bulbosum, Saccharum officinarum*) and self-pollinated species (e.g. wheat, barley, oats, *Setaria italica*). Thus, both structural and functional aspects of flowers must be studied and comprehended for proper appraisal of the pollination mechanisms.

1.2.2 Modes of Reproduction

The multicellular plant, as all living systems, has a restricted life span. To conserve the life of the organism, alternation of generations is required by reproduction. Moreover, successful maintenance depends on increase in numbers, on adaptation, as well as on colonization and dispersal of populations. Reproduction of the multicellular plant is achieved asexually or sexually. Asexual reproduction occurs readily in higher plants but not at all in higher animals. Mitotic cell divisions and establishment of asexual propagules and formation of seeds without a sexual process provide for alternation of successive sporophyte generations by asexual reproduction and rapid increase of adapted populations in a stable environment. Meiotic cell division, sexual differentiation and syngamy provide for heteromorphic alternation of gametophyte-sporophyte generations by sexual reproduction and proper adaptation in changing environments. Pollination mechanisms have a decisive role in sexual reproduction in plants. To put this role in the proper perspective, sexual and asexual modes of reproduction in plants will be discussed next.

1.2.2.1 Sexual Reproduction

Sexual reproduction is the prevalent mode of propagation in higher plants. The biological significance of sexual reproduction in diploid and polyploid plants lies in pooling of genetic information carried by the individuals of an interbreeding population. Phenotypic uniformity must not be sacrificed by storage of a large amount of genetic variability in such a population.

1.2.2.1.1 Spatial Separation of Sex Organs as an Outbreeding Device

The reproductive organs of the spermatophytes, the flowers, exhibit almost infinite morphological and physiological expressions. Floral adaptations are concerned mainly with efficient pollination and regulation of level of outbreeding. One way of adjustment for evolutionary proper hybridity in plant populations is *spatial separation* of sexual organs within or between plants. In normal bisexual plants female and male reproductive organs and gametes are produced in the same individual. In normally unisexual plants we find a separation of sexes (sexual dimorphisms) and only one kind of gamete (female or male) is produced in each individual. There exist also plant species in which different individuals are either bisexual or unisexual. The ratio and expression of the sexual flower types of polymorphic plants may be modified.

1.2.2.1.2 Basic Concepts of Structural Differentiation

From the considerations stressed above a thorough understanding of the sexual reproduction in plants is a prerequisite for sensible utilization of the pollination mechanism for breeding and improvement of crop productivity. We shall, therefore, focus our attention on the differentiation of structure and function of the flower, the medium of sexual reproduction in angiosperms. Some fundamental considerations of differentiation of multicellular structures should, nevertheless, be dealt with first. We shall keep in mind that at the level of complexity in which we are dealing the structure and function of an elaborate organ, such as the flower, the direct approach to search for the processes which ultimately lead from the genetic information stored in the cells to a functional three-dimensional structure, is far beyond our aim. At the level of our present understanding this approach is useful at only very low levels of differentiation (e. g. the composition and form of viral coat proteins; c. f. POGLAZOR, 1973; HARRISON et al., 1973). In these simple systems the causal relationship between DNA base sequences and the polypeptides which assemble in a predictable way to form the ultimate viral coat is a rather fruitful and interesting research objective. Nevertheless, it is also by deducing from the many genes involved in the construction and assembly of subunits of such primitive forms that we can infer the immense complexity of the morphogenetic processes leading to a multicellular structure. We shall, therefore, keep in mind that what we mean by investigating the differentiation in multicellular structures, such as flowers, is merely to seek an understanding as to how specific lines of morphogenetic events, out of two or more potential ones, are actually taking place. To illustrate this point let us take a simulation, far removed from our system, but which bears the essence of the above consideration. Consider a railroad system toy, with fixed ramifications of rails, permitting a train to be sent to a variety of destinations. The role of the student of differentiation can then be exemplified by a child devoting himself to learning the ways of diverting his train along some of the fixed routes. He may familiarize himself with rail shifts and their function and effects on the destiny of the train, but will not be concerned with the layout of the rails or with the propulsion of the train.

1.2.2.1.3 Models of Floral Differentiation

The initiation of floral development and the modification of floral structure attracted many students of differentiation. Taking earlier studies and their interpretation into account, WARDLOW (1957) treated the floral meristem as a reaction system which passes through an irreversible sequence of phases. WARDLOW'S analysis regarded the floral bud as a shortened, determinate variant of a shoot tip in which the apex was converted into a receptacle and the foliar initials into floral members (calyx, corolla, androecium and gynoecium). He then followed the lines of thought earlier formulated by MATHER (1948) and applied them to the examination of floral ontogeny. For his consideration, WARDLOW divided the apex into several zones, giving the 'sub-distal region' a prime regulatory role: "In the sub-distal reaction system we may envisage the 'flower inducing substance' (or substances) not merely as affecting an initial change in the working system, but as initiating an orderly sequence of changes in its activity, partly by evoking successively the action of specific genes and partly as a result of physical correlation." We shall see more clearly what was meant by WARDLOW when we turn to the model of regulation of sex expression in flower proposed by HESLOP-HARRISON (1963, 1972). HESLOP-HARRISON extended MATHER'S and WARDLOW'S concepts, and by taking into consideration the current knowledge on gene action and regulation as well as the available experimental results obtained by him and by others, he arrived at a detailed model. This model illustrates the events at a terminal floral bud, but is also applicable to axillary buds (Fig. 1.2). Our starting point will be the vegetative, uncommitted, apex induced to transform into a floral meristem. It then undergoes a scheduled program of growth in seriation, producing lateral determinate members which are either sterile or spore-producing: the sterile producing sepals and petals (or tepals) and the sporophylls differentiating either as stamens or as carpels. Experimental evidence indicates that during a certain early period, these lateral initials entertain a 'plastic' phase and that only at the latters' termination is the member completely autonomous in its irreversible fate to become a periant or an essential floral member. Before and during the 'plastic' phase the primordium is under the control imposed on it by the surrounding tissue and its spatial location relative to other primordia as well as to the apex, is determinative to its final ontogeny. Thus, the floral meristem generates sequentially primordia

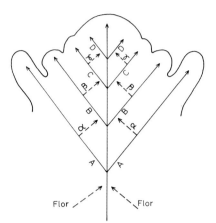

Fig. 1.2. A model for a differentiating floral apex according to HESLOP-HARRISON (1963). Floral induction *(Flor)* causes the shift from vegetative growth to flowering. In cells of the lower floral member gene complex A is activated and inducer α is produced. The latter flows to the next floral member activating gene complex B etc.

for the lateral members and sends them to their prospective final form in a fixed order (Fig. 1.3).

The above is a schematic model which attains an idealized completion in the formation of a bisexual flower such as that of *Helleborus*. By appropriate regulation of the number and character of the lateral members, a vast variety of final floral structures can be visualized. According to this model we may regard the female flower of cucumber as resulting from an early suppression of further development of anther initials (Fig. 1.4) and the female flower of hemp as resulting from a direct 'passover' from periant initials to carpel initials, lacking any vestiges of stamen initials.

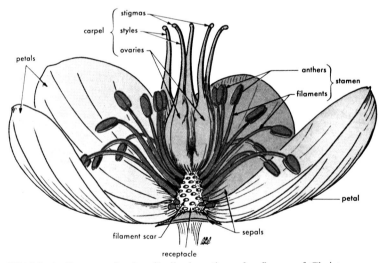

Fig. 1.3. A diagram of a longitudinal section of a flower of Christmas rose *(Helleborus)*. The perianth consists of two similar whorls; there are numerous stamens arranged in a spiral on a cone-shaped receptacle. Five separate carpels form the central whorl of floral parts. From WEIER et al. (1974)

In the foregoing discussion several points were intentionally omitted. These should now be considered. They concern arguments in favor of a complete separation of cell lineage between vegetative (e. g. leaves) and reproductive (e. g. carpels) tissues, which are based on histological observations. This apparently obligatory 'division of labor' would speak against the 'plasticity' of shoot apices and their projected primordia to become either vegetative shoots and leaves, respectively, or to be shifted into floral members. However, enough evidence has been provided in recent years about the retainment of totipotency. The production of whole plants from a single leaf mesophyll (TAKEBE et al., 1971) and from leaf epidermal cells (e. g. THI DIEN and TRAN THANH, 1974) are two of several indications of such totipotency.

Another point concerns the evolutionary origin of the unisexual flowers in diclinous plants. Two possibilities were put forward (see HESLOP-HARRISON, 1957). All unisexual flowers could have a primary bisexual origin (i. e., descending from a preangiosperm with bisexual strobili) or two kinds of unisexual flowers coexist: those of a primary bisexual origin and those from a primary unisexual origin (i. e., descending from a preangiosperm with unisexual strobili). However, for our present considerations this is of no importance since experimental evidence shows that even in the candidates of the later group, e.g. hemp flowers, both stamens and carpels can occur in the same flower under certain conditions (HESLOP-HARRISON, 1958).

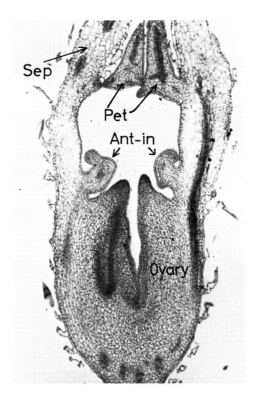

Fig. 1.4. Section through an in vitro cultured cucumber floral bud. Anther initials (*Ant-in*) were arrested while ovary did develop. A female flower would ultimately result by completion of ovary, sepals (*Sep*) and petals (*Pet*) differentiation

The last point is the question whether or not in fact the genetic information needed to construct all floral members exists potentially in every floral bud, including those borne on unisexual plants. This question seems to require a definite answer because only if it is answered in the affirmative way may we assume that the final expression of the floral bud can be fully conditioned by regulatory processes. Unfortunately, we cannot furnish a decisive answer to this question. Even though there are cases which appear to be rather clear, indicating a lack of genetic information for the construction of one of the essential members, we should be careful in our reasoning. This is exemplified in *Melandrium* (WESTERGAARD, 1958) where the genes for stamen formation were assigned to a specific section of chromosome Y and are thus apparently absent from female plants. Nevertheless, there are arguments for the claim that the genetic information for the construction of stamens is not entirely lacking in female *Melandrium* plants. To stress this point we may divert to mammals where the genes responsible for the construction of the testes are assigned to the Y chromosome, and thus should be absent in the XX females. Still the case is not finally stettled. Gonad primordia in human embryos first appear as genital ridges about 5 weeks after conception, and for one more week they remain in an indifferent state. Only then the sex cords appear in male embryos, while in female embryos the 'resting' period lasts longer. MITTWOCH (1969, 1973) suggested that the role of the Y chromosome is merely a regulatory one: in XY embryos cell division is facilitated, especially at the gonad primordium, inducing testes differentiation, while the slower mitotic activity in the XX embryos retards further gonad development and thereby enables the differentiation of the ovaries. Although this specific problem has not yet been settled, there is a warning for us: unidirectional differentiation in higher organisms, which seems to be caused by lack of structural information, may actually be based on regulatory mechanisms.

1.2.2.1.4 Definitions of Sex Types in Flowering Plants

Plants with only staminate or only pistillate flowers are commonly termed male and female plants, respectively. We shall not change this widely used designation. Nevertheless, the reader should be aware that from a strict botanical point of view this terminology is questionable. Strictly speaking, 'male' is the organism which produces the usually motile male gametes, and female is the organism producing the usually less motile or sessile female gametes. These roles are attributed in the class Angiospermae, to the gametophytes. Gametophytes in this class are always unisexual, and should thus be designated as 'male' or 'female'. But, while bearing in mind these 'true' definitions, we should consider the actual relation between the gametophyte and the sporophyte in flowering plants.

This relation imposes a control of the autotrophic sporophyte over the heterotrophic gametophyte; thus, the sex differentiation of the latter depends on factors effective in the sporophyte. To put it more clearly, the sex expression of the gametophyte is imposed by the kind of sporophyll in which it develops: in a stamen, a male gametophyte will result; in a carpel, a female gemetophyte will result. Therefore, as rightly put forward by HESLOP-HARRISON (1972), the heterangy of the sporophyte can be regarded itself as a form of sexual dimorphism, and by extending this reasoning we may term *staminate* and *pistillate* structures as *male* and *female*, respectively.

The terminology and classification of sex in plants can be quite confusing. Moreover, there is no general agreement in either the classification of sex types or in the symbols assigned to them. We shall avoid here a critical review of sex terminology in flowering plants, and not claim that our usage has the most sound botanical basis. The following classification is merely intended to avoid confusion, and still to include most of the existing types of sex expression in plants.

We should first keep in mind that one may speak about sex expression in an individual *flower*, in an individual *plant*, or in a *group* of plants (e.g. a species or a cultivar). The terminology should make it clear which of the above three we have in mind.

1. Sex expression of the individual flower
   Hermaphrodite-flower having both stamens and carpel (or carpels)
   Staminate (or androecious)-flower having only stamens but no carpels
   Pistillate (or gynoecious, carpillate)-flower having only carpels but not stamens

2. Sex expression of the individual plant
   Hermaphrodite-plant with only hermaphrodite flowers
   Monoecious-plant with both staminate and pistillate flowers
   Androecious (or male)-plant with only staminate flowers
   Gynoecious (or female)-plant with only pistillate flowers
   Andromonoecious-plant with both hermaphrodite and staminate flowers
   Gynomonoecious-plant with both hermaphrodite and pistillate flowers
   Trimonoecious-plant with all three types of flowers: hermaphrodite, staminate and pistillate.

Basic flower types in individual plants are shown in Fig. 1.5.

3. Sex expression of groups of plants

Here the term 'groups' is not clearly defined. In the most common case, this would mean the taxon or the species as a whole; but it may also comprise a natural isolated subspecies population of plants. In cultivated crops the group may be formed during the breeding process and thus kept as such under culture or breeding conditions.

   Hermaphrodite—a group consisting of only hermaphrodite plants
   Monoecious—a group consisting of only monoecious plants
   Dioecious—a group consisting of androecious and gynoecious plants
   Androdioecious—a group consisting of hermaphrodite and androecious plants
   Gynodioecious—a group consisting of hermaphrodite and gynoecious plants.

This list is obviously not complete, as the possible theoretical combinations of plant sex types is almost unlimited. Some of the types not listed here do

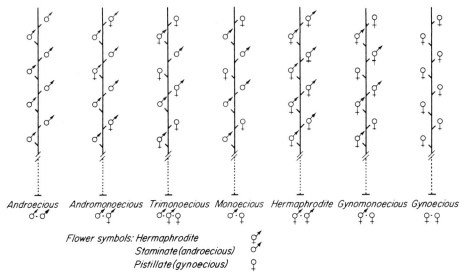

Fig. 1.5. A scheme for the classification of sex in individual plants. The vertical lines represent individual plants or sections of plants i.e. a plant may be first androecious and then (e.g. in the next flowering season as in *Arisaema*) become gynoecious. Flower and sex type symbol are given as used in this book

occur in either natural or artificially bred populations but they are rather rare and could be described without a previously defined term.

Terminology does not help us in rather common cases where either the flower or the plant cannot be included strictly under one designation: an androecious plant may have some flowers with degenerated stamens; moreover, as we shall see, temporary changes in sex expression can be imposed experimentally, e.g. a gynoecious plant can be induced to develop male flowers.

In the classification of sex in individual plants, we neglected the patterns of floral distribution. Plants may be classified under the same terminology (i.e., monoecious, andromonoecious or gynomoecious) and still vary vastly in their sex expression. We shall not go into such details here but will exemplify this situation by mentioning three, out of many more, patterns in monoecious plants:

1. Staminate and pistillate flowers are distributed throughout the plant—e.g. many cucumber cultivars.

2. Staminate and pistillate flowers are grouped in male and female inflorescences, respectively—e.g. corn.

3. At any specific time the plant bears only one type of flowers—e.g. *Arisaema japonica*, having first only staminate flowers and in subsequent seasons only female flowers. Populations of such plants may erroneously be classified as dioecious by a careless observer, and in fact this has happened (MAEKAWA, 1924).

In conclusion we should bear in mind that this terminology is intended only to shorten our communication and by no means should it be regarded as a description of sex expression. Having thus defined the terminology regarding the sex of flowers, plants and groups of plants, we should always be careful to express ourselves clearly. That is, we should speak about a *staminate* flower, a *pistillate* plant and a *dioecious* population (or group).

1.2.2.1.5 Temporal Separation of Sex Organs as an Outbreeding Device

Until now we have considered spatial separation of sex organs as an outbreeding device. A second device for the adjustment of evolutionary proper hybridity in plant populations is *temporal separation* of the activity of sexual organs. In bisexual plants, such separation ensures that a plant or an individual flower is unable to shed and receive pollen at the same time, and thus becomes dichogamous. Bisexual plants, shedding and receiving pollen at the same time, are termed homogamous. Homogamy does not necessarily result in self-fertilization ("autogamy"). Morphological, as well as various physiological barriers to autogamy, may exist in the hermaphrodite flower ("hercogamy"). The structure of homomorphic flowers may facilitate cross-fertilization. Flowers may also be of more than one morphological type ("heteromorphy"); such a condition usually influences breeding behavior. Functional failure of self-pollination due to mechanical or physiological barriers ("self-incompatability") as well as the absence of functional male gametes ("male sterility"), are important outbreeding devices of hermaphrodite flowers. Outbreeding devices will be dealt with in Chapter 3, and inbreeding devices, such as functional autogamy and cleistogamy, will be discussed in Chapter 2. The various modifications of dichogamy and homogamy of hermaphrodite flowers are shown in Table 1.1.

In monoecious plants, dichogamy may be due to the pattern of floral distribution (see Section 1.2.2.1.4). Thus, we must distinguish between *plant dichogamy* in a monoecious plant, and single *flower dichogamy* in a hermaphrodite flower.

Table 1.1. Classification of hermaphrodite flower modifications

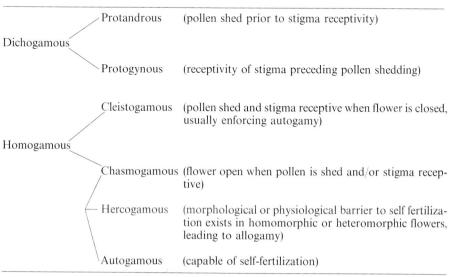

Dichogamous	Protandrous	(pollen shed prior to stigma receptivity)
	Protogynous	(receptivity of stigma preceding pollen shedding)
Homogamous	Cleistogamous	(pollen shed and stigma receptive when flower is closed, usually enforcing autogamy)
	Chasmogamous	(flower open when pollen is shed and/or stigma receptive)
	Hercogamous	(morphological or physiological barrier to self fertilization exists in homomorphic or heteromorphic flowers, leading to allogamy)
	Autogamous	(capable of self-fertilization)

Furthermore, dichogamy may exist also in dioecious species such as in the date palm *(group dichogamy)*.

1.2.2.2 Asexual Forms of Reproduction

Asexual reproduction affords continuity and increase in populations of higher plants by means of natural or artificial propagules. These propagules have the capacity of regeneration of the whole plant by successive mitotic divisions. Asexual propagules may be located outside or within the floral regions of the plant.

1.2.2.2.1 Asexual Propagules outside the Floral Region

Among the natural propagules many herbacious perennials feature dormant fleshy organs—roots, stems or leaves, specialized for food storage during cold or dry seasons, which function in multiplication. The organs bear buds to produce new shoots in the next season (see review by WEBER, 1967). Thus, we find modified fleshy roots in the sweet potato or in *Dahlia*, modified fleshy stems such as bulbs in *Narcissus* or in *Allium*, corms in *Gladiolus* or in *Colchicum*, tubers in the Irish potato or in *Caladium*, rhizomes in *Iris* or in *Phragmites*, and modified leaf structures such as leaf bulbils in *Bryophyllum* or in *Cardamine*. In some cases basal side shoots develop into runners which root and produce new plants (e. g. in strawberries and *Saxifraga*).

Asexual plant propagation in agriculture is based on the natural propagules mentioned or on artificial propagules. Such propagation is required in many cases: to increase the proportion of female plants in the population (e. g. dioecious date palm), clonal maintenance of self-incompatible varieties (e.g. avocado), fixation of heterozygosity and hybrid vigor (e. g. roses), multiplication of material grown under conditions in which plants do not flower (e. g. sweet potatoes), and production of pathogen-free clones (e. g. carnations).

Artificial propagules of various kinds are employed in vegetative plant propagation (see review by HARTMANN and KESTER, 1968). Many horticultural crops are propagated by stem, leaf or root cuttings endowed with a potential shoot system. Shoots and roots are initiated by buds already present, or by the development of primordia via meristematic cells. An additional method of vegetative propagation is "layering", utilizing the development of roots on stems still attached to the parent plant. The rooted stem, after being detached, becomes a new plant. Layering sometimes occurs in nature (e. g. raspberries). Another important method of vegetative propagation consists in joining parts of plants together ("grafting"). The united parts, being symbiotic, behave like one plant. Multiplication of the desired type is achieved by subdivision into many parts. Such parts may be very small, often consisting of a small piece of bark and wood and a single bud ("budding"). Finally, aseptic tissue culture techniques become practical in the propagation of horticultural plants (see recent discussion and review by NICKELL and HEINZ, 1973; MURASHIGE, 1974). Excised shoot tips ("meristem culture"), callus and even single cells are utilized. The large-scale multiplication of a desired individual plant achieved by these techniques, and the fact that sometimes desired clones can be freed of pathogens by culturing terminal growing points, make the so-called "micropropagation" increasingly attractive for agricultural purposes.

1.2.2.2.2 Asexual Propagules within the Floral Region

Apomixis, i.e., replacement of sexual by asexual reproduction, is the evolutionary result of a tendency to reduce genetic recombination. Flowers, designed for efficiency in the sexual process, underwent an evolutionary adjustment either by producing parthenogenetic seed ("agamospermy") or by producing propagules instead, or in the addition of normal flowers ("vivipary"). Agamospermy may result in *diploid* parthenogenesis by: (1) formation of an unreduced embryo sac from nucellar or integument cells ("apospory") or from archesporal cells ("diplospory"); or (2) directly from the sporophyte tissue ("adventitious embryony"). In the broad sense, agamospermy may include also haploid parthenogenesis by the development of a haploid embryo from the egg ("gynogenesis") or microspore ("androgenesis").

Haploid parthenogenesis may be utilized in breeding procedures, e.g. for the production of homozygous parents (see reviews by HORN, 1972; NITCH and NORRELL, 1973).

In vivipary, bulbils may be produced as accessory formations in the floral axils or branches (e.g. *Agave*) or instead of flowers (e.g. *Allium*). The bulbils as well as the transformation of flowers into vegetative shoots (as in *Poa*) propagate the diploid female sporophyte. Table 1.2 summarizes the different modes of apomictic reproduction. For detailed discussions of apomixis see STEBBINS (1941), GUSTAFSSON (1946, 1947a, b), NYGREN (1954), RUTISHAUSER (1967).

Apomicts may be obligate, in which case heterozygotic genotypes are preserved at the cost of evolutionary flexibility. More often, apomixis is facultative, in which case apomictic and sexual reproduction coexist. Pollination may or may not be required for the production of functional parthenogenetic seed. Apomictic seeds are of agricultural importance by (1) providing for uniformity in seed propagation of rootstocks (e.g. in *Citrus*, where the sexual seedling is eliminated

Table 1.2. Modes of apomictic reproduction

Apomixis		
Agamospermy (parthenogenetic seed)		Vivipary
Gynogenesis Androgenesis	Apospory Diplospory Adventitious Embryony	Vegetative Proliferation Production of Bulbils, etc.
female or male haploid gametophyte Haploid Parthenogenesis ↓ New Haploid Sporophyte	diploid female gametophyte Diploid Parthenogenesis ↓ New Diploid Sporophyte	female somatic tissue ↓ Diploid Sporophyte

by weakness; CAMERON and SOOST, 1953; or in aposporic mango seed); (2) for true-breeding F_1 hybrids (e. g. in *Poa*, where up to 85% of the seeds are apomicts; BRITTINGHAM, 1943); and (3) for removal of viruses from clones of vegetatively propagated plants (e. g. *Citrus*; CAMERON et al., 1959).

1.2.2.3 Distribution of Modes of Reproduction among Cultivated Plants

As mentioned before, propagation of many economic plants is by *asexual* reproduction. Fruit, nut and ornamental trees, shrubs and woody vines are mostly clonally propagated, as is sugar cane, agave and pineapple. However, also seasonal crops, such as Irish potato, sweet potato, onion, garlic, strawberry, artichoke, and many flower crops are clonally propagated by asexual propagules outside the floral region or by aseptic micropropagation. In these crops pollination may be of importance only in the breeding of new cultivars and for proper fruit set. Pollination may become important in some cases of apomictic reproduction where seed development depends on the stimulation by pollination ("pseudogamy"). Such stimulation is often by the secondary effect of fertilization required for endosperm development. Stimulation of apomictic seed development occurs in *Citrus* spp., *Eugenia* spp. (adventitious embryony), and in *Parthenium argentatum* (guayule), *Rubus* spp. (raspberry), *Pyrus* spp. (apple), some *Poa* spp. (blue grass), as well as a result of interspecific pollination (e. g. in *Brassica* species crosses). In other agricultural plants no stimulation by pollination is required for agamospermy; *Opuntia* spp. (prickly pear or tuna-adventitious embryony), *Paspalum* spp. and *Poa* spp., *Simondsia sinensis*, *Capparis* spp. and in some vegetable crops (e. g. *Taraxacum officinale*—Huang-hua tsai). The possibility of genetic manipulation of agamospermy as indicated in sorghum and other Gramineae (see TALIA-FERRO and BASHAW, 1966; HANNA et al. 1970) as well as in some flower crops of the Compositae (e. g. *Centaurea cyanus* = cornflower), opens up the way for stable maintenance and multiplication of superior hybrid cultivars of field, garden and forage crops. For vivipary no stimulation by pollination has been reported for cultivated plants, but genetic control is indicated (see MCCOLLUM, 1974). Vivipary is found among *Agave* spp., *Allium cepa* var. *viviparum* (top onion), *Allium sativum* (garlic), *Allium fistulosum* (Welch onion), *Poa bulbosa*, *Festuca* spp. and others.

The distribution of *sexual* forms of reproduction in about 120,000 species in the phanerogamic flora has been cataloged (YAMPOLSKY and YAMPOLSKY, 1922): 86% of the species were bisexual (72% hermaphrodite, 7% monoecious and 7% gynomonoecious, andromonoecious or androgynomonoecious), 4% of the species were unisexual (dioecious), and 10% were both unisexual and bisexual (gynodioecious, androdioecious or trioecious). Unisexual forms, predominating in animals, are relatively scarce in plants. Gene-flow is enhanced and inbreeding is restricted by unisexual forms. Plants, as a consequence of being essentially stationary with only passive mobile stages of pollen and seed dispersal, have a restricted size of breeding population. Flexible, efficient and economic breeding systems have evolved in plants to adapt to a stable or changing environment. Bisexuality, coupled with isolation mechanisms of various types, provides greater flexibility, efficiency and economy in gene-flow than does unisexuality. Sexually polymorphic species, which include both unisexual and bisexual individuals, feature intermediate compromises between flexibility in hybridity and economy in gamete waste.

Among cultivated plants few are sexually polymorphic; in many of the dioecious crop plants hermaphroditic or monoecious individuals are produced regularly or occasionally in proportions depending on genic and environmental factors. Dioecy in economic plants has been reported mainly among perennials (Table 1.3); these plants are more or less obligately outcrossing, depending on the number and self-compatibility of hermaphroditic or monoecious individuals in the population.

Rather wide differences may be observed in the breeding behavior of bisexual plants. Such differences are caused by genotypical variation within the species and by the ecological conditions imposed on the potentially interbreeding population. Nevertheless, bisexual cultivars (as well as cultivars not entirely unisexual) may be grouped by their mode of pollination under standard conditions, i. e., normal conditions of agricultural practice. Thus, we may classify species or cultivars as belonging to inbreeders or outbreeders. Table 1.4 lists largely self-pol-

Table 1.3. Partial list of largely dioecious economic species[a]

			Main pollen vector
Fruit and forest trees	Carob	*Ceratonia siliqua*	wind
	Date palm	*Phoenix dactylifera*	wind
	Fig	*Ficus carica*	wasp
	Mulberry	*Morus alba, M. nigra*	wind
	Muscadine grape	*Vitis rotundifolia*	insect
	Papaya	*Carica papaya*	wind and insect
	Persimmon	*Diospyros* spp.	insect
	Pistacio	*Pistacia vera*	wind
	Pili nut	*Canarium ovatum*	insects
	Ash	*Fraxinus* spp.	wind
	Box elder	*Acer negundo*	insect
	European holly	*Ilex aquifolium*	insect
	Poplar	*Populus* spp.	wind
	Sea buckthorn	*Hippophae rhamnoides*	wind
	Varnish tree	*Rhus vernicifera*	insect
	Willow	*Salix* spp.	insect
Range plants	Buffalo grass	*Buchloe dactyloides*	wind
	Saltbush	*Atriplex* spp.	wind
	Texas blue grass	*Poa arachnifera*	wind
Vegetable crops	Asparagus	*Asparagus officinalis*	insects
	Spinach	*Spinacea oleracea*	wind
	Yam	*Dioscorea* spp.	?
Condiment plants	American bittersweet	*Celastrus scandens*	?
	Black pepper	*Piper nigrum*	rain
	Hop	*Humulus lupulus*	wind
Industrial crops	Hemp	*Cannabis sativa*	wind
	Jojoba bean	*Simmondsia californica*	?

[a] Nomenclature consistent with UPHOF (1968).

Table 1.4. Partial list of normally inbreeding cultivated plants[a]

Family Botanical name Common name	Main sex types[b]	Flower modifications[c]	Natural cross-pollination percentage and predominant agent[d]	Isolation standard (Registered seed) (m)[e]
Compositae				
Bellis perennis Daisy	☿			
Callistephus chinensis China aster	☿		10 (flies)	
Chichorium endivia Endive	☿ ☿♀		15 (flies)	10
Lactuca sativa Lettuce	☿	partly cleistogamic	1–6 (flies)	10
Graminaceae				
1. Cereals				
Avena spp. Oat	☿	upper fl. usually imperfect. largely cleistogamic	data variable max.: 10 (wind)	0
Hordeum vulgare Barley	☿	many cultivars cleistogamic	as above (wind)	0
Barley hybrids	☿+♀(ms)	as above		200
Oryza sativa Rice	☿	some cv cleistogamic varies climatically		3
Panicum miliaceum Proso Millet	☿		>10 (wind)	400
Sorghum vulgare Sorghum	☿			300 (400 from *Sorghum sudanense*)
Triticum spp. Triticale Wheat	☿	largely cleistogamic	data variable max.: 6 (wind)	0
Wheat hybrids	☿+♀(ms)	as above		200
2. Forage grasses				
Agropyron trachycaulum Slender wheatgrass	☿			0
Bromus cartharticus Rescue grass	☿			0
Bromus marginatus Mountain bromegrass	☿	mostly cleistogamous		0
Bromus mollis Soft brome	☿			0
Chloris gayana Rhodes grass	☿	mostly cleistogamous		
Elymus spp. Wild rye	☿			0
Eragrostis trichoides Sand love grass	☿			0

Table 1.4.

Family
  Botanical name
    Common name
	Flower characters		Natural cross-
			pollination
			percentage and
			predominant
			agent[d]
				Isolation
				standard
				(Registered
				seed)
				(m)[e]

	Main sex types[b]	Flower modifications[c]		
Hordeum jubatum Foxtail barley	☿			0
Lolium temulentum Annual Rye grass	☿			
Setaria italica Foxtail millet	☿			0
Sorghum vulgare var. *sudanensis* Sudan grass	☿			0
Stipa spp. Needle grass	☿			0

Leguminosaceae
  1. Seed and Vegetable
    Legumes				

Arachis hypogaea Peanut	☿			0
Cicer arietinum Chickpea	☿			not specified
Glycine max Soybean	☿			0
Lens culinaris Lentil	☿			not specified
Phaseolus aureus Mung bean	☿			0
Phaseolus lunatus Lima bean	☿		0–80 (bees)	45
Phaseolus vulgaris Common bean	☿		1–8 (bees)	45
Pisum sativum Pea	☿		some cv up to 25	0
Vicia faba Broad bean	☿		> 30 (bees)	
2. Forage and other Legumes				
Crotalaria spp. Crotalaria	☿			
Lathyrus cicera Garousse	☿			
Lathyrus sativus Grass peavine	☿			
Lathyrus tingitanus Tangier pea	☿			
Lespedeza spp. Lespedeza	☿	partly cleistogamic	dependent on proportions of chasmoga- mous flowers chasmogamous, flowers exhibit 70 (bees)	3

Table 1.4.

Family Botanical name Common name	Flower characters		Natural cross-pollination percentage and predominant agent[d]	Isolation standard (Registered seed) (m)[e]
	Main sex types[b]	Flower modifications[c]		
Lupinus albus	☿		10 (bees)	
Lupinus luteus	☿		10–25 (bees)	
Lupinus perennis	☿			
Lupine				
Medicago hispida	☿			
Bur clover				
Medicago lupulina	☿			
Black medic				
Melilotus dentata	☿			
Melilotus indica	☿			
Sweet clover				
Onobrychis vicifolia	☿			90
Sainfoin (Esparcette)				
Phaseolus mungo	☿			
Urd bean				
Trifolium fragiferum	☿			90
Trifolium glomeratum	☿			90
Trifolium procumbens	☿			90
Trifolium subterraneum	☿			90
Clover				
Vicia benghalensis	☿			3
Vicia pannonica	☿			3
Vicia sativa	☿			3
Vetch				
Linaceae				
Linum usitatissimum	☿		3 (bees)	0
Flax				
Malvaceae				
1. Fiber crops				
Gossypium arboreum	☿	protandrous, but staminal column facilitates self-pollination	2 (insects)	400
Gossypium barbadense	☿		5–10 (insects)	400
Gossypium herbaceum	☿		5–20 (insects)	400
Gossypium hirsutum	☿		5–40 (insects)	400
Cotton				
Hibiscus cannabinus	☿		2–45 (bees)	
Kenaf				
2. Vegetables				
Hibiscus esculentus	☿		5–20 (insects, hummingbirds)	400
Okra				
Hibiscus sabdariffa	☿			
Roselle				
Rosaceae				
Prunus armeniaca	☿		some cultivars self-incompatible	
Apricot				
Prunus persica			self-incompatible	
Peach and Nectarine	☿			

Table 1.4.

Family
  Botanical name
    Common name
	Flower characters		Natural cross-pollination percentage and predominant agent[d]	Isolation standard (Registered seed) (m)[e]

	Main sex types[b]	Flower modifications[c]		
Rutaceae				
Citrus spp. Citrus	⚥	many cv apomictic		
Pedaliaceae				
Sesamum indicum Sesame	⚥	protandrous	about 5, some cv up to 65 (bees)	acc. to cv 180–360
Solanaceae				
Capsicum annuum	⚥		5–10 (bees and thrips)	30
Capsicum frutescens Pepper	⚥		7–36 (bees and thrips)	
Lycopersicon esculentum Tomato	⚥	protogyneous, but pendant flower and anther cone facilitate self-pollination	<2 (solitary bees and thrips)	30
Nicotiana rustica	⚥			
Nicotiana tabacum Tobacco			2–3 (hummingbirds, bees)	50
Solanum melongena Eggplant	⚥		7 (insects)	
Solanum tuberosum Potato	⚥	many cv produce non-functional pollen		
Umbelliferae				
Apium graveolens Celery	⚥		30 (insects)	
Pastinaca sativa Parsnip	⚥	protandrous	30 (insects)	
Vitaceae				
Vitis vinifera Grape	⚥	some cv partly or fully self-incompatible		

[a] Nomenclature largely consistent with UPHOF (1968). Table compiled from various sources (e. g. SPECTOR, 1956; FRYXELL, 1957; ALLARD, 1960; JOHNSON, 1962).
[b] For definition of symbols, see Fig. 1.5.
[c] For definitions, see Table 1.1. cv = cultivar, fl. = flower.
[d] Percentage reported under average conditions with own pollen competition.
[e] Standards set mainly by the Association of Official Seed-Certifying Agencies, 1973 for registered seed.

linated economic plants. Most of these feature less than 5% cross-pollination under average agricultural conditions; a few of them consistently show a certain amount of natural cross pollination (NCP), which is normally less frequent than self pollination. An indication of the degree of crossbreeding may serve reported ranges of NCP and the specific standard levels of isolation requirements for certification of seeds by the U.S. Association of Seed-Certifying Agencies. Standard levels are aimed at the maintenance of specific genetic purity and identity and graded to classes of seed. Defined standards for the third seed class ("Registered" seed, which is the progeny of "Breeder" or "Foundation" seed) are given in Tables 1.4 and 1.5, for self-pollinated and cross-pollinated cultivated plants, respectively. Some relevant details of pollination ecology are also given in the tables.

Table 1.5. Partial list of normally outbreeding, bisexual, cultivated plants[a]

Family Botanical name Common name	Flower characters		Compatibility[d]	Main pollen vector[e]	Natural cross-pollination percentage[f]	Isolation standard (Registered Seed) (m)[g]
	Main sex types[b]	Flower modifications[c]				
Annacardiaceae *Mangifera indica* Mango	♂·☿	some pseudo-gamic apomixis	S.I. or S.C.	flies		
Annonaceae *Annona cherimola* Cherimoya	☿	strongly F–PG		beetles		
Bromeliaceae *Ananas comosus* Pineapple	☿		S.I.	humming-birds		
Chenopodiaceae *Beta vulgaris* Sugar beet Fodder beet Red beet Swiss chard	☿	F–PA	S.C. +S.I. S.I.	wind		
Chenopodium quinoa Quinoa	☿			wind		
Compositae						
1. Oil crops *Carthamus tinctoria* Safflower	☿		S.C.	bees	5–90	
Helianthus annus Sunflower		F–PA	variable S.I.	bees	20–75	800
2. Vegetable and garden crops						

Table 1.5.

Family Botanical name Common name	Flower characters		Compatibility[d]	Main pollen vector[e]	Natural cross-pollination percentage[f]	Isolation standard (Registered Seed) (m)[g]
	Main sex types[b]	Flower modifications[c]				

Chrysanthemum spp.
  Chrysanthemum
  Pyrethrum
	☿		S.I.	various insects		

Cichorium intybus Chichory	☿		S.I.			
Dahlia rosea Garden dahlia	☿		S.I.			
Helianthus tuberosus Jerusalem artichoke	☿					
Convolvulaceae						
Ipomea batatas Sweet potato	☿	F–PG	S.I.	bees		
Corylaceae						
Corylus spp. Hazelnut (filbert)	♂·♀	dicho-gamous	S.I.	wind		
Cruciferaceae						
Brassica spp. Broccoli	☿					
Brussels sprouts			S.I.	bees		
Kale			S.I.	bees		
Kohlrabi			S.I.	bees		
Mustard			S.I. or S.C.	bees		
Chinese cabbage			S.I.	bees		
Cabbage			S.I.	bees		
Collard			S.I.	bees		
Colza			S.I.	bees		
Cauliflower			S.I.	bees		
Rape			S.C.	bees	>10	40
Rutabaga			S.C.	bees	>25	
Turnip			S.I.	bees		
Crambe maritima Sea kale	☿		S.I.	bees		200
Raphanus sativus Radish	☿		S.I. or S.C.	bees	>85	
Cucurbitaceae						
Citrullus vulgaris Watermelon	♂·☿ or ♂·♀		S.C.	bees		800
Cucumis melo Melon	♂·☿ or ♂·♀	P–PA	S.C.	bees	0–100	
Cucumis sativus Cucumber	♂·♀ or ☿	P–PA	S.C.	bees	70	

Table 1.5.

Family Botanical name Common name	Flower characters		Compatibility[d]	Main pollen vector[e]	Natural cross-pollination percentage[f]	Isolation standard (Registered Seed) (m)[g]
	Main sex types[b]	Flower modifications[c]				
Cucurbita spp. Pumpkin Winter squash Marrow Gourd	♂·♀		S.C.	bees		
Euphorbiaceae *Aleurites montana* Tung tree	♂·♀	predominanting male or female individuals. P–PG		insects		
Hevea brasiliensis Rubber tree	♂·♀		S.I. or S.C.	midges		
Manihot esculenta Cassava, manioc	♂·♀	P–PG		insects		
Ricinus communis Castor bean	♂·♀	P–PG		wind		
Fagaceae *Castanea* spp. Chestnut	♂·♀		S.I.	insects		
Quercus spp. Oak	♂·♀	Plant dichogamous		wind		
Graminaceae						
1. Cereals *Secale cereale* Rye	⚥		S.I.	wind		200
Zea mays Corn	♂·♀			wind		200
2. Forage grasses *Agropyron cristatum* *Agropyron desertorum* *Agropyron elongatum* *Agropyron intermedium* *Agropyron repens* *Agropyron smithii* *Agropyron trichophorum* Wheat grasses	⚥		S.I.	wind		90
Agrostis alba Redtop	⚥		S.I.	wind		90

Table 1.5.

Family Botanical name Common name	Flower characters		Compatibility[d]	Main pollen vector[e]	Natural cross-pollination percentage[f]	Isolation standard (Registered Seed) (m)[g]
	Main sex types[b]	Flower modifications[c]				
Andropogon furcatus *Andropogon scoparius* Blue stem	⚥			wind		90
Arrhenatherum avenaceum Tall oat grass	⚥			wind		10
Bouteloua curtipendula *Bouteloua gracilis* Grama	⚥	some forms apomictic				90
Bromus erectus *Bromus inermis*	⚥ ⚥		S.I. S.I. or S.C.	wind		90
Bromus pumpellianus Brome grass	⚥					
Cynodon dactylon Bermuda grass	⚥			wind		90
Dactylis glomerata Orchard grass	⚥		S.I. or S.C.			
Festuca elatior *Festuca rubra* Fescue	⚥	facultative apomictic	S.I. or S.C.			
Lolium italicum *Lolium perenne* Ryegrass	⚥		S.I.	wind		
Panicum virgatum Switchgrass	⚥	facultative apomictic		wind		90
Paspalum notatum Bahia grass	⚥	facultative apomictic	S.I.			500
Pennisetum glaucum Pearl millet	⚥		S.I. or S.C.			
Phalaris arundinacea Red canary grass	⚥		S.I.	wind		90
Phleum pratense Timothy	⚥		S.I.	wind		90
Iridaceae						
Freesia spp. Freesia	⚥		S.I.	bees		
Gladiolus spp. Gladiolus	⚥		S.C.			
Iris spp. Iris	⚥		S.C.			
Juglandaceae						
Carya pecan Pecan	♂·♀	variable plant dichogamous	S.C.	wind		

Table 1.5.

Family
  Botanical name
  Common name
	Flower characters		Compatibility[d]	Main pollen vector[e]	Natural cross-pollination percentage[f]	Isolation standard (Registered Seed) (m)[g]

	Main sex types[b]	Flower modifications[c]				
Juglans regia Walnut	♂·♀	P–PA	S.C.	wind		
Labiaceae						
Mentha piperita Peppermint	⚥			flies		
Origanum vulgare Marjoran	⚥			flies		
Rosmarinus officinalis Rosemary	⚥			bees		
Lauraceae						
Persea spp. Avocado	⚥	P–PG	S.I.	bees		
Leguminosaceae						
1. Seed and vegetable legume						
Cajanus indicus Pigeon pea	⚥		S.C.	honey bees	13–65	
Phaseolus coccineus Scarlet runner	⚥		S.C.	honey and bumble bees	>30	
2. Forage legumes						
Lotus corniculatus Trefoil	⚥		S.I.	honey bees	>90	
Medicago sativa Alfalfa	⚥		S.I. or S.C.	honey bees	>80	90
Trifolium alexandrinum *Trifolium hybridum* *Trifolium pratense* *Trifolium repens* Clover	⚥		mostly S.I.	bees	>90	
Liliaceae						
Allium cepa Onion	⚥	P–PA	S.C.	bees	93	800
Allium porrum Leek	⚥			bees		
Allium schoenoprasum Chive	⚥		S.I.	bees		
Lilium spp. Lily	⚥		S.I.			
Tulipa spp. Tulip	⚥		mostly S.I.			

Table 1.5.

Family / Botanical name / Common name	Flower characters		Compatibility[d]	Main pollen vector[e]	Natural cross-pollination percentage[f]	Isolation standard (Registered Seed) (m)[g]
	Main sex types[b]	Flower modifications[c]				
Myrtaceae						
Eucalyptus spp. Eucalyptus, gum	☿	P–PA	S.I.	bees		
Feijoa sellowiana Feijoa	☿		S.I. or S.C.	bees		
Psidium guajava Guava	☿		S.C.	bees	35	
Oleaceae						
Fraxinus spp. Ash	♂·♀			wind		
Olea europaea Olive	♂·☿		some cv S.I.	wind		
Palmaceae						
Cocos nucifera Coco palm	♂·♀	P–PA		wind and insects	acc. to cv	
Elaeis guineensis Oil palm	♂·♀					
Pinaceae						
Abies spp. Fir	♂·♀			wind		
Larix spp. Larch	♂·♀			wind		
Picea spp. Spruce	♂·♀		S.C.	wind		
Pinus spp. Pine	♂·♀			wind		
Pseudotsuga menziesii Douglas fir	♂·♀		S.I.			
Piperaceae						
Piper nigrum Black pepper	☿ or ♂·♀ or ♂+♀	P–PG		rain		
Proteaceae						
Grevillea robusta Silk oak	☿					
Macadamia ternifolia Macadamia nut	♂		partial S.C.			
Rosaceae						
Fragaria spp. Strawberry	☿ or ☿·♀ or ♀ (ms)	F–PG	S.C.	bees		

Table 1.5.

Family Botanical name Common name	Flower characters		Compatibility[d]	Main pollen vector[e]	Natural cross-pollination percentage[f]	Isolation standard (Registered Seed) (m)[g]
	Main sex types[b]	Flower modifications[c]				
Prunus spp.	☿			bees		
Almond			S.I.			
Cherry			S.I.			
Plum			S.I. or S.C.			
Pyrus spp.	☿			bees		
Apple			S.C. or S.I.			
Pear			S.C. or S.I.			
Rosa spp.	☿		S.I.			
Rose				bees		
Rubus						
Blackberry	♂ ♀		S.I.			
Raspberry	☿		S.C.			
Rubiaceae						
Coffea arabica	☿		S.C.	insects and wind	7–90	
Coffee						
Solanaceae						
Atropa belladonna	☿	fl. dichogamous				
Belladonna						
Petunia hybrida	☿		largely S.I.			
Petunia						
Sterculiaceae						
Theobroma cacao	☿		S.I. or S.C.	midges and ants	> 30	
Cacao						
Theaceae						
Camellia sinensis	☿		S.I.	insects		
Tea						
Umbelliferae						
Daucus carota	☿	F–PA	S.C.	bees		
Carrot						
Petroselium hortense	☿					
Parseley						

[a] Nomenclature largely consistent with UPHOF, 1968. Table compiled from various sources (e.g. SPECTOR, 1956; FRYXELL, 1957; ALLARD, 1960; JOHNSON, 1962).

[b] For definition of symbols, see Fig. 1.5.

[c] For definitions see Table 1.1. cv = cultivar; fl. = flower; F–PG = flower protogyneous; P–PG = plant protogyneous; F–PA = flower protandrous; P–PA = plant protandrous.

[d] S.I. = self-incompatible; S.C. = self-compatible.

[e] Most common and effective pollinating agent.

[f] Reported percentage under average conditions with own pollen competition.

[g] Standards set mainly by AOSCA, 1973.

1.3 Ecology and Dynamics of Pollination

Two dispersal units exist in higher plants: the pollen grain which is a microspore with a resistant outer cover, and the seed, constituting an arrested stage of a young sporophyte. In contrast to animals and most lower plants, dispersal units of higher plants are nonmotile. Moreover, if we disregard motion provided by anthers or stigmas on touch (e.g. "tripping" in legumes), or by the release of tensile strength of certain fruit structures, all energy required for dispersal after the release of pollen or seed as discrete units must be provided by external agents. Vectors for dispersion may be abiotic or biotic. Resistance to dry conditions is required to insure the survival of pollen and seed of terrestrial plants during the mobile phase of dispersion. The biological function of pollen and seed as dispersal units depends on germination and growth in a proper ecological niche. For seeds this niche is relatively broad and gravity provides for a consistent tendency to reach this niche, although longevity in semidormant conditions may be required to bridge periods of unfavorable milieu at that niche. Pollen function, on the other hand, depends on a much restricted and much more specific ecological niche—the compatible and receptive surface of the ovule. It follows that economical transfer of pollen requires extreme precision. Longevity of pollen is much less important than precision of transfer. Thus, the evolution of pollination mechanisms must be viewed in the light of functional efficiency of pollination syndromes, built of coadapted specificities of flowers and pollen vectors. For a recent and detailed discussion of pollination syndromes and ecology, the reader is referred to the books by FAEGRI and VAN DER PIJL (1971) and PROCTER and YEO (1973).

The basic breeding behavior of higher plants depends on the pollination syndrome. STEBBINS (1970) discussed the evolutionary aspects of these syndromes and concluded that "the diverse floral structures and pollination mechanisms in angiosperms represent a series of adaptive radiations to different pollen vectors and different ways of becoming adapted to the same vector". However, in cultivated plants in particular, superimposed on the normal pollination syndromes are a variety of interacting factors modifying the basic pollination dynamics and breeding behavior (Fig. 1.6).

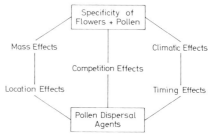

Fig. 1.6. Factors in pollination dynamics of cultivated plants

1.3.1 Specificity of Flowers and Pollen

The adaptation to different pollen vectors is expressed by *structural, spectral* and *olfactorial* floral specificities. The functional relation of the diverse characters to the vectors can be attributed to the type of the vector—whether *involuntary*, as are abiotic and unspecialized biotic vectors, or *voluntary*, i.e. specialized animal vectors.

Adaptation to voluntary specialized vectors is based on a multiformity of stimuli (providing attractiveness, as well as on structural and proportional *morphology*) providing for efficient vector approach, pollen reception and pollen deposition. Specialized animal vectors are stimulated by the provision of food and shelter and perhaps by sexual attraction. Food comes mainly in the form of pollen and nectar. The flower may serve as a refuge or as a breeding site for insect pollinators (e.g. for *Blastophaga* wasps in *Ficus*). Insect behavior resembling "pseudocopulation" of males has been observed with some orchid flowers imitating the female. Identification of, and discrimination between flowers by animal vectors is based on odor or visual attraction. Visual attraction is due to color, shape, texture, location and movement of flowers. FAEGRI and VAN DER PIJL (1971) summarized some of the "harmonic" relations between flower morphology, flower color and the voluntary pollination vector (Table 1.6).

Table 1.6. "Harmonic" relations between pollinators and blossoms (after FAEGRI and VAN DER PIJL, 1971)

Structural blossom class	Pollinator	Color preference (human perceptive light)
Dish bowl	Beetles	Brown
Bell beaker	Wasps	Drab
	Flies	
Brush	Bats	White
	Bees	Yellow
Gullet		Blue
	Moths	
Flag	Butterflies	Red
Tube	Birds	Green

Color preference is largely subject to the spectral range of the pollinator's light perception (in bees, between 300 and 650 nm—thus including UV and excluding red), whereas blossom class preference relates to easy approach of the vector.

Adaptation to abiotic and involuntary, unspecialized animal vectors is based mainly on the physical characteristics of the pollen, the exposure of the pollen to the dispersal agent, the pollen-catching capability of the stigmatic surface, and a high ratio of male to female gamete production. Physical characteristics of pollen are boyancy, adherence and stickiness. Flowers usually are small, with perianths insignificant or lacking.

1.3.2 Pollen-Dispersal Agents

Pollination, in most cases, is not carried out exclusively by one single agent. The origin of flower specificities of the species must be related to the most efficient pollen-dispersal agent present in the region it is evolving. The process of evolutionary floral modification may be retarded by the presence of secondary dispersal agents (STEBBINS, 1970). Furthermore, biotic pollen-dispersal agents may undergo parallel evolutionary adaptations to structural, spectral and olfactorial flower specificities (BAKER and HURD, 1968). Joint pollen vectors can be all biotic or both biotic and abiotic. Cooperating biotic vectors may be all specialized to different degrees or both specialized and unspecialized. However, normally we find one of the vectors predominant. Gymnosperms feature primarily abiotic pollination and biotic pollination is a derived condition. On the other hand, in angiosperms biotic pollination is most common and abiotic pollination is considered a derived condition.

1.3.2.1 Biotic Vectors

Anthophily can be found among a large number of insects, and some small vertebrates are also anthophilous (Table 1.7).

Table 1.7. Biotic pollination syndromes in cultivated plants[a]

Vectors		Syndrome term	Crops: Examples of families or species
Order	Common name		
Coleoptera	beetles	Cantharophily	Cucurbits, pyrethrum, mango
Hymenoptera	sawflies	Symphytophily	Umbelliferae, Roseceae, yellow-flowered Compositae
	wasps	Vespophily	Figs, gooseberry, alfalfa, carrot
	ants	Formicophily	Umbelliferae
	bees	Melittophily	numerous—see Table 1.5
Diptera	flies	Myophily	Onion, carrot, blackberry
Lepidoptera	butterflies	Psychophily	Rubus, aster, clovers
Lepidoptera	moths	Palaenophily	Clovers, Dianthus, tobacco
Neognathae	birds	Ornithophily	Pineapple, eucalyptus, strelitzia
Chiroptera	bats	Chiropterophily	Kapok, banana, agave

[a] see also Table 1.5.

Not all anthophilous animals are effective pollinators. Mutualistic systems of flowers and biotic pollen vectors are governed by energy needs and stimulus of the vector, and by the precision and efficiency of pollen transfer. In other words, anthophilous animals of pollination syndromes relate to specific ecological conditions providing for an optimal energy budget. Thus, we may contrast the low energy system of the ant-pollination syndrome, present primarily with plants featuring small inconspicuous flowers, offering limited amounts of nectar in hot and dry habitats (HICKMAN, 1974), to the high energy system of the bird-pollination syndrome found primarily with tropical plants having conspicuous, vividly colored flowers offering large amounts of nectar per flower (GRANT and GRANT, 1968).

Insects are the predominant biotic pollen-dispersion agents; primitive insects, like beetles, developed before flowering plants. Since plant products form the major food of most insects, unspecialized entomophily was established early in the evolution of flowering plants. Specialized pollinators, like bees, evolved concurrently with flowering plants and developed more and more the ability to perceive and discriminate between floral specificities. Flower discrimination is important, since efficient pollination depends on successive flower visits and flower type constancy of the pollinator. The incentive for such a type of pollinator activity is based on profitability. The amount of reward per flower must be sufficient to justify a visit, but not be so extensive as to limit successive flower visits. Figure 1.7 illustrates "the point of view of bees" for profitability of foraging for pollen as conceived by DOULL (1970). Similar regulatory models could be made for other stimuli and other specialized pollinators.

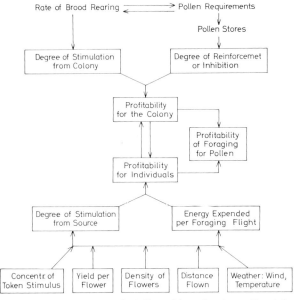

Fig. 1.7. A model for profitability of foraging for pollen (after DOULL, 1970)

Efficient insect pollination is of utmost importance in fruit or seed production of crops like safflower, sunflower, alfalfa, many clovers, crucifers, carrot, onion, cucurbits, and self-incompatible fruit trees. Among insect-pollinated forest trees, maple, chestnut, eucalyptus, willow, linden, magnolia and catalpa should be mentioned. Thorough insect pollination at an appropriate time is very often a prerequisite for agricultural success. Early, uniform and concentrated yields, but also quality of fruits (e. g. strawberry, melon), depend on full fertilization of an optimal number of ovules with the help of abundant vector activity.

Entomophily in cultivated plants has been subject to extensive research activity. For a thorough review of insect pollination in crops, the reader is referred to a recent work of reference by FREE (1970). Entomophilous crops are characterized by large or grouped flowers and conspicuous perianths; often, petals are colorful and with nectar guide marks; nectaries and scent are often present; pollen grains are relatively large (75–150 microns), sticky and oily, often in tetrads, with an ornamented surface, and are not carried by air currents. Although flower constancy of insects would be encouraged by hermaphrodite flowers, some entomophilic crops are monoecious (e. g. squash and cucumber) or even dioecious (e. g. fig and asparagus).

Bees (Apidae) in general, and social bees in particular, are the most important pollinators for cultivated plants. Bees are the most specialized pollen vectors and depend almost entirely on the pollen and nectar of flowers. The mutual adjustment of flower phenology and dynamics of bee populations has been well documented not only for honeybees (see FREE, 1967), but also for bumble bees (see MACIOR, 1970).

Melittophily is a very efficient pollination system and the most important one for cultivated plants. A honeybee may visit 100 flowers per trip and carry an estimated load of 5 million pollen grains (about 20 mg). The same bee may make 5 to 10 trips a day involving visits to different species, but remaining flower-constant in every single trip. An average honeybee hive may be responsible for 4 million trips a year, and gathering about 2 kg of pollen. Honeybee populations can be artificially managed much easier than other bees and hence are of utmost importance as agricultural pollinators. Bumble bees may be even more efficient pollinators than honeybees, since they work faster and for longer hours, and carry greater loads of pollen. The occurrence of solitary bees is often temporary and localized because it is regulated by the "flower calender" of the adapted host plant. Cross-pollination vectors for the production of uncontaminated seed in enclosures is often needed. In large cages or greenhouses, honeybees can be used, but confinement of honeybee colonies in small enclosures is difficult to manage. Bumble bees or even solitary bees (e. g. the leaf cutter bee, *Megachile rotundata*) may work better in small enclosures.

The management of insects other than bees is sometimes of importance in agriculture. Thus, only the introduction of the appropriate *Blastophaga* wasps from the Mediterranean area to California made it possible to grow Calimyrna figs there. Cross-pollination in enclosures of onion, carrots and brassicas has been facilitated by artificially bred flies, in particular blow flies. Ants may be proposed as artificial cross-pollination vectors in enclosures or other isolations for plants like bell peppers, legumes, carrots and cacao.

Very few cultivated plants are pollinated primarily by vertebrate vectors although vertebrates, and in particular flying vertebrates such as birds and bats, constitute primary pollinators for about 20% of the tropical flora. A large number of agricultural pests serve as involuntary biotic pollen vectors and often become of importance in pollination; these may include thrips, aphids, grasshoppers, moths and even squirrels, rats and other mammals.

1.3.2.2 Abiotic Vectors

The three abiotic agents for pollen transfer are gravity, air movement and water movement. Gravity transfer is confined to one dimension, and hence ordinarily results in autogamy. Pollen dispersal by air movement is three-dimensional and known as "wind-pollination" or "anemophily." Pollen dispersal by water movement may be two-dimensional, when taking place on water surfaces, or three-dimensional, when mediated by volumes of water or by raindrops; it is known as "water-pollination" or "hydrophily". All abiotic pollination mechanisms are random in nature, but structural adaptations of flowers and pollen and of plant populations have evolved to increase the efficiency of the random pollen-transfer process.

Anemophily is based on an indiscriminate, inefficient dispersal mechanism, and hence requires very large amounts of pollen to insure proper pollination. If a stigmatic surface is one square millimeter, then 1 million pollen grains distributed evenly over a area of 1 square meter are required for reasonable success in fertilizing a single ovule. The potency of the mechanism is further modified by the caprice of air movement, depletion of pollen grains in the atmosphere by filtration or washout by raindrops, and by the steep gradient in pollen dispersion density in relation to distance from the pollen source. Prerequisite for anemophily are conditions conducive to pollen transport toward appropriate distances. Such conditions may be present in open and dry habitats. Consequently, anemophily is frequently found in deciduous forests, in prairies, and in savannas—but not in tropical rain forests. The mechanism is independent of the occurrence and behavior of biotic vectors and thus may have been established as a response to the absence of biotic pollinating agents. Among wind-pollinated forest trees are gymnosperms (fir, lark, spruce, pine, etc.) and several deciduous trees (ash, oak, poplar, birch). Many nut trees are wind-pollinated (hazelnut, pecan, walnut). Wind-pollination is the dominant pollination mechanism in Gramineae, beet, castor bean, olive and date. Evolutionary and environmental considerations of wind-pollination have been presented by WHITEHEAD (1969).

If we disregard inconsistencies of atmospheric circulation and pollen density gradients, efficiency of anemophily may be expressed by the rough equation:

$$\text{effective pollen grains} = \text{total pollen output} \times \frac{\text{stigmatic surface}}{\text{total pollen filtration area}}$$

It follows that if total pollen output per ovule is increased, efficiency is increased. As a matter of fact, anemophilous plants are very prolific pollen producers: a floret of rye produces more than 50,000 pollen grains, a hazelnut catkin nearly 4 million pollen grains, and a ragweed something of the order of 8 billion pollen grains per day. However, gametic waste as compensation for inefficiency of the pollination mechanism is evolution-wise not a desirable feature. Other compensating devices have evolved in anemophiles taking advantage of the aerodynamics of pollen transport and capture ability of stigmatic surfaces. Variables that control dispersal of pollen through and above the vegetation, as well as the pollen filtration by obstacles, and pollen deposition on receptive stigmas, have been studied in detail (see GREGORY, 1961; TAUBER, 1967). For a discussion of the effective atmospheric pollen depletion by rain (the "rain-scavenging process"), see MCDONALD (1962).

Several types of adjustment of pollen-releasing and -collecting structures account for increased potency of wind-pollination. Enlarged, ornated, featherlike stigmas (e. g. in grasses) function as effective pollen filtration grids and the number of ovules per stigma is one or a few (e. g. hazelnut, grasses). The inert pollen filtration area is decreased in anemophiles by reduced perianths, exposure of flowers above or outside the leaf canopy, flowering before leaves come out, and reduction of leaf surfaces. Several pollen-releasing adjustments, such as anther exposure by long filament, explosive pollen dispersal into a strong wind (e. g. in pistacio and castor beans), and regulation of pollen dispersal by arresting devices adapted to proper wind force (e. g. in grasses and gymnosperms), decrease pollen waste. Anther dehiscence ordinarily is regulated to coincide with dry, relatively hot weather, conductive to effective wind and proper pollen germination on the stigma (e. g., in ragweed; see PAYNE, 1963).

Since the concentration of airborne pollen in the neighborhood of dehiscent stamens is very high, stigmas close to the stamens would be covered profusely with their own pollen; this is not only a waste of pollen but also decreases the opportunity to achieve cross pollination. Thus, wind-pollinated plants tend to have the sexes in separate flowers (dioecious or monoecious plants) or are dichogamous (hermaphrodite plants).

In contrast to the sticky, ornamented pollen of entomophilous species, pollen of anemophiles has a smooth, dry surface. Thus, grains are dispersed singly and not in groups. Buoyancy of pollen depends also on size and weight. Pollen comes generally in sizes between 2.5 micron (Ficus purmila) and 250 micron (squash) in diameter. Windborne pollen is usually smaller than pollen carried by insects—in angiosperms between 10 and 25 microns and in the conifers between 30 and 60 microns. The optimal size depends on a compromise between the dispersal range of pollen and the trapping efficiency of stigma. In this respect, adjustment such as airsacs in windborne pollen grains can be considered a solution for increasing size for trapping efficiency without loss of buoyancy.

Water pollination is relative rare. In aquatic plants, surface pollination, i. e., movement of pollen or pollen-bearing structures on the water surface, is found in some dioecious plants (Elodea, Vallisneria). Pollen dispersed by means of volumes of water (submerged hydrophily) is less common. In terrestrial plants, among them a few cultivated ones, hydrophily is due to rain. A prerequisite for rain pollination is water-repellent or water-insensitive pollen. In anemones (Ranunculus), rain pollination leads to autogamy by filling the erect perianth with water raising the floating pollen grains to the level of the stigma. In other plants, raindrop splash and wind may achieve allogamy. Thus, in black pepper (Piper nigrum) rain breaks up the glutinous masses of pollen in flowers lacking perianths, and the stigmas are constructed to strain pollen from water (MARTIN and GREGORY, 1962). The breeding behavior of black pepper is modified by protogyny in hermaphrodite cultivars or by monoecy and dioecy found in other cultivars.

1.3.3 Timing and Climatic Factors in Pollination Dynamics

Climatic variables such as wind, rain, air humidity, temperature, light intensity and spectral quality, are the triggers for the timing of flowering in plants and for pollen vector activity; their time constancy or change, the abrupt or gradual variability and the diurnal and seasonal periodicity, are the major factors determining the effective flowering period.

Flowering in plants may be continuous, seasonal or gregarious. The inherent flower-bearing characteristics depend on the plant-specific location of flowering buds. Buds may be located terminally or laterally on the main or on the side axis of the plant. Thus flowers are either terminal and on the main axis in plants such as wheat or pineapple, or terminal and on the side axis in plants such as in apple and determinate tomatoes. Flowers are either lateral and on the main axis in plants such as in coconut and date palm, or lateral and on the side axis in plants such as cucurbits and almond. The initiation and development of flowering buds, as well as the anthers of flowers, are subject to various environmental variables. Among the important variables influencing timing of flowering are light (photoperiod, light intensity and quality), temperature (thermoperiod, temperature extremes), moisture supply (soil moisture level and regime, atmospheric relative humidity) and nutrient supply (in particular nitrogen). Agricultural practices such as pruning, training, grafting, planting density, and chemical treatments (growth regulators) influence flowering.

Continuity of flowering can be due to constancy of environmental conditions (constancy of day length at equatorial regions) or insensitivity to environmental fluctuations. Most commonly we can recognize seasons of flowering which can be related by fluctuations of the various variables mentioned. Flower-inducing conditions in perennials may exist only once a year, when the dominant factors such as a certain sequence of light and temperature occur just once or twice a year in response to equal daylength and temperature in spring and fall. Cyclic or alternating flowering may occur as a consequence of growth flushes or overbearing (e.g. in fruit trees). Gregarious flowering, terminal in the life cycle of a plant, is common in quite a number of perennial plants (e.g. sisal, bamboo). Most annuals go through a vegetative stage of growth before conditions are right to induce single or successive flowering periods.

Cross-pollination can be achieved only by synchronization of flowering of the pollen donor and pollen receptor. The chronological adjustment of pollen shedding and stigma receptivity is termed "nicking" by plant breeders and hybrid seed producers. Nicking depends on a variety of timing devices (Table 1.8). Exact coordination of pollen shedding and stigma receptivity is facilitated by the prolongation of the effective flowering period. A sequence of events must be synchronized to achieve pollination: flowers must be open→anthers must

Table 1.8. Timing effects in pollination dynamics

Anthesis

Flowering period
Diurnal timing
Persistence in single flowers

Genetic dichogamy

Interval of sex separation
Diurnal separation of sexes
Overlap in maturity of sexes

Anther dehiscence and pollen supply

Duration of dehiscence in single flowers
Duration of pollen discharge in single flowers
Diurnal timing of pollen release
Pollen longevity

Pollen vector activity

Seasonal activity
Calendar overlap of competitive flowers
Diurnal activity in correspondence with stimulus presentation

Stigma receptivity

Persistence in single flowers
Diurnal timing

release pollen → pollen vectors must be active → stigma must be receptive → pollen must be capable of germination and fertilization.

Anthesis brings about exposure of anthers and stigmas to pollen vectors. The *flowering period* is the length of time a population of flowers maintains exposed anthers and stigmas. Flowers ordinarily open and close at definite hours. Such *diurnal timing* of anthesis ("floral clock") may be synchronized with pollen vector activity: the bat-pollinated kapok tree *(Ceiba pentandra)* flowers for 15 h during the night (between 1700 and 0800 h), whereas the insect-pollinated night-blooming cactus, *Selenicereus grandiflorus*, flowers for only 6 h (between 2000 and 0200 h). The flowers of most crop plants open during the early morning and stay open for various lengths of time (e. g. castor bean—4 h, lettuce—5 h, potato—8 h, cotton—10 h, cabbage—14 h). The exact hour of flower opening is subject to weather conditions and sometimes also to the age of the flower. Thus, in *Vicia faba* fresh flowers open at 1600 h, one-day-old flowers at 1300 h, and 2-day-old flowers at 1100 h (SYNGE, 1947). In the dichogamous coffee plant rainfall triggers flower opening on alternate days only.

Persistence of a single flower varies with species. Single flowers may persist for only one day ("ephemeral flowers"), such as in portulaca and many anemophilous plants. Single flowers of poppy, flax and raspberry fade after 2 days, others persist for a few days (pepper, cotton, apples). In a few plants flowers persist for longer periods (in cranberry about 20 days and in some orchids up to 80 days). In general, flower persistence is shorter in plants producing abundant pollen (anemophiles, poppy, portulaca) than in plants producing moderate or small amounts of pollen (insect-pollinated plants with relatively small numbers of anthers per flower).

Genetic dichogamy, based on temporal separation in the function of sexes, is another factor in pollination dynamics. Temporal separation is often found reinforced by spatial separation of the sexes, such as by dioecy, monoecy or heteromorphy. The significance of the *interval of sex separation*, i.e., the time interval between the maturation of the two sexes, depends not only on the length of the interval but also on the sequence (protandry or protogyny). In protandrous plants such as cucurbits, coconut, cherimoya and gerbera, the interval of about 5 days may easily be bridged by *overlap of maturity* in consecutive flowers or by pollen longevity. Protogyny found in many anemophilous forest trees, hemp and date palm pays for cross-fertilization by the loss of the first flowers. In agricultural practice pollen must sometimes be stored for a whole year to assure fruit set in protogynous plants (e.g. date palm). Dichogamy is sometimes due to *diurnal separation of sexes*. Thus, the pollen-stigma maturation schedule in the hermaphrodite cacao flower, based on a 12-h interval, insures cross-pollination between plants with matching maturation schedules only. A similar situation is encountered in the protogynous avocado (see Fig. 1.8); varieties are classified according to schedule A (flowers open on the morning of the first day, then close, and reopen in the afternoon of the following day) or schedule B (flowers open in the afternoon of the first day, then close, and reopen on the following morning).

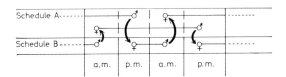

Fig. 1.8. Stigma-pollen maturation schedule and pollination in Avocado

Chronological adjustment of *anther dehiscence and pollen supply* to vector activity and stigma receptivity are an additional requirement for effective pollination. The *duration of dehiscence* in single flowers differs. Anthers in individual flowers may all dehisce simultaneously in the closed bud (as in most legumes) or at the time the flowers open (as in black currant, cabbage, pumpkin, corn). Anthers in individual flowers dehisce successively within 2 days in radish, within about 3 days in cherry, plum, apricot and apple, within about 7 days in pear, and up to 9 days in raspberry (PERCIVAL, 1955). Pollen is released mostly passively, through openings such as splits along the longitudinal grooves of the anther, transverse cracks, pores near the end of the locules, or apertures on one side of the anther which are closed by a trap door in wet weather—as in some tropical plants. Pollen shedding from the anthers is regulated by restrictions imposed by the openings and by the maturity gradients which exist between the anther lobes and along the main axis of the individual pollen locule.

Temperature and relative humidity regulate the *duration of pollen discharge* in single flowers. Thus, in castor beans a raceme may shed pollen for 16 days between 400 and 2000 h, but pollen release is restricted to temperatures higher than 24 °C and to relative humidity lower than 70% (MEINDERS and JONES, 1950). *Diurnal timing of pollen release* has a species-specific pattern. PERCIVAL (1955) grouped pollen presentation according to the daily period. In most crop plants the peak period of pollen release coincides with the peak period of pollen vector activity. The bulk of pollen is released early in the morning in crops such as sunflower, cabbage, radish and mustard; release until noon is found in wheat, corn, black currant and strawberry. Pollen is shed more or less uniformly throughout the daytime in crops such as onion, leek, raspberry and watermelon. Pollen shedding is concentrated in the afternoon in crops such as broad bean, apple, plum and white clover and in some cultivars of pumpkin. In the bat-pollinated kapok, pollen is presented mainly during the night.

Pollen survival under natural conditions is essential for a certain period of time: from pollen maturation through pollen release, dispersal and germination on a receptive stigma, to fertilization. During the mobile phase of dispersion pollen is exposed to quite severe environmental conditions, but usually this phase is only a matter of minutes or hours. In dichogamous crops the interval of sex separation must be bridged by *pollen longevity*. The life span of pollen of different species under natural conditions varies from a few hours to more than a year. Pollen generally survives for only a few hours under natural conditions in guava (about 4 h), in cacao (about 12 h) and in corn, wheat, barley and other grasses (up to 24 h). In deciduous fruits and alfalfa, pollen generally remains viable for a number of days and in some wind-pollinated trees (e.g. pines and dates) pollen can be stored viable for a year or more under natural conditions. Investigations on pollen physiology (see reviews by VISSER, 1955; JOHRI and VASIL, 1961; LINSKENS, 1964) revealed that longevity depends on variables such as atmospheric humidity, temperature, air composition and air pressure, as well as on pollen vigor—which might be modified by plant nutrition, viruses and other pathogens.

BREWBAKER (1967), summarizing physiological differences between binucleate and trinucleate pollen grains, observed the general tendency that trinucleate pollen loses its viability much faster than binucleate pollen. Trinucleate pollen is also difficult to germinate in vitro. It may be assumed that the second mitotic division deprives the pollen grain of sufficient reserves for longevity and germination. However, changes in metabolism associated with germination and longevity of trinucleate pollen, such as of corn and wheat, are not very well understood (GOSS, 1968). Among the cultivated plants, trinucleate pollen is found in the Caricaceae (papaya), Caryophyllaceae (dianthus), Chenopodiaceae (beet, spinach, quinoa), Cruciferaceae (cabbage, radish, etc.), Linaceae (flax), Umbelliferae (carrot, celery) and Graminaceae (wheat, corn, oat, barley, sorghum, range grasses).

Drought resistance by pollen appears to be an adaptation to the conditions prevailing during its dispersion phase. Nevertheless, no consistent differences can be found between drought resistance of anemophilic and entomophilic pollens. Most pollens retain viability longer at low relative humidities (0–40%) than at higher ones, but usually not below a distinct, critical level which is species-specific. As a general rule, low humidity is decidedly harmful to trinucleate pollen, such as that of grasses.

Temperatures above the optimum for the growth of the species, in all cases decrease pollen longevity. Viability is extended progressively with a decrease in temperature. Pollen grains of many crops can be stored viable for quite long periods at $4°-5°C$ if properly dried under vacuum and maintained in sealed containers. (Those remaining viable for more than one year are beet, corn, alfalfa, tobacco, cacao, clover, pine, coconut, potato and pea.) No damage is usually found if air-dried binucleate pollen is stored at deep-freezer temperatures $(-10°--35°C)$; under these conditions pollen such as that of tomato, potato, petunia, grape, apple and pine has been kept viable routinely for more than one year. Storage at uncontrolled temperature in sealed containers after freeze-drying appears to be a convenient method for long-term storage and for shipping pollen, provided it is properly rehydrated before use under species-specific conditions (see KING, 1965). Indications are that ultimate prolongation of the pollen life span, required for germplasm preservation in "pollen banks", will probably be achieved by the proper combination of conditions in storage: extremely low temperatures, low humidity, atmosphere enriched by CO_2 or reduced air pressure. Pollen preparation for storage and proper thawing and rehydration after storage will have to be determined for each species.

Sufficient *pollen vector activity* at the right place and time is required for effective pollination. Atmospheric conditions modulate abiotic vectors' activity seasonally and diurnally. Saturation of flowers with a strong and effective biotic vector is modified by quite a number of factors. Climatic factors condition seasonal and diurnal variations in the presentation of stimuli (such as nectar) for anthophilous animals (PERCIVAL, 1955). Climate also influences directly the physical ability of anthophilous animals, limiting or extending pollinating activity in degree and distance. Floral preference and constancy of the pollinator relate not only to discrimination between flowers, but also to the calendar overlap of competitive flowers and quantity of flowers present in the effective foraging area of the pollinator (FREE, 1970). The population size of pollinators obviously depends on the succession of flowers of different species ("floral calendar").

Compatible and receptive stigmas are a requisite for pollination. Persistence of *stigma receptivity* in individual flowers varies between species and lasts from a few hours up to one month. Emasculated flowers of self-pollinated crops (e. g. sesame, cotton, wheat, tomato) generally show brief persistence of receptivity, although optimal environmental conditions may extend receptivity somewhat. Insect-pollinated crops, as a rule, display receptivity for a number of days (e. g. guava, apple, sweet potato, onion), but crops belonging to the crucifers and cucurbits are exceptions to the rule with a stigma receptivity lasting a few hours only. The tendency in wind-pollinated crops is to maintain stigma receptivity for a week or more (e. g. date, beet, pecan). In most plants the best pollination is achieved on the day of flower anthesis, although stigmas may be fully receptive already some hours before anthesis (e. g. in cabbage, cucumber, and many self-fertilizing grasses). Often we encounter a diurnal variation in stigma receptivity. Thus, in hot weather crops such as watermelon and sorghum, morning pollination sets much better than afternoon pollination. In crops releasing pollen all day long, stigma receptivity is usually higher in the morning and evening than at noon. In dichogamous sweet potato, stigma receptivity is restricted to the late afternoon (1800–2000 h), as is the case with the type B avocado cultivars mentioned before.

The effective flowering period of each individual flower is of practical relevance, in particular for self-fertilizers. Thus, synchronization of the function of the two sexes in the individual flower must be achieved by coincidence of maturity or life span overlap. For cross fertilizers, the effective flowering period depends on the potentially interbreeding population of flowers. Developmental gradients provide for successive flowering within plants, and ecological or genetic differences extend the flowering season in the interbeeding plant population to spread the effective flowering period.

Breeding new cultivars, cultivar maintenance and proper pollination for crops where seed set is required, hang upon synchronization of the male and female functions. Matching the effective flowering period of the male and female parent must often be achieved by *artificial nicking procedures*. Thus, it is common practice in the production of hybrid seed to achieve pollination between early and late-flowering parents by adjustment of planting dates (for onions, see ATKIN and DAVIS, 1954). Artificial nicking is achieved by pollen storage in breeding programs, cultivar maintenance and date production. Interplanting of cultivars with pollinizer plants is practiced for diurnal and seasonal nicking. Such pollinizers are required for dichogamous, self-incompatible, sexually female or male sterile cultivar clones (avocado, tangelo, apple, plum, pear, cherry, Muscadine grape, fig, papaya, persimmon and pistacio) as well as for gynoecious cultivars (cucumbers). Some techniques used to insure nicking in plant breeding and seed production will be mentioned later on in the book in connection with the crops concerned.

1.3.4 Location and Mass Effects in Pollination Ecology

We have mentioned the artificial adjustment of pollination for agricultural or breeding purposes by control of nicking and manipulation of vector activity. The manipulation of climate for the adjustment of pollination is restricted to physical interventions such as windbreaks, greenhouses and other enclosures which are often economically not feasible. Location and mass effects in pollination ecology, on the other hand, provide many practical means of handling cross pollination.

Distance between pollen donor and acceptor is the major element affecting horizontal pollen dispersion. Generally, pollen dispersion curves follow leptokurtic distribution, i. e., they feature a higher concentration around the pollen source than would be expected with a corresponding normal distribution. Thus, progressive increase in distance becomes rapidly less effective in pollination. Leptokurtic distribution of pollen migration is typical both for abiotic and biotic dispersal, but methods of pollen dispersal modify the *direction* of pollen migration. Pollen dispersed by wind is subject to unpredictable variations in wind direction and velocity and is distributed in a downwind direction. On the other hand, biotic vectors move independently of one another in a statistically predictable manner, equally in all directions within a species-specific foraging area. The systematic, often flower-constant pollen transfer may be modified only by unequal dispersion of the plant species concerned or climatic conditions limiting normal activity of the vector. Data on horizontal pollen dispersion have been compiled by WOLFENBARGER (1973).

Pollen dispersion distances are best presented in terms of a curve for crossing frequency. Such a curve has a negative slope with its steepness decreasing as the distance increases. Experimental results (BATEMAN, 1947b, c) indicated that mathematical formulae for actual biotic and abiotic pollen dispersion curves are unexpectedly similar. Logarithmic transformation of crossing frequency data yield a straight line regression where the intercept (a) is the logarithm of the crossing frequency at the pollen source, and the slope (b) is equal to 0.4343 the coefficient of rate of decrease of frequency with distance (Table 1.9).

Table 1.9. Approximation of actual pollen dispersion curve (after BATEMAN, 1947b, c; WRIGHT, 1962)

$\log F = \log F_0 - 0.4343\,kD$
$\quad\quad = $ logarithmic statement of
$F = F_0 e^{-kD}$
F and F_0 are the frequencies at distances D and O, respectively
$\quad$ e = base of natural logarithms $\cong 2.718$
$\quad$ k = slope of straight line obtained by plotting log frequency over distance.
$\quad$ Logarithms used are to base 10.
$\quad$ The logarithmic statement of $F = F_0 e^{-kD}$ is similar to the generalized straight line formula, $Y = a + bX$. The intercept, a, is the logarithm of the source frequence $(=\log F_0)$, and the slope, b, is equal to $0.4343\,k$. The intercept and slope can be calculated by ordinary regression formulas as follows:

$$b = 0.4343\,k = \frac{\sum (wXY) - \dfrac{\sum(wY)\sum(wY)}{\sum w}}{\sum(wX^2) - \dfrac{[\sum(wX)]^2}{\sum w}}$$

$$a = \log F_0 = \frac{\sum(wY) - b[\sum(wX)]}{\sum w}$$

X = distance
Y = log frequency
w = weight = frequency

The use of the frequency (F) as the weight (w) to be allotted to a given observation was suggested by Sewall Wright. It is based upon the fact that frequency at any one point probably follows a binomial distribution, for which the variance $(=\,^1/\text{reliability})$ varies with frequency.

Barriers of various types limit pollen dispersion. We have already mentioned the filtration of pollen from the air stream by physical barriers such as are imposed by topography, interception by vegetation, or rain. For abiotic dispersion, the height of the pollen-bearing flowers and pollen buoyancy determine interception by the soil surface. Biotic pollen vectors may be limited by physical barriers in free movement or arrested by deterrents. Apart from physical barriers, pollen is diluted by the *mass* of the plant species. The plant's own pollen may compete with pollen from a different plant occupying stigma space, or may even precede fertilizing an ovule (if compatible). The mass of the plant species will decrease biotic pollen migration distance, because a biotic pollen vector can carry only a limited amount of pollen. Data given by FRYXELL (1956) illustrate the relative importance of varietal mass and distance in specifying the population breeding structure of cotton.

1.3.5 Competition Effects in Pollination Dynamics

The pollen-carrying capacity of the wind may be considered unlimited, and thus one type of pollen does not compete with another. On the other hand, the amount of pollen of one variety carried by a biotic-dispersal agent can increase only at the expense of another variety. Thus, biotic pollination becomes a function of the effective population size of the vector, the mass of the plant species, the foraging area size of the vector, the distribution of the vector, and vector preference for alternative species and shorter foraging trips. Competition during the spore-dispersal phase may be followed by competition between genotypically different pollen grains after the deposition on the receptive surface of the ovule. Such competition could depend on differential longevity, germination or growth of unlike pollen varieties, and on selective fertilization—the latter being an outcome of structural or physiological compatibility barriers imposed by interaction of the genotypes of the pollen parent, the pollen grain, the ovule parent and the ovule.

1.3.6 Determination of the Natural Cross-Pollination Rate (NCP)

The basic breeding behavior of species is a consequence of spatial, temporal or physiological separation of sexual organs. Pollination ecology and dynamics modify the basic breeding behavior and thus impose difficulties in measurement. In spite of the difficulties, we must arrive at a discriminate estimate of the NCP of a species under set environmental conditions if we attempt to restrict or promote NCP artificially.

The definition of NCP has sometimes been based simply on the physical interfloral transfer of pollen (KING and BROOKS, 1947). From the genetic point of view, cross-pollination is meaningful only if it involves crosses between unlike genotypes. Thus, interfloral-intraplant pollination, and even interplant-intraclonal pollination (e.g. in fruit orchards), are obviously genetically equivalent to self pollination. The term self pollination could be extended even further to interplant-intravarietal pollination, in which the participating gene pool is restricted. The extent of genetically significant cross-pollination, i.e., allogamy, is a function of (1) numerical relations of genotypes within a nicking and interbreeding population of flowers; (2) reproductive potential of the different flower genotypes; and (3) efficiency of the pollen vector.

Since fertilization is a mutually exclusive event, differences in gene frequency between the pollen pool and the zygotes may occur: pollen genotypes may differ in speed of germination and growth, egg cells closer to the stigma may be fertilized earlier than those located farther away, and seed maturation in the fruit may not be uniform.

Inferences on the rate of cross-pollination are made from the frequency of genetically marked individuals in the progeny. A cultivar, as a natural population of a plant species, may be composed of a number or different genotypes. If such a cultivar is seed-propagated, uniform distribution of the genotypes throughout the field is expected. This is a result of thorough mixing of the seed in agricultural practice. Estimates of the rate of cross-pollination will be different if based on progeny of individual flowers, single plants, certain genotypes within the population, or the whole interbreeding population. Furthermore, different results will be obtained from identical populations grown in different environments. Particularly large differences between estimates can be expected with normally inbreeding plants and plants with a high seed number per flower, where low-frequency cross-pollination events may produce differentials in the progeny of fruits. Thus, a minority of fruits may show a high rate of cross-pollination, whereas the average of the whole population may be exceedingly low.

In plant breeding and seed production of hybrid and synthetic cultivars, we are interested in the rate of allogamy *within* the population in a certain field, i.e., *intrapopulation crossing*. On the other hand, in seed production or ordinary (uncontrolled or "open-pollinated") cultivars, we are concerned in particular with undesirable cross-pollination *between* cultivars, i. e., *interpopulation crossing*. The artificial intrapopulation promotion of cross-pollination, as well as the interpopulation restriction of cross-pollination, depend on the basic natural breeding behavior of the species.

FRYXELL (1957) gives a diagrammatic representation of the basic reproductive methods of species by locating the species within an equilateral "reproductive triangle". We might extend such representation to include also the average sexual generation time and thus describe more fully the evolutionary significance of the breeding system of a species (Fig. 1.9).

Basic assumptions made in all methods for the estimation of the NCP are: monogenic inheritance and full penetrance of marker genes, equal survival of zygotes and of seed, equal seed dispersal, and equal seedling establishment and survival in the experimental progeny of the tested species population. For a quantitative estimate of outcrossing, different phenotypes may be used in progeny tests of populations or of specific phenotypes.

Bearing in mind the limitations and assumptions mentioned, the NCP may be estimated on the basis of gametic and/or zygotic frequencies. Zygotic frequencies in a random interbreeding population follow the Hardy-Weinberg equilibrium. A partially inbreeding population may be subdivided for convenience into a random interbreeding segment following the Hardy-Weinberg equilibrium, and into an inbreeding segment, subject to the probability that an individual

Fig. 1.9. The "breeding system square"

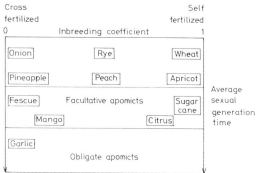

carries two alleles at a given gene locus derived from the same ancestor through a single generation, i.e., dependent on F, the coefficient of inbreeding (Table 1.10). The inbreeding coefficient, F, relates to the outcrossing frequency, λ in that $F = \dfrac{(1-\lambda)}{(1+\lambda)}$ and $\lambda = \dfrac{(1-F)}{(1+F)}$ (NEI and SYAKUDO, 1958).

Table 1.10. Subdivision of a partially inbred population

Zygotic		Random breeding segment:	Inbreeding segment:	Total frequency of genotypes in partially inbred population
frequency	genotype	$1-F$	F	
D	AA	p^2	p	$(1-F)p^2 + Fp = p^2 + Fpq$
H	Aa	$2pq$	—	$(1-F)(2pq) = 2pq - 2Fpq$
R	aa	q^2	q	$(1-F)q^2 + Fq = q^2 + Fpq$

Outcrossing estimation methods make use of observed deviations in zygotic frequencies of experimental progenies from the Hardy-Weinberg equilibrium. Different progenies may be used to obtain estimators for nonrandom mating as in the following examples:

1.3.6.1 Progeny Testing of Dominants

Assumptions: 1. Population at equilibrium.
2. No selection.
3. Cross-fertilization random with respect to all genotypes.

Procedure: 1. Count dominant and recessive phenotypes of population to obtain $(D+H)$ and R.
2. Progeny-test dominant phenotypes to determine proportions of heterozygotes to homozygotes.

3. Set the expected frequencies of heterozygotes and recessives equal to observed frequencies:

observed   *expected*

$$H = 2pq - 2pqF \qquad (1)$$
$$\text{and } R = q^2 + pqF \qquad (2)$$

4. Solve simultaneously for F and p as follows:

$$p = D + \frac{H}{2}$$

$$F = \frac{[(4D \times R) - H^2]}{(2D + H)(H + 2R)}$$

5. Calculate outcrossing frequency $\lambda = \dfrac{(1-F)}{(1+F)}$

1.3.6.2 Progeny Testing of Recessives

Assumptions: 1. Gene frequency in pollen pool is the same as in the zygotes that produce the pollen.
2. Outcrossing estimate applies equally to all genotypes in the population.

Procedure: 1. Determine zygotic proportions (see 1.3.6.1) and calculate dominant gene-frequency (p).
2. Progeny-test recessives to determine frequency of heterozygotes (DR).
3. Set expected frequencies of heterozygotes (function of dominant gene frequency, p, in the pollen pool and outcrossing frequency) equal to observed frequency:

observed   *expected*

$$DR = \lambda p \qquad (3)$$

4. Solve for outcrossing frequency λ.

1.3.6.3 Progeny Testing of Heterozygotes

Assumptions: 1. Gene frequency in pollen pool equals frequency in the zygotes that produce pollen.
2. Cross-fertilization random with respect to all genotypes.
3. Outcrossing estimate applies equally to all genotypes.

Procedure: 1. Determine zygotic proportions in the population (see 1.3.6.1) and calculate dominant gene frequence, p.
2. Progeny-test heterozygotes to determine frequency of recessives.
3. Set expected frequency of recessives in the progeny of heterozygotes equal to observed frequency:

observed   *expected*

$$RH = \frac{1+\lambda-2\lambda p}{4} \tag{4}$$

4. Solve for outcrossing frequency λ.

1.3.6.4 Progeny Testing of Recessives and Heterozygotes

Assumptions: 1. Recessives and heterozygotes have similar outcrossing frequencies.
2. *No* assumptions are made concerning gametic or zygotic frequencies.

Procedure: 1. Progeny-test recessives to determine frequency of heterozygotes (DR).
2. Progeny-test heterozygotes to determine frequency of recessives (RH).
3. Equate expected and observed frequencies of DR and RH as follows:

observed   *expected*

$$DR = \lambda p \tag{3}$$

$$RH = \frac{1+\lambda-2\lambda p}{4} \tag{4}$$

4. Solve simultaneously for p and λ.

VASEK (1968) reviewed a variety of outcrossing estimation methods and concluded that estimations based on recessives in the progeny of heterozygotes are not efficient, in particular if the frequency of the alleles approaches equality or outcrossing is low. Errors in estimation may also be considerable, if the frequency of the heterozygotes themselves is used. Consequently, the best estimators for outcrossing are frequencies of recessives and of heterozygotes in the progeny of recessives, which are observed directly. Estimates for outcrossing can be based on expected and observed frequencies in any two of the four basic estimators $(H, R, DR$ and $RH)$ solved simultaneously for p and λ in Eqs. (1), (2), (3), (4). If two or more independent markers can be utilized in the estimations of outcrossing, assumptions may be minimized. Thus, assumptions need not be made for selection and equilibrium in the population, and gene frequencies in the pollen pool and zygotes producing that pollen pool must not be assumed to be equal.

1.3.7 Artificial Control of Outcrossing

In plant breeding and seed production fields, cross-pollination must be controlled. Enclosures, emasculation of flowers or plants, and hand pollination normally provide the ultimate control of pollination required in plant breeding. The various factors involved in pollination dynamics have to be considered in the artificial control of open pollination. Effective *isolation of seed crops* is required to avoid interpopulation crossing and resulting contamination of cultivars. On the other hand, *promotion of cross-pollination*, i. e., facilitation of intrapopulation crossing, is often necessary to produce hybrid seed or to secure wide crosses in a breeding program.

In seed production, isolation standards have been established voluntarily by seed producers or by public agencies in many countries. Such standards take into consideration the breeding behavior of species, barriers and cultivar mass (see GRIFFITHS, 1956; Association of Official Seed Certifying Agencies, 1973). Thus, inbreeding, natural barriers, border rows, border-row removal, and large fields reduce the set standards for isolation distances. Cross-pollination may be artificially facilitated by choice of a vector-saturated location, planting design, downwind location of pollen acceptor, ratio between pollen donor and acceptor, proper adjustment of nicking, and artificial manipulation of vector activity.

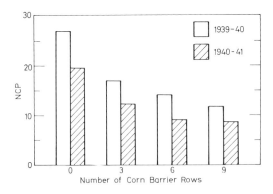

Fig. 1.10. Percentage of natural crossing (NCP) in cotton as averages of four replications by barrier condition and year (after POLE et al., 1944)

Intentionally planted border barrier-rows are useful as physical barriers intercepting pollen from outside the seed production field. The efficiency of such border barrier-rows may be illustrated by the data of POPE et al. (1944) for the NCP in cotton protected by corn barrier-rows (Fig. 1.10). The height and density of border barrier-rows appear to be important characteristics for efficient interception of wind- and insect dispersed pollen.

Isolation distance may also be modified by border rows consisting of the same variety ("guard rows"), or of the pollen parent in a hybrid seed production field. Such border rows impose dilution of foreign pollen, due to the addition of mass to the cultivar (BATEMAN, 1947a). The reduction in contamination as related to mass and border rows is brought out in the certification standards set by the AOSCA (1973), indicating reduction of distance from contaminants with the increase of border row number, size of seed production field, and removal of "guard rows". Thus, for instance, the required minimum isolation distance for sweet corn is 200 m from the contaminating source, plus four pollen-parent border rows when the field to be inspected is 4 ha or less in size. However, this distance may be decreased by 4.5 m for each increment of 1.6 ha in the size of the field, to a maximum of 16 ha. The distance may be further decreased by 12 m for each additional border row, to a maximum of 16 pollenparent border rows. These border rows can not be used for maintenance of the pollen parent.

The isolating efficiency of guard rows was demonstrated for cotton by GREEN and JONES (1953). If such rows are discarded, the isolating potency of the discarded rows may surpass border rows of a different species such as corn. Consequently, minimum isolation distances for many crops have been set lower for the cases when guard rows are removed. For example, isolation distances for cross-pollinated grasses have been set as follows: 50 m without guard strip removal; 30 m with guard strip removal of 2.7 m; 23 m with guard strip removal of 4.5 m (AOSCA, 1973). Utilization of guard or barrier rows is widely applied in seed production, and alternate strip planting of different seed crops is the most common design used.

Promotion of intrapopulation crossing for hybrid seed production, using male-sterile, self-incompatible or emasculated ovule parents, may be achieved by choosing a vector-saturated location or by manipulating biotic vector activity artificially. Locating the ovule parent plants downwind from the pollen parent may improve the required crossing both in wind-pollinated crops (for outbreeders such as corn, see BATEMAN, 1947b; for inbreeders, e. g. wheat, see DE VRIES, 1974) and in insect-pollinated crops (for outbreeders such as onions, see SPARKS and BINKLEY, 1946; for inbreeders, e. g. lima beans, see ALLARD, 1954). Planting design is of major importance in facilitating intrapopulation crossing. The most efficient alternation of rows, strips or plants of pollen and seed parents depends essentially on the steepness of the pollen dispersal curve (see Section 1.3.4). The ratio of pollen donor to acceptor plants in the field must be set by considering concurrently the percentage of seed set, area of pollen parent lost for seed production, seed quality achieved, and economics of production. These, and additional aspects connected with the promoting of intrapopulation crossing will be taken up in more detail later on.

Controlled cross-fertilization in plant breeding is normally achieved by hand pollination. Incompatibility barriers in wide crosses may bring about minimal success in hybridization and a very large number of pollinations may result in only a few viable seeds. In such cases, hand pollination is not practical.

A good strategy for overcoming difficulties in wide crosses would be to try
to force natural introgression by putting self-incompatible, male-sterile or emasculated single seed parents into a field of pollen donors with the hope of successful
NCP (KATZNELSON, 1971). The pollen-saturated condition of the seed parent
provides for maximal opportunity of the cross, and no hand work is required
to achieve the desired cross.

Chapter 2. Autogamy

2.1 Evolutionary Aspects of Autogamy

2.1.1 Strategies for Adjustment of Recombination

Higher plants are characterized by bisexuality—a prerequisite to autogamy—and by species which are broken up into populations of remarkably small effective breeding size even if distributed continuously (WRIGHT, 1946; EHRLICH and RAVEN, 1969). Such genetic discontinuity is an outcome of the random, but leptokurtic form of gene dispersal by pollen and seed, and of disruptive selection encouraging polymorphism (MATHER, 1955) in extremely localized environments. Plant population differentiation over distances of less than 30 m and in less than 40 years has been demonstrated experimentally (SNAYDON, 1970). Local differentiation of outbreeding species is maintained by selection in spite of foreign pollen contamination as high as 50–60 % at the site (JAIN and BRADSHAW, 1966). Therefore, the adoption of self-fertilization by many plants as an evolutionary strategy must be regarded mainly as a response to certain conditions of existence such as present in pioneer habitats, where initial reproduction must be assured and a rapid build-up of uniform population may be advantageous (STEBBINS, 1957; BAKER, 1959; ROLLINS, 1967; LEVIN, 1972). As a matter of fact, the majority of the most successful noncultivated colonizers are predominantly self-pollinated annual species (ALLARD, 1965). Cultivated plants have been selected for genetic constancy and reliable fertility since the beginning of plant domestication. Selection for genetic constancy and stable fertility mean restriction of genetic recombination and independence of the vaguaries of pollination vectors in the cultivated field; these are facilitated by self-pollination. Actually, the evolution of many self-pollinated cultivated plants can be traced to outbreeding ancestors. Such evolution most likely arose through a gradual adjustment of the mating system (loss of self-incompatibility, structural adjustment of flowers) as well as through restriction of the effective breeding population.

Although pollination mechanisms are among the most effective means for the attunement of recombination, many other factors modulate the recombination rate. Gene flow may be restricted or extended by pollen vectors and seed dispersal mechanisms. The breeding system may be adapted by the length of generation time, by chromosome number, and by the extent of structural hybridity of chromosome segments. Selective values may be varied by specific habitats favoring or rejecting recombinant genotypes. Thus, the evolution of the mating systems must be considered within the frame of quite a variety of forces active in species differentiation.

2.1.2 Origin of Mating Systems in Higher Plants

Although the most probable ancestors of the angiosperms appear to be the seed ferns, the phylogenetic relationship between the predominantly anemophilous gymnosperms and the predominantly entomophilous angiosperms is still obscure (CRONQUIST, 1968). Furthermore, phylogeneticists are still divided on the sexual potency of the primitive angiosperm flower. Ontogenetic observations do not negate a polyphyletic origin of sexual potency (HESLOP-HARRISON, 1972), but the fossil record available and the very special features of the angiosperm flower (i.e., enclosure of the ovule, shortening of the floral axis, presence of a perianth, etc.) support the notion of bisexual potency in the cantharophilic primitive angiosperm flower (BAKER, 1963; CRONQUIST, 1968; FAEGRI and VAN DER PIJL, 1971; GOTTSBERGER, 1974).

STEBBINS (1970) considers adaptation of pollination mechanisms in the angiosperms as governed by a number of evolutionary principles. The principle of the "most effective pollinator" suggests coordinated adaptation of the hermaphrodite flower and insect pollination. The efficiency of entomophily in the hermaphrodite flower is based on the deposition of pollen on the stigma while the insect is gathering pollen for food at the same visit. On the other hand, insects do not have incentives to visit pistillate flowers devoid of pollen (and other attractants), and, in view of gamete waste in anemophily, the hermaphrodite flower may have become a necessity. The significance of the coadaptation of anthophilous insects and hermaphrodite flowers is well illustrated by the problems in hybrid seed production using male sterile mutants as female parents of normally entomophilous crops. This subject will be discussed later on.

The pollination efficiency of entomophily in hermaphrodite flowers promotes the chance of self-pollination. Reduced hybridity in the plant population as a result of self-pollination may become disadvantageous in evolution. BAKER (1963) suggested that the presence of a style, and of pollen tube growth through it, provided an evolutionary opportunity for the control of fertilization by sieving of pollen tubes. Apart from self-incompatibility, barriers to self-pollination by temporal or spatial separation of the sex organs may have served as another evolutionary strategy for the adjustment of the hybridity optimum.

A reversal in evolutionary trends of mating systems may occur when environmental conditions elicit them (STEBBINS, 1970). Examples of such reversals are redifferentiation of bisexual potency toward unisexual potency, and shifts between insect pollination and wind pollination. The mulberry family (Moraceae) furnishes a good example of multiple reversal of trends. The bisexual flowers of the primitive Moraceae are entomophilous. A shift to unisexual flowers is exemplified by the entomophilous monoecious jack fruit (*Artocarpus heterophyllus*). The monoecious breadfruit tree (*Artocarpus communis*) growing at the fringe of the tropical regions in monsoon countries, developed the deciduous habit and, in the absence of the flies and beetles pollinating the jack-fruit, became anemophilous. Unisexual potency of plants developed in the entomophilous dioecious mulberry (*Morus* spp.) and in the anemophilous dioecious hemp *(Cannabis sativa)* and hops *(Humulus lupulus)*. The entomophilous fig has unisexual or bisexual flowers.

Many genera of the grasses *(Hordeum, Elymus, Agropyron, Festuca, Poa, Bromus)*, of the legumes *(Medicago, Trifolium, Phaseolus)*, and of the Solanaceae *(Solanum, Lycopersicon, Nicotiana)*, include both inbreeding and outbreeding species. The inbreeding species are usually more specialized than the outbreeding ones and are most likely derived from cross-fertilizing ancestors (STEBBINS, 1957). The shift to autogamy usually requires a breakdown in hercogamy or dichogamy upon which obligate cross-breeding is based (see Chapter 3). Self-compatibility may have evolved through loss mutation of genes at the self-incompatibility locus (LEWIS, 1954), by breakdown of self-incompatibility in derived polyploids (LEWIS, 1943; GRANT, 1956), or by fixation of self-compatible, monomorphic recombinants produced by "illegitimate crosses" in heteromorphic species (BAKER, 1966). Adjustments in structure, position and size of anthers, stigmata and perianths are often associated with the change of the breeding system. Figure 2.1 presents one possible outline of the hypothetical phylogeny of angiosperm mating systems.

2.1.3 Variation in Autogamous Populations

Although predominant self-fertilization reduces the capacity for genetic recombination, its adoption is by no means a prelude to evolutionary extinction. This is evident from the success of some autogamous groups outlasting their cross-fertilizing relatives (STEBBINS, 1957) and the persistence of genetic variability in inbreeding populations (ALLARD et al., 1968). Restriction of genetic recombination by inbreeding can be compensated for by features such as short sexual generation time, high chromosome numbers, absence of structural hybridity of chromosome segments, wide hybrid fertility, and long-range seed dispersal (GRANT, 1958). However, persistence of genetic variability in inbreeding plants is most probably due to local differentiation in ecological niches which maintains massive storage of genetic variability (ALLARD et al., 1968).

Successful inbreeding and outbreeding species are little different in total genetic variability. Inbreeders are more likely to be variable between families, whereas outbreeders tend to vary within families. Natural populations of inbreeders as well as most autogamous cultivars are made up of different, mostly homozygous genotypes (JOHANNSEN, 1926; KNOWLES, 1943; JAIN and MARSHALL, 1967; KANNENBERG and ALLARD, 1967). Furthermore, the experience of plant breeders indicates a certain residual heterozygosity in almost every autogamous cultivar and response to directional selection of progeny of individual plants often yield improved cultivar types. Within-family variability has actually been demonstrated in experimental populations of beans (see ALLARD et al., 1968) and in wild oats (IMAM and ALLARD, 1965).

Genetic polymorphism in self-pollinating species is of utmost importance in plant breeding methodology. Maintenance of variability and directional changes (ALLARD et al., 1972), in spite of the prevalent autogamy, must be taken into consideration. The evolutionary response of an inbreeding population is not only dependent on natural or artificial selection, but also on the original genetic make-up and on small, environmentally modulated changes in breeding systems.

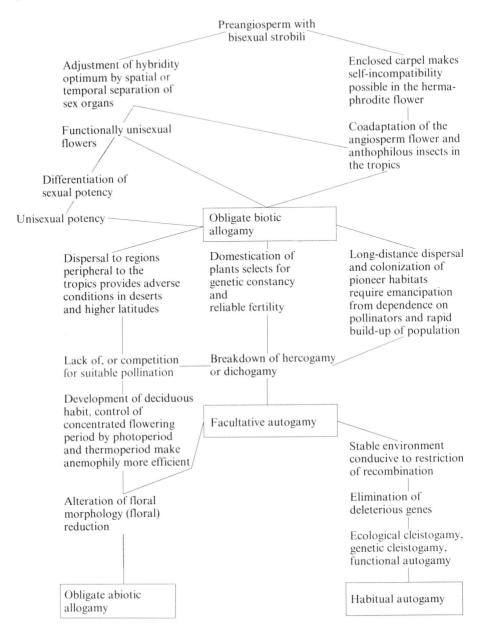

Fig. 2.1. Hypothetical phylogeny of angiosperm mating systems (assuming a monophyletic origin from a preangiosperm featuring bisexual strobili).

The reproductive values of genotypes cannot be deduced from their performance in pure culture, because it is modified in plant associations by frequency-dependent competition. Thus it has been demonstrated that heterozygotes at very low frequency in a population may produce about three times as many progeny as homozygotes in the same population (HARDING et al., 1967), and with a decreasing frequency in the population the selective value of heterozygotes in a mixture may become higher (WORKMAN and ALLARD, 1964).

Competitive ability and yield under natural selection have been studied in populations synthesized by mechanical mixtures of a number of pure lines. HARLAN and MARTINI (1938) first demonstrated rapid changes in mixture of barley cultivars in response to contrasting climate. Subsequent experiments with small grains showed repeatedly that one pure line usually becomes rapidly predominant in such populations (SUNESON and WIEBE, 1942; SUNESON, 1949). However, competition ability of a constituent line in a mixture is not necessarily related to its yielding ability, as illustrated in an equiproportional mixture of five rice cultivars (JENNINGS and DE JESUS, 1968). The taller cultivars were most competitive but low-yielding in pure stand, whereas the dwarf cultivars were less competitive but high-yielding in pure stand. In populations of autogamous plants derived from intercrossing different genotypes, directional and stabilizing responses to both natural and artificial selection are usually experienced (see ALLARD et al., 1968).

Normal residual heterozygosity in inbreeding crop cultivars and a frequency-dependent heterozygote advantage signify the importance of maintaining a panmictic gene pool and a proper level of hybridity in an inbreeding crop. Introduction of male-sterile mutants into autogamous populations has been proposed as a means to enforce outbreeding artificially (see JENSEN, 1970). JAIN and SUNESON (1966) and JAIN (1969) studied the fate of barley populations into which various proportions of male-sterile mutants were introduced. Comparison of an "outbreeding series" (derived from a population in which high proportions of male steriles were included) with an "inbreeding series" (derived from a population in which low proportions of male steriles were included) indicated increased within- and between-family variability but no improvement of population fitness in the population derived from the "outbreeding series". From these and other experiments it becomes clear that sufficient variability is normally being released in autogamous crops at the normal low levels of outcrossing.

Increased between-family variability, as provided by multiline mixtures, appears to enhance yield in some cases. Multiline mixtures of small grains and soybeans, assembled on the basis of pure stand yields and interactions between components, produced increased yields—apparently through efficient use of different environmental niches in the field (SIMMONDS, 1962; JENSEN, 1965; BRIM and SHUTZ, 1968). Multiline mixtures are thought to be more disease tolerant than pure lines (SUNESON, 1960; LEONARD, 1969), but studies with grain sorghum indicate that only substantial advantages in terms of yield stability or disease tolerance justify multiline blends (MARSHALL and ALLARD, 1974).

Recurrent cycles of enforced outbreeding and concurrent artificial selection—instead of relying on natural selection—may prove to be a useful breeding strategy in autogamous crops. FREY (1967) has shown that mass selection in oats is as useful in inbreeding as in outbreeding systems. The usefulness of mass selection even for characters of low heritability, has been indicated in enforced outbreeding populations of tobacco (MATZINGER and WERNSMAN, 1968), wheat and barley (REDDEN and JENSEN, 1974) and soybean (HANSON et al., 1967). Hence, again it is evident that the total genetic variability in autogamous plant populations is quite significant.

2.2 Mechanism of Autogamy

Breakdown of hercogamy, dichogamy or self-incompatibility must precede the shift of the breeding system from obligate outcrossing toward autogamy. A number of stages are clearly involved in such a shift. The first step toward the development of autogamy is characterized by *late anthesis "emergency pollina-tion"* securing fertilization in the case cross-pollination did not materialize. The mechanism of such pollination could be based on late anthesis changes in the position of own pollen and stigmatic surface (e. g. by bending of style, stamens or corolla, or closing of the corolla), incidence of overlap in style receptivity and pollen shedding in dichogamous plants, or weakening of self-incompatibility mechanisms (e. g. by decrease of impeding stigmatic substances). A further stage toward autogamy may involve accidental self-pollination supplementary to cross-pollination. *Mixed cross- and self-pollination* is a result of a breakdown of the outbreeding mechanism not replaced as yet by definite mechanisms of autogamy. The initial moves toward an active device for self-pollination are structural and positional changes in the hermaphrodite flower resulting in *functional auto-gamy*. A further step toward complete autogamy is *preanthesis self-pollination* ("bud pollination"), which make chasmogamy functionally redundant. Finally, flowers need not open at all to insure complete self-fertilization; such a condition is referred to as *cleistogamy*.

2.2.1 Cleistogamy

Cleistogamy is quite a common phenomenon in cultivated plants, but rarely obligatory in a species. The frequent presence of transitional flower types indicates that cleistogamy is a condition derived from chasmogamy. Flowers closing during unfavorable weather (ecological cleistogamy) may have given rise to the per-manently closed condition. Cleistogamic flowers are characterized by a reduction in number and size of floral parts such as stamens, and by modifications of the perianth. For that reason, UPHOF (1938), summarizing the literature on the subject, considered cleistogamous flowers essentially retarded, precociously functional forms of chasmogamous flowers. On the other hand, evidence is at hand of cases indicating genetic control of cleistogamy (such as in sorghum and rice). Thus, cleistogamy may be a product of genotype-environmental interac-tion (MAHESHWARI, 1962).

Ecological cleistogamy may be a device to safeguard pollination under unfavor-able conditions: the presence or absence of cleistogams is conditioned by such environmental factors as drought, heat, cold, shade, nutrition, submersion or burying of flowers, etc. (UPHOF, 1938). Thus, a cleisto-chasmogamous type of floral dimorphism may become seasonal. Weed sorrel *(Oxalis acetosella)* or sweet violet *(Viola oderata)*, for instance, if overgrown by other vegetation later in the season, becomes cleistogamous. In lespedeza *(Lespedeza stipulacea)*, cleistogamy is prevalent in apetalous flowers at 21 °C, whereas chasmogamy is predominant in petalous flowers at 27 °C (HANSON, 1943). Cleisto-chasmogamic floral dimorphism often exists in grasses. Panicles of rescue grass *(Bromus cathar- ticus)* produce many cleistogamous florets in dry and hot weather. Under stress conditions panicles of California brome *(Bromus carinatus)* and of sorghum will produce chasmogamic florets at the base and center, and cleistogamic florets at the extremes of the panicle (HARLAN, 1945; DOGGETT, 1970). High humidity will delay spikelet opening in "Japonica" rice or even inhibit opening completely and cause cleistogamic pollination (NAGAI, 1959).

Differential fertility of the two flower types has been reported among cleisto- chasmogams. MADGE (1929) states that chasmogamic flowers of sweet violet *(V. oderata)* do not function and seeds are produced only from cleistogamous flowers. BEATTIE (1969) describes in detail two phases of flowering in *Viola* species: the first cleistogamic and fertile and the second chasmogamic, which normally does not result in crossing (possibly because of protandry). An interesting example of the involvement of cleistogamy in the adjustment of recombination is provided by Carolina *(Lithospermum caroliniense)*. In this plant a heteromor- phic incompatibility system (see Chapter 3) is complemented by self-compatibility of cleistogamic flowers. Cleistogamic flowers produce more seeds than chasmoga- mic flowers and thus more selfed seeds are produced than one would expect on the basis of the number of cleistogamic plants in a population (LEVIN, 1972a). On the other hand, in many grasses, cleistogamic florets found at the extremes of inflorescences often are undernourished and hardly produce seed.

Constitutional cleistogamy has been reported in a number of grasses. In annual fescue *(Festuca microstachys)*, in "Sathi" rice cultivars of Uttar Pradesh in India, in the "Nunaba" sorghum cultivar of West Africa and *Sorghum papyr- escens (membranaceum)*, as well as in some barley cultivars, cleistogamy is con- trolled genetically. Panicles of annual fescue and of "Sathi" rice are pollinated within the boot-leaf sheaths (CHANDRARATHNA, 1964; KANNENBERG and ALLARD, 1967). Glumes of cleistogamic sorghums are closed or rolled (AYYANGAR and PONNAIYA, 1939; BOWDEN and NEVE, 1953). The inheritance of cleistogamy in *S. papyrescens* depends on the action of two recessive genes controlling rolling in of the edges of lower floral and upper involucral glumes (gx) and controlling a papery glume (py). According to AYYANGAR and PONNAIYA (1939), the concur- rent presence of gx and py results in cleistogamy. Cleistogamic barley varieties are characterized by poorly developed lodicules (BERGAL and CLEMENCET, 1962). *Poa chapmaniana* is considered a cleistogamic mutant of the annual blue grass, *Poa annua* (WEATHERWAX, 1929). In certain South American plantain species *(Plantago* spp.) hermaphrodite plants are normally cleistogamous and upon selfing produce only cleistogamic progeny. A cross between hermaphrodite and chasmo- gamous unisexual male plants yields only chasmogamous male progeny. Thus, cleistogamy and chasmogamy may be considered sex-linked in these plantain species (SCHÜRHOFF, 1924).

The *mechanism of cleistogamy* is based on nicking within the closed flower; it is a very economical pollination device. In *F. microstachys* only one of three anthers develops and it produces only a few pollen grains. Pollination is assured by enclosure of the single dehiscent anther in the stigmatic hairs (KANNENBERG and ALLARD, 1967). In other cases, such as in sweet violet *(V. oderata)*, wood sorrel *(O. acetossella)* and lespedeza *(L. stipulata)*, pollination is "cleistantheric", i.e., anthers normally do not dehisce, and pollen germinates within the closed anther locule wall (MADGE, 1929; HANSON, 1943, 1953; BEATTIE, 1969). Figure 2.2 illustrates floral morphology in cleistantheric-pollinated lespedeza.

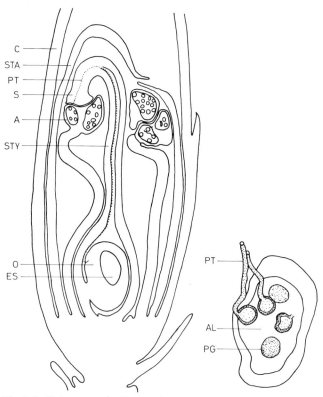

Fig. 2.2. Cleistogamy in Korean lespedeza.
Left: Median longitudinal section of cleistogamous flower at time of fertilization. C: calyx, A: anther, STA: standard, PT: pollen tube, S: stigma, STY: style, O: ovule, ES: embryosac ($\times 74$, after HANSON, 1953)
Right: Section of anther of apetalous flower showing penetration of pollen tubes through anther wall ("cleistanthery"). PT: pollen tube, AL: anther locule, PG: pollen grain ($\times 325$ after HANSON, 1943)

2.2.2 Chasmogamic Selfing

Preanthesis self-pollination ("bud pollination") is common in a number of cultivated legumes, although their flowers are of conspicuous entomophilous type. In the vetches *(Vicia angustifolia* and *V. sativa)* anthers dehisce in the bud and self-fertilization takes place before or at about anthesis (LECHNER, 1959).

Self-pollination in the bud occurs also in the common bean *(Phaseolus vulgaris)* and the mung bean *(Phaseolus aureus)*, although fertilization may often be delayed until after anthesis because of protandry (PURSEGLOVE, 1968). In field and garden peas *(Pisum sativum)* anthers generally dehisce in the bud and pollen accumulates at the tip of the keel (HERTZSCH, 1959). Pollen becomes deposited on the stigma with the growth of the pistil and self-fertilization is assured. A similar mechanism of preanthesis self-pollination occurs also in the peanut *(Arachis hypogaea)*, pigeon pea *(Cajanus cajan)*, soybean *(Glycine max)*, and some other legumes. Examples of preanthesis self-pollination in other families are: the garden cress, *Lepidium sativum (Cruciferae)*, and some grasses, e. g. rye grass *(Lolium temulentum)*, wheat grass *(Agropyron trachycaulum)*, and brome grass *(B. carinatus)* (SMITH, 1944; HANSON and CARNAHAN, 1956).

Grass flowers also provide good illustrations of active devices for self-pollination. *Functional autogamy* in wheat and barley is based on the maturation of pollen within the anthers before the florets open. When the lodicules swell the lemma and palea are pushed apart, and the two stigmata spread outward concurrently with the extension of the staminal filaments. Anthers begin to dehisce from the top while the filaments lengthen rapidly and pollen falls within the flower and becomes lodged in the feathery stigma. Normally the anthers are pushed out of the floret, but the extent of anther extension is variable, dependent on cultivar and on environmental conditions. JOPPA et al. (1968) report on 72 % anther extrusion for spring wheat as compared with 22 % for durum wheat, and on a positive correlation between anther extension and the amount of pollen shed. Pollen-shedding ability is of prime importance in hybrid seed production. For a detailed review of flowering biology of wheat see DE VRIES (1971), and for barley see BERGAL and CLEMENCET (1962). In oats (*Avena* spp.) and rice *(Oryza sativa)*, anther dehiscence atop the stigma coincides or precedes spikelet opening (BONNETT, 1961; CHANDRARATHNA, 1964), thus facilitating selfing.

Examples of devices affording functional autogamy are given in Fig. 2.3. In sesame *(Sesamum indicum)*, pollen is shed concurrently with the anthesis and the bifid stigma opens and becomes covered with pollen (WEISS, 1971). In many autogamous plants, anthers are grouped at the time of dehiscence near or above the stigma, increasing the chance of self-pollination [e. g. in self-pollinated legumes such as chickpea *(Cicer arietinum)* or soybean *(G. max)*, and self-pollinated crucifers such as Thale cress *(Arabidopsis thaliana)*]. The efficiency of selfing due to anther location close to the stigma is often enforced by a variety of mechanisms. Anthers may be open and stigmata may be receptive even before the flower unfolds completely (e. g. in *Nicotiana* spp.; GOODSPEED, 1961). In flax *(Linum usitatissimum)*, dehiscent anthers are located above the five erect styles touching the stigmata as the flower opens. Afterwards anthers often fall together to form a cap over the stigmata (DILLMAN, 1938). In the Compositae, pollen is shed inside the anther tube and the hairy developing stigma pushes the pollen mass outside like a piston and then it opens up outwards; thus, if self-compatible, such as lettuce, endive and aster, self-pollination is facilitated. In lettuce, obligatory self-fertilization is further enforced by a very short duration of anthesis (JONES, 1927). Self-fertilization in the cultivated tomato

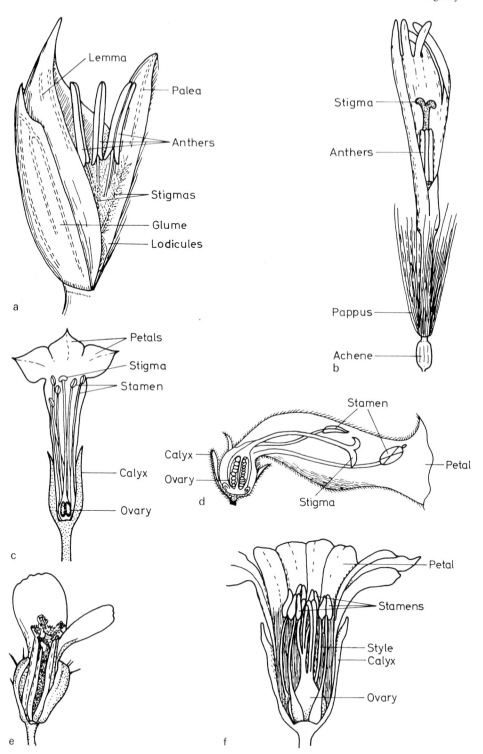

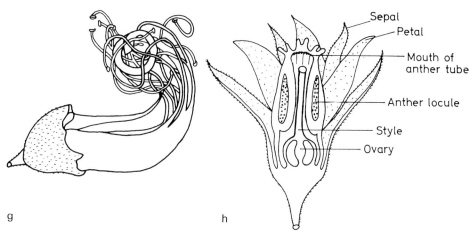

Fig. 2.3 a–h. Examples of devices affording functional autogamy. (a) Elongation of pollen-shedding stamens—wheat (Triticum), (b) Piston action of stigma in anther tube—lettuce *(Lactuca sativa)*, (c) Corolla unfolding while anthers shed pollen—tabacco *(Nicotiana tabacum)*, (d) Unfolding stigma lobes brush pollen—sesame *(Sesamum indicum)*, (e) Grouping of stamens above stigma—thale cress *(Arabidopsis thaliana)*, (f) clustering of anthers over style—flax *(Linum usitatissimum)*, (g) Twisting of stamens and style—common bean *(Phaseolus vulgaris)*, (h) Stigma enclosed in anther tube—tomato *(Lycopersicon esculentum)*

is favored by the position of the receptive stigma within the cone of the anthers and the normal pendant condition of the flower. Varietal differences in the degree of stigma exertion (RICK and DEMPSEY, 1969) are of major importance in tomato fruit setting and crossing techniques. The inserted stigma position affords better self-pollination and fruit set, whereas the exerted stigma position aids in artificial crossing and hybrid seed production utilizing male-sterile mutants as female parents.

A good example of partial breakdown of outbreeding mechanisms in mellitophilous legumes is the broad bean *(V. faba)*. Such partial breakdown results in *mixed cross- and self-pollination*. Pistil and anthers are liberated through a slit in the top of the keel under the weight of the alighting bumble bee, and pollen is discharged and received by the vector but also deposited on the pistil. Thus, although differential pollen germination and growth exist, the absence of a strong barrier to self-pollination results in about 30 % NCP (SOPER, 1952; ROWLANDS, 1958; BRANDENBURG, 1961; PURSEGLOVE, 1968). In peppers *(Capsicum annuum* and *Capsicum frutescens)*, mixed cross- and self-pollination is probably due to variation in protogyny based on the relative delay in anther dehiscence. Vector saturation (bees, thrips, ants), climate and genotype appear to alter the proportion of selfing (unpublished observations). Cross-pollination for a number of genotypes may range from 9 to 32 % under pollen-saturated conditions (ODLAND and PORTER, 1941). Cotton is another important crop featuring mixed cross- and self-pollination. NCP in cotton is subject to planting design, vector population and cotton species. Although upland cotton *(Gossypium hirsutum)* generally has shorter styles than extra-long-staple cotton *(Gossypium barba-*

dense), and the protruding stigmata are surrounded by erect stamens in intimate contact with the stigma, some cultivars show over 40 % NCP, as compared with 5 to 10 % for *G. barbadense* (SIMPSON, 1954; SIMPSON and DUNCAN, 1956; MEREDITH and BRIDGE, 1973).

There are numerous examples of *"late anthers emergency self-pollination"* due to a weakening of the outbreeding mechanisms. Among normally protogynous plants, dichogamy may be bridged in the absence of foreign pollen (e. g. in *Gerbera viridifolia*). Among normally self-incompatible crops, incompatibility may weaken in the old flower (e. g. in crucifers such as *Sinapsis alba* or *Brassica campestris*) and self-pollination will be successful. Crops featuring selective fertilization by foreign pollen competing with own pollen will be self-pollinated by slow-growing own pollen late in anthesis when foreign pollen is absent [e. g. in certain sunflower *(Helianthus annuus)* and raspberry *(Rubus idaeus)* cultivars].

2.3 Management of Pollination in Autogamous Crops

The control of parentage, as required for breeding purposes and for hybrid seed production, may call for techniques of emasculation of hermaphrodite flowers, pollen collection, pollen storage, artificial pollination, isolation and marking of flowers, and selective harvesting of fruits or seed. The natural adjustment to self-pollination makes aspects of emasculation, pollen collection and artificial cross-pollination most important for the control of parentage in autogamous crops.

2.3.1 Emasculation

Mechanical removal of microsporophylls, differential destruction of spores, and induction of male impotence are the means by which self-pollination can be prevented in autogams. Choice and efficient use of the different means depend on various factors such as flower morphology and size, timing of anther dehiscence, requirements for the reliability of the method, and economic considerations.

2.3.1.1 Mechanical Removal of Microsporophylls

The removal of microsporophylls may involve deposal of pollen grains, anthers, separate stamens, coalesced staminal columns, tubes, cones, sheaths, or even perianths with adnated stamens. Generally, it is advisable not to remove any part of the flower unless required and to do as little mutilation as necessary. Often slitting or clipping petals, sepals, glumes, etc., is preferable to removing these parts entirely.

For flowers large enough to permit manual emasculation, the *excision of unopened anthers* is the most common method used. The larger the anther and the better separated from the rest of the floral parts, the easier is the work. Hence, excision is usually delayed until shortly before dehiscence. Pollen in such anthers is already mature and extreme care must be taken to keep instruments and fingers free of pollen released from the anthers accidentally injured during the operation. To avoid unwanted contamination it is further advisable occasionally to dip instruments and fingers in alcohol or water, and to dry or wipe them clean.

The estimation of the proper stage for hand emasculation must be based on developmental criteria specific for each crop and cultivar. Since in most autogams anthers dehisce and stigmata are receptive before or at anthesis, emasculation must precede the opening of the flower. The change of color at the tip of the closed corolla is used as the criterion for the proper emasculation stage in plants such as flax, tobacco and tomato. In other plants the size of the flower bud, protrusion of petals from the calyx, or location of the flower, serves as a reference or interpolation of the prospective anthesis or anther dehiscence. Thus, cotton is emasculated one day before anthesis. Some barley cultivars must be emasculated when the ear is still deep within the boot leaf (POPE, 1944), while in other barley cultivars the proper stage is identified by separation of the edges of the boot leaf by the expanding spike, and for some cultivars the spike should be entirely out of the boot leaf.

Every feasible step must be taken to minimize the danger of self-pollination during the emasculation step. The operation is preferably carried out at the time of day when the danger of anther dehiscence is the least, i.e., generally, when humidity is high. Dehiscence may in some cases be prevented or delayed by artificially maintaining high humidity, e.g. by covering panicles of sorghum with plastic bags (SCHERTZ and CLARK, 1967), or by artificial fog—delaying dehiscence in *Paspalum* (BURTON, 1948) and other plants. The undehisced anthers of many sorghum cultivars can be successfully and fully detached from their filaments by repeated vigorous shaking of the panicle within a plastic bag, leaving the open florets essentially emasculated and ready for cross-pollination (A. BLUM, pers. comm.). Adjustment of the time of emasculation to the diurnal variations in stigma receptivity is another approach to reduce the danger of selfing.

Excision of unopened anthers from still closed flowers is often a delicate and time-consuming operation. JORDAN (1957) gives a detailed description of the use of warm air to force rice spikelet opening and the extrusion of unopened anthers. The interior of a vacuum flask is heated with warm water (43°C) and emptied. The bottle is inverted over the panicle from which mature spikelets have been cut off, and held for a few minutes with the fingers across the mouth, thus closing it. After a short time several spikelets open and stamens hanging out can be easily removed with a pair of forceps. Unfortunately, attempts in other plants to accelerate anther extrusion without anther dehiscence have been unsuccessful. Hence, the microsporophylls must be excised from closed flowers.

In relatively large flowers petals may be torn appart or slit and anthers plucked off with the fingers (e.g. tobacco) or with a pair of forceps (e.g. bell pepper). In other cases the corolla or even the calyx must be cut off with scissors or the fingernails to facilitate excision of stamens. Stamens may be

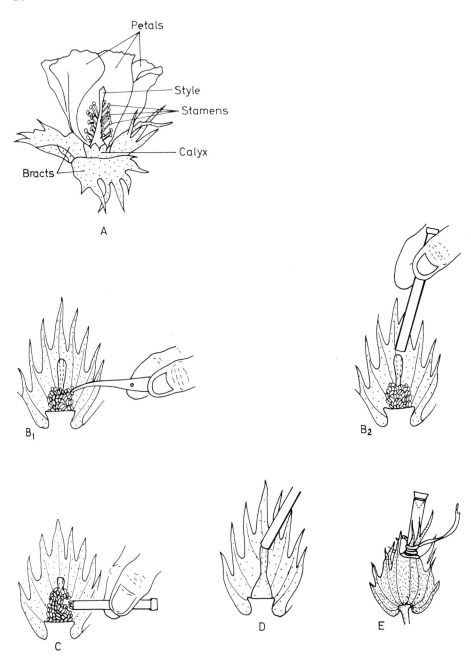

Fig. 2.4. Crossing technique for cotton (partly after POEHLMAN, 1959 modified.) (A) Flower morphology, (B₁) Emasculation by forceps, (B₂) Emasculation by tube, (C) Ripe anther collection in isolation straw, (D) Pollination with straw, (E) Tying and fastening of straw to bracts

removed by a pair of forceps, dissecting needle, or special devices adjusted
to the flower type. In cotton (DOAK, 1939) and other plants featuring staminal
columns, a proper size straw tube may be forced over the column to remove
the anthers and isolate the stigma from its own pollen (Fig. 2.4). Simultaneous
excision of stamens and glume apices is possible in rice when, at a certain
developmental stage, the anther level within the spikelet exceeds the level of
the stigmata (see Fig. 2.5).

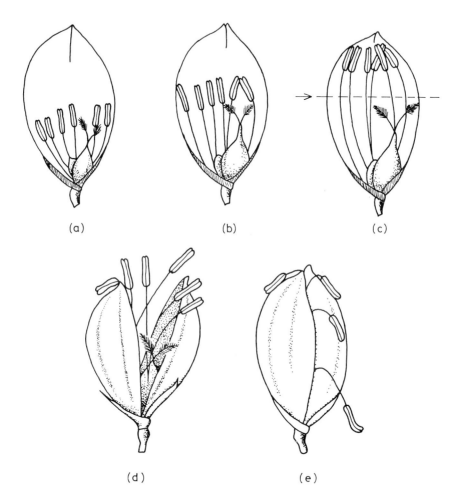

(a)   (b)   (c)

(d)   (e)

Fig. 2.5. Blooming and emasculation of the rice spikelet (modified after COYAUD, 1950).
(a), (b) and (c) stages in the elongation of staminal filaments; (d) spikelet opening; (e)
spikelet closure. Broken line indicates simultaneous excission of stamen and glume apices

In wheat, barley and other grasses the level of undehisced anthers within
the florets relative to the stigmata does not permit clipping off the anthers

without cutting part of the stigmata. However, florets may be clipped with scissors just below the anther tips without damaging the stigmata. Such mutilation of the anthers, called "scissor emasculation", is in some cases quite reliable, and does not require removal of the rest of the anthers when carried out several days before anthesis. The failure of the lower portion of the anthers to form viable pollen after the removal of the top portions is probably due to the destruction of the pollen by desiccation (WELLS and CAFFEY, 1956). Scissor emasculation is less time consuming than the more common method of removal of anthers or anther remnants with a small forceps after clipping off the tip of the floret to gain access to the anthers (POPE, 1944; JENNINGS et al., 1964). The latter procedure is the much more reliable one, and, therefore, preferred in practice. In some cases emasculation of grass flowers can be accomplished without clipping florets. The glumes may be opened by applying pressure with a pointed instrument (e.g. a pencil point) and rolling unbroken anthers out from the lemma and palea (HARRIS, 1955).

In small-flowered species, such as flax and soybean, it is common to remove the corolla and even the sepals to facilitate the tedious operation of stamen excision by fine-pointed forceps. Suction produced by small vacuum devices has been used for the effective emasculation of small-flowered legumes (KIRK, 1930; STEVENSON and KIRK, 1935) and also for a rapid emasculation rate in rice (CHANDRARATHNA, 1964). Anthers are aspirated off the opened flowers. In large-flowered legumes (e.g. common beans, see Fig. 2.6) it is sufficient to open only one side of the bud by careful removal of one half of the keel and exsection of the anthers (BUISHAND, 1956; BOLING et al., 1961; SINGH and MALHOTRA, 1975). In certain sympetalous flowers in which stamens are adnate to the corolla, the entire corolla may be pulled off together with the stamens. This may be done, for instance, using small scissors with a triangular notch (BARRETT and ARISUMI, 1952; CARVALHO and MONACO, 1969). In the cultivated tomato the stamens are coalesced to form an anther cone and filaments are adnated at their base to the corolla. Since joint vascular bundles lead to the corolla and anther cone at the base of the petals (Fig. 2.7), the base of two petals can be grasped by a straight forceps with flattened tips and a gentle pull can remove and slip the combined corolla and undehisced anther cone over the style (BARRONS and LUCAS, 1942; LAPUSHNER and FRANKEL, 1967).

In a small-flowered species it is extremely difficult to emasculate by forceps, needles or even aspiration. Removal of pollen from stamens and stigmata can be achieved by fine jets of water, or mist spraying. This procedure has been found quite efficient in the emasculation of protandreous flowers such as those of lettuce (RYDER and JONSON, 1974).

2.3.1.2 Male Gametocide

Emasculation may be based upon differentials in the sensitivity of sporophylls to unfavorable conditions. The androecium is in general more tender than the gynoecium, and microsporocytes in the pollen sac are less protected by the

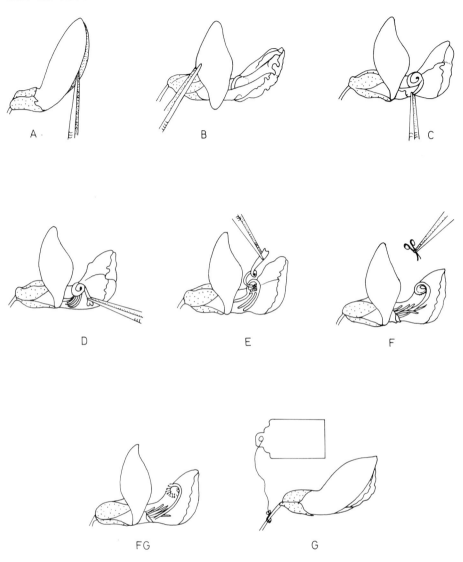

Fig. 2.6. Crossing technique for beans. (A) the standard is detached and (B) unfolded; (C) the left-hand wing has been removed, and the keel is being removed piecemeal (D and E); (F) the stamens have all been removed; (FG) the desired pollen is rubbed into the stigma; (G) the flower is folded and provided with a number (redrawn after BUISHAND, 1956)

anther walls than megasporocytes by the wall of the ovary and the protective layers of the ovule (integuments). Furthermore, the most sensitive stage in micro-sporogenesis (usually meiosis) often does not coincide with the most sensitive stage in megasporogenesis.

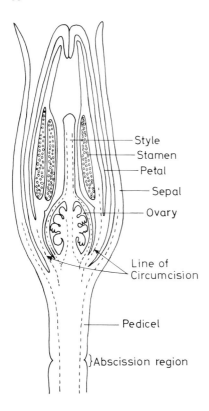

Fig. 2.7. Diagrammatic longisection of a tomato flower prior to anthesis i.e. proper stage for emasculation. Vascular bundles indicated by broken lines

— Style
— Stamen
— Petal
— Sepal
— Ovary

Line of Circumcision

Pedicel

}Abscission region

A variety of chemicals, mainly growth retardants, have been tested for selective gametocidal action after differentiation of the sporogeneous tissue. Ethyl alcohol has been found efficient for emasculating open, tripped alfalfa flower buds (TYSDAL and GARL, 1940). Spray applications of sodium 2,3 dichloroisobutyrate (DCIB) cause male sterility after 2 weeks in cotton (EATON, 1957) and tomato (MOORE, 1959, 1964), and after 4–5 weeks in clovers and beet (WIT, 1960). Male sterility was induced after 2 weeks by sodium dichloroacetate in snapdragon (KHO and DE BRUYN, 1962), by maleic hydrazide in tomato (REHM, 1952), by sodium 1-(P-chlorophenyl)-1,2-dihydro-4,6-dimethyl-2-oxonicotinate in wheat (JAN et al., 1974) and by other growth retardants.

Although male sterility can be induced by chemicals without serious harm to female fertility, no chemical gametocides are yet used in practice. The main difficulty in the use of chemical gametocides is correct timing of application. Proper timing depends not only on the exact developmental stage in microsporogenesis, but also on absorption, movement and persistence of the chemicals. Furthermore, not all flower buds in a plant develop synchronously and repeated treatments, often harmful to the plant or female fertility, are usually required.

To overcome these difficulties a number of growth substances have been tried to modify early differentiation of the tissues of the androecium, or to modify sex expression rather than to kill microsporocytes. This subject will be taken up in Chapter 3 of this book. In addition to the modification of differentiation or sex expression, some growth substances may cause abnormalities in microsporogenesis. Thus, 2-chloroethylphosphonic acid, often promoting changes toward femaleness when applied early in the development of the plant, will, when applied during meiosis, also cause male sterility by inducing abnormal additional nuclear divisions in pollen grains of wheat (ROWELL and MILLER, 1971; BENNETT and HUGHES, 1972), rice (PERETZ et al., 1973) and other plants (MACDONALD and GRANT, 1974). Another growth regulator, gibberellin, will act sometimes as a male gametocide (e. g. in onion—VAN DER MEER and VAN BENNEKOM, 1973; or sweet corn—NELSON and ROSSMAN, 1958). Functional interference to self-pollination can also be achieved by gibberellin-induced heterostyly, in which the stigma extends beyond the stamens before pollen shedding (e. g. in tomato—BUKOVAC and HONMA, 1967).

A somewhat obsolete method of emasculation is employment of temperature as a selective gametocide. Immersion in hot water has been used to emasculate flowers at or after meiosis. Water temperatures selectively lethal to pollen of sorghum (QUINBY and MARTIN, 1954), rice (CHANDRARATHNA, 1964) and other grasses (KELLER, 1952) range between 42° and 48°C and immersion periods range from 1 to 10 min, depending on the species. The differential sensitivity of the androecinum and gynoecium of wheat (RUDORF and JOB, 1934) to low temperatures can be used for emasculation (SUNESON, 1937). Immersion of rice panicles for 10 min in cold water (0°–4°C) has also been found in certain cases to be an effective emasculation method (CHANDRARATHNA, 1964).

2.3.1.3 Circumvention of Emasculation Requirements

Hybridization of autogamous plants without emasculation may be achieved by preanther-dehiscence pollination or by "genetic emasculation", i. e., utilization of male sterility. The utilization of male sterility will be dealt with in Chapter 3 of this book.

Preanther-dehiscence pollination is possible when the stigma receptivity precedes pollen shedding sufficiently to allow pollination and pollen germination before anther dehiscence. This method is useful in particular for small-flowered species in which the expenditure in labor and the damage to the flowers make hand emasculation inadvisable. Such a situation exists in the chickpea. In the small papilionaceous flowers of this crop, nine stamens with fused filaments are placed ventrally around the ovary, leaving the tenth stamen free, situated dorsally. Anthers deposit pollen on the stigma within the keel when they grow past the style. Hand emasculation is very inefficient and laborious (VAN DER MAESEN, 1972). Flower buds may be opened carefully with forceps and pollinated when the petals are just starting to come out of the calyx. At this stage anthers reach only about half the height of the pistil, and upon pollination fruit set reaches normal proportions. RETIG (1971) found that two-thirds of the seed were crosses, as indicated by marker genes. It is obvious that such hybridization without emasculation must make use of marker genes to identify the crossed progeny.

2.3.2 Controlled Pollination

2.3.2.1 Isolation

To avoid contamination by unwanted pollen, flowers prepared for hybridization
as well as flowers serving as pollen sources must be isolated. Isolation in strict
autogamous plants is less critical than in plants showing some NCP. In entomophilous flowers, absence or exclusion of vectors by screens, nets, etc., may suffice
for isolation, if care is taken to avoid contamination during the manipulation
of flowers and pollen. For anemophilous flowers, paper, plastic or tightly woven
cloth enclosures must be used and care be taken not only to avoid accidental
contamination during manipulation but also by airborne pollen during emasculation, pollen collection, and hybridization.

Isolation may be accomplished by enclosure of stigmata, single flowers, inflorescences, single plants or groups of plants. Isolation of large single entomophilous
flowers may be accomplished by simply tying or clamping petals and/or sepals
in order to exclude pollen vectors. With plants having entomophilous flowers
there are generally no aeration or temperature problems in growing plants
in screened enclosures. For small plants, cages covered with nylon stockings
are convenient. Larger cages may be built of fabric bags supported by a collapsible
skeleton of galvanized wire rings. Such bags can be suspended by a string
closing the top of the bag and tied to a wire or structure above the bag (see
KOBABE, 1965). Anemophilous flowers or inflorescences (e. g. grasses) are usually
isolated by paper, vegetable parchment, glassine, cellophane or plastic bags
(e. g. synthetic sausage casings). If whole plants or groups of plants are to be
isolated, filtered air must be supplied to maintain the suitable environment.
Polyethylene bags may be inflated by a continuous stream of filtered air escaping
through a second filter in the wall of the bag (KNIGHT, 1966; ENGLAND, 1972).
Such bags may be fitted with a suitable access for manipulation which can
be closed airtight, e. g. with screw caps (HARRIES, 1972). Likewise, positive pressure
(inflated) plastic greenhouses exclude airborne pollen if the inflating air is filtered.

2.3.2.2 Pollen Collection and Storage

In many cases pollen is functional already before anther dehiscence and, if
convenient, may be extracted from the closed anther by a dissecting needle
or spear-headed scalpel and transferred directly to the stigma. Anthers can
be surface-sterilized, thus making isolation of the pollen donor unnecessary.
Similarly, artificial pollination can be accomplished by *direct transfer* of pollen
from dehisced anthers of the pollen donor to the stigma. The stigma may be
touched by the anther or whole flower of the pollen donor, or pollen may
be applied by means of a transfer tool (forceps, needle blackened toothpick,
camel brush, pencil point, dental spoon, etc.). Direct transfer of pollen may
be very convenient for small numbers of hybridizations, but when used on
a large scale it is more time consuming and requires a larger number of nicked

pollen donor flowers than the use of pollen previously extracted by a pollen-collection device. The amount of pollen collected can also be stretched by diluents (see STANLEY and LINSKENS, 1974).

Pollen collection means induction of, or waiting for anther dehiscence and gathering the pollen before it is dispersed. If anthers are collected undehisced, surface-sterilized, and induced to dehisce under controlled conditions, isolation of the pollen donor plant is not required (PETRU et al., 1964). Inflorescences, flowers or anthers may be removed from the pollen donor a few days before anther dehiscence and allowed to mature under controlled environmental condition in well-aerated trays, canvas bags, sausage casing, etc. Pollen may be shaken off the dehisced anthers (e.g. WRIGHT, 1962; MIRAVELLE, 1964) to settle on the bottom of the tray, on paper, sheet glass, a watch glass, beaker, vial, etc., or be separated from the anthers by the electrostatic charge between the pollen and the dry polyethylene wall of the container while shaking. Certain entomophilous flowers may be rubbed on a 16 mesh/in screen to strip the anthers from their filaments. Anthers fall through the mesh. After maturation and dehiscence, anthers are transferred onto a 60 mesh/in screen placed over a sheet of window glass, and rubbed. This frees the pollen, which falls through the mesh onto the glass. Pollen can be scraped off with a razor blade into the storage container (BARRETT and ARISUMI, 1952). A common practice for the anemophilous grasses (e.g. sorghum, corn) is to shake pollen donor inflorescences upon dehiscence within their paper or plastic isolation bags in order to dislodge the pollen, which will become attached to the wall of the bag or dispersed in the air within the bag. The bag is then put over the female or emasculated inflorescence to accomplish the simultaneous pollination of many flowers.

Efficient and large-scale separation of pollen from dehisced anthers may require mechanical devices to dislodge and separate the pollen. A large variety of shaking, vacuum, sieving and electrostatic instruments are used to fit the particularities of different flowers and pollen. Pollen-collecting apparatus designed for use with tomatoes are good examples of the application of the *vibration principle and gravity separation* (Fig. 2.8). Portable vibrators may be used to extract pollen from flowers on the plant. Pollen may be collected in glass tubes (COTTRELL-DORMER, 1945), gelatine capsules (MARTIN, 1960; CRILL et al., 1970) or small cups (MARCHESI, 1970). However, stationary extraction of previously collected flowers is more efficient (LAPUSHNER and FRANKEL, 1967) and stationary shaking devices are used in large-scale hybrid seed production by seed companies (e.g. DOROSSIEV, 1962).

Vacuum aspiration of undehisced anthers (KING, 1955) or of pollen from dehisced anthers is a common method in pollen collection. There are no difficulties in separating anthers from the air stream, but separation of pollen from the air stream must be accomplished by filtration, centrifugal force, or air draft-buoyancy gradients (Fig. 2.9). A filter trap may be made of a fine mesh sieve, nylon cloth or fritted glass (BARRETT, 1969). Filters get easily blocked by accumulating pollen. Pollen separation utilizing centrifugal force does not have this shortcoming. Centrifugal separation can be accomplished by "cyclone-type" separators (LINDEN, 1949; OGAWA and ENGLISH, 1955) or "inertia-type" traps (IZHAR et al., 1975). If necessary, the differential buoyancy of aspirated pollen and

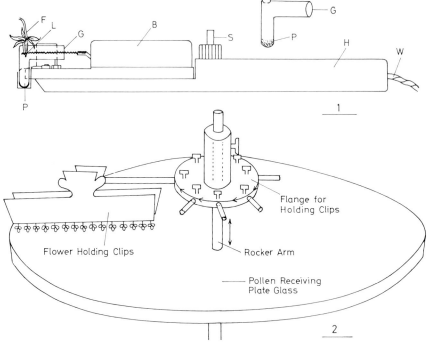

Fig. 2.8. Examples of vibratory pollen collectors. (1) *Portable vibrator*; B: Buzzer, L: Vibrator loop, F: Position of flower during pollen extraction, S: Switch, H: Handle, W: Wire to battery, G: Glass pollen chamber, P: Pollen collection area. (2) *Stationary vibrator*

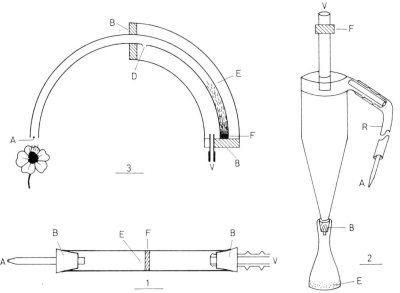

Fig. 2.9. Examples of vacuum aspiration pollen collectors. (1) Fritted glass filter trap type (BARRETT, 1969), (2) Cyclone separator type (after OGAWA and ENGLISH, 1955), (3) Inertia trap type (after IZHAR et al., 1975), A: intake, B: rubber stopper, C: cyclone, D: aspirating aperture, E: pollen collection area, F: filter, V: suction source, R: flexible rubber tubing

dust may be used to fractionate the functional pollen from foreign matter and dead pollen grains.

Separation of dry, nonsticky pollen from dehisced anthers and from fresh or vacuum-dried and crushed flowers can be achieved by sieving. The uniform size and shape of the pollen is a good basis for *screen separation.*

For maximum yield of highly viable pollen, collection should take place under optimal growing conditions for pollen donor plants. Pollen, being generally more sensitive than the female gametophyte, can often be used for pollination under marginal conditions if collected and stored under favorable conditions. Thus, the season and flexibility in large-scale hand hybridization schedules can be extended by pollen storage. Proper conditions for pollen storage and pollen viability tests have been outlined in detail by STANLEY and LINSKENS (1974). Pollen longevity is discussed in Chapter 1.3.3 of this book.

2.3.2.3 Pollen Transfer Methods

2.3.2.3.1 Utilization of Natural Pollen Vectors

Hand pollination is a time-consuming operation and in commercial hybrid seed production usually not economical. Utilization of natural pollen vectors for controlled pollination saves the work of pollen collection and hand hybridization. However, prerequisites for such controlled pollination are numerous: seed parents must be emasculated or self-sterile; pollen parents must be "nicking"; the pollen supply must be abundant; the environment must be vector-saturated; and discrimination of the seed parent by the pollen vector must be absent. In autogamous plants, which do not feature barriers to self-pollination, seed parents must always be emasculated mechanically, chemically or genetically. Seed and pollen parents may be grown in pairs, in small or large groups of plants, under vector-saturated enclosures; or be isolated by distance from potential contaminating sources. Potted and emasculated seed parent plants may be put into a pollen parent field to produce hybrid seed.

Good illustrations of problems associated with pollen transfer in autogamous, but inherent anemophilous plants, are wheat and rice. After the discovery of genetic male sterility and fertility-restoring genes, cross-pollination remains the main limiting factor in economical F_1 hybrid seed production. Genetic floral characteristics of pollen parents promoting cross-pollination include large and extruded anthers with profuse pollen production, ear or panicle level above that of the seed parent, and proper nicking (DE VRIES, 1972, 1973; VIRMANI and ATHWAL, 1973, 1974). Generally, awnless and tall wheats release more pollen than awned and semidwarf wheats (OLSON, 1966). Seed parents should feature long stigmata, a high degree of stigma exertion, and proper flowering pattern and duration. Since pollen dispersal is inversely proportional to the distance from the pollen source and seed set decreases rapidly beyond a distance of 3 m (see e. g. KHAN et al., 1973), the planting design is of extreme importance. The distance from the pollen source and consequent width of the male-sterile strips, the overall ratio of pollinator to male-sterile plants in the field, and the upwind location of the pollen parent, influence the extent of seed set on the male-sterile plants. In experiments by DE VRIES (1974), a 2-m male-sterile strip and a 1:2 ratio of pollinator to male-sterile plants yielded the highest quantity of hybrid wheat seed (Fig. 2.10). For the production of hybrid sorghum seed, a 1:3 ratio of a succession of 12 male-sterile seed parent rows and four pollen-parent rows is customary, indicating a broader dispersal of sorghum pollen (QUINBY and SCHERTZ, 1970).

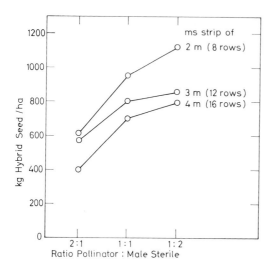

Fig. 2.10. Hybrid wheat seed yield (kg/ha to be expected at pollinator: male sterile ratios 2:1, 1:1 and 1:2 and male sterile strip widths of 2, 3 and 4 m). Sowing rate ms seeds: 22 seeds/m (after DE VRIES, 1974)

With autogamous, inherently entomophilous plants the genetic floral characteristics promoting cross-pollination in the open field are those making them attractive or accessible to the vector. Thus, e. g. in flax, the degree of floral opening (disk-form flower) of the pollen and male-sterile seed parents is important (COMSTOCK, 1965; DUBEY and SINGH, 1966), and in male-sterile tomato, stigma exertion beyond the mouth of the anther tube is naturally a prerequisite to insect-mediated crossing (see RICK, 1950). Genetic variation in attractiveness, based on quality or quantity of the vector or of the pollen offered by the crop and the individual flower, may be exploited or must be avoided as in the case of extra-floral nectaries in cotton, distracting the vector's attention away from the flowers.

The absence of pollen in the male-sterile seed parent may cause difficulties in utilizing *voluntary* pollen vectors for hybridization. If nectar is secreted in both the pollen and male-sterile seed parent (e. g. in cotton, pepper, flax, tobacco, field beans), voluntary vector activity could be based on nectar gathering (see, e. g. BOND and HAWKINS, 1967). However, if no nectar is secreted (e. g. in tomato), voluntary vectors will soon ignore male-sterile flowers and restrict their activity to pollen-gathering from pollen-fertile flowers only. Thus, we can employ *involuntary* pollen vectors (e. g. thrips, beetles or wind) or increase artificially the frequency of "trial and error" visits of voluntary pollen vectors. To mislead pollen-gathering bees, seed of the two parents may be mixed or be planted in close alternate rows so that discrimination between male-sterile and fertile plants before entering the flower becomes arduous. If pollen-parent plants cannot be roughed from the field or harvested separately, hybrid seed can be sorted from male parent seed if genetically marked by color or size (e. g. in onions, DAVIS, 1966; in field beans, BOND and HAWKINS, 1967; in flax, BARNES et al., 1960), or hybrid plants can be selected on the basis of seedling marker genes (e. g. in peppers, CH. SHIFRISS, pers. comm.).

Number, location and condition of honey bee colonies or of other potential pollen vectors decisively influence efficiency of crossing. For best results, vectors could be conditioned or trained to the target crop. Conditioning includes moving bee hives to the crop not before flowering starts, and to replace colonies which start foraging on crops other than the target (FREE, 1970). In some cases it is possible to direct and train bees for a particular crop by "scent feeding" (VON FRISCH, 1947, 1967). A number of techniques have been advocated to increase pollination activity and efficiency of honey bees: pollen trapping (LIN-DAUER, 1952), sugar syrup feeding (FREE, 1965), and pollen-dispensing devices ("pollen inserts"), hopefully charging outgoing foragers with the proper pollen (TOWNSEND et al., 1958; GRIGGS and IWAKIRI, 1960; JAYCOX and OWEN, 1965).

Pollen may also be dispensed in the open field or within isolation enclosures, and transferred by natural vectors by placing detached inflorescences ("bouquets") of the pollen parent near the seed parent. This method has been used since the early domestication of dioecious (or functional dioecious) fruit trees to insure proper pollination in the anemophilous date (OUDEJANS, 1969) and the entomophilous fig (FREE, 1970); it is also used to a certain extent in deciduous tree orchards (LÖTTER, 1960). With autogamous anemophilous plants, detached inflorescences are commonly used in the so-called "approach crossing" techniques. In this technique, used in wheat (CURTIS and CROY, 1958), barley (ROSENQUIST, 1927), oats (MCDANIEL et al., 1967) and rice (ERICKSON, 1970), the pollen donor inflorescence is put slightly above the emasculated inflorescence and bagged or enclosed in opposite ends of a plastic sleeve. The cut ends of detached inflorescences are sometimes placed in water containers to achieve normal pollen shedding. Handling can be facilitated by growing plants in pots and several inflorescences can be enclosed under the same isolation bag or cage.

Approach crossing with entomophilous plants in enclosures is used in plant breeding to save hand pollination, avoid contamination in the open field, and speed up the breeding program (by utilizing out-of-season time to grow additional generations in greenhouses, etc.). Problems arising with the confinement of the insect vector are usually the only ones in the controlled hybridization of allogamous plants (see e. g. KRAAI, 1954). Bumble bees work well in confinement (e. g. with broad beans; BERTHELEM, 1966); honey bees are easier to obtain, but may require larger enclosures than bumble bees (FREE, 1970); flies are used extensively for hybridization in small isolation cages of myophilous crops such as onions (JONES and EMSWELLER, 1934), carrots (BORTHWICK and EMS-WELLER, 1933) and Brassicaceae (WIERING, 1969). Caging autogamous, inherently entomophilous crops with pollination insects which have been cleansed first of contaminating pollen, appears to work satisfactorily in crops with a slight NCP. Thus, male-sterile cotton produced 84 % of the seed of the maintainer line when caged with honey bees in a relative large cage (MOFFETT and STITH, 1972), and good seed set was obtained on single male sterile pepper plants caged with fertile maintainer and houseflies in small screen enclosures (unpublished observation). Lack of attraction and timing of stigma receptivity make attempts to utilize insects for the pollination of many complete selfers, such as tomatoes, soybean and lettuce, problematic. Approach crossing of fruit trees can be facilitated by grafting scions of one variety into an older, isolated tree of a second variety using the scion as the female parent (see e. g. BERGH and STOREY, 1964).

2.3.2.3.2 Forced Pollination

Forced pollination may be carried out by hand or compressed air devices. Large scale application of anemophilous pollen can be made easily by compressed air dusters. The difficulties posed by placing male inflorescences into high date palms and the waste of pollen in such procedure brought about the use of a variety of ground level equipment (OUDEJANS, 1969; VIS et al., 1971) and fixed wing aircraft or helicopters (BROWN, 1966) for the application of pollen. In hybridization programs with inherently anemophilous plants, flowers of the seed parent can be pollinated by pollen dispersed into the isolation bag with the help of polyethylene puffer bottles, glass tubes and blowers, injection syringes, insecticide pressure sprayers or dusters and the like (see e. g. BARRETT and ARIZUMI, 1952; BÉNARD and MALINGRAUX, 1965; TAMMES and WHITEHEAD, 1969).

For economical use and even distribution, pollen dilution by inert substances such as wheat or cornflour, kaolin, carbola, magnesia or other talcs, lycopodium powder, casein, egg albumin, nutshell flour, ground cellulose, glass beads, dead pollen and other diluents have been tried. Basically, diluent's particles should not stick to the pollen, not lump, be of similar size and weight as the pollen grains, not be hygroscopic and be inert to the pollen and stigma. The degree of pollen dilution must be determined in each case and depends on pollen viability, stigmatic surface, number of seeds per fruit, climatic factor, method of application and physical characteristics of the pollen and diluent. Low pollen filtration ability of stigmata and often high numbers of seed per fruit common in entomophilous plants, decrease effectiveness of forced, overall pollen dispersal. Nevertheless, in certain cases, increased fruit set has been claimed as a result of forced pollination by diluted pollen of apples, grapes and other fruit trees (OVERLEY and BULLOCK, 1947; PFEIFFER, 1948; GARDNER, 1966).

Economy of pollen use, provided by pollen dilution, is of special importance in artificial hybridization programs and in the production of high value hybrid seed (e. g. tomato, petunia) when pollen must be applied to each individual stigma. The pollen may be applied to the stigma by transfer tools such as mentioned above or, preferably, directly from a pollen collection or storage container in the form of a vial, or J shape or straight glass tubes (Fig. 2.11).

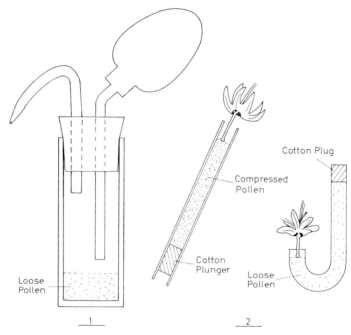

Fig. 2.11. Pollen containers used directly as pollen application tools. (1) Puffer bottle for anemophilous plants. (2) Glass tubes for entomophilous plants

The use of the latter is less wasteful of pollen and less time consuming than the use of intermediary tools transferring pollen from the pollen container to the stigma. The flow chart in Fig. 2.12 outlines methods in manual pollen transfer.

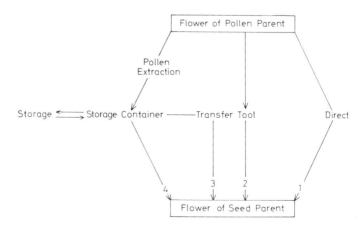

Fig. 2.12. Chart of manual pollen transfer methods. 1: Flower→Flower, 2: Flower→transfer tool→flower, 3: Flower→storage container→transfer tool→flower, 4: Flower→storage container→flower

Marking of artificially pollinated flowers or flower clusters is necessary when the seed parent is pollen fertile. Removal of flowers or fruits from uncontrolled pollination, though sometimes required to obtain optimal fruit set and seed count from the controlled pollination, is usually not reliable enough to assure harvest of pure hybrid seed. Flowers can be marked by clipping part of the calyx, by paint, pieces of plastic covered wire, tags and the like.

Chapter 3. Allogamy

3.1 Sexual Reproduction—Structures and Functions

In this chapter we shall turn our attention to the flower. We shall look into some of the structures and functions of the anther and the male gametophyte as well as of the ovule and the female gametophyte. We shall concentrate on those features which will help in a better understanding of sex expression, self-incompatibility and male sterility. For more information and references, the reader should consult such specialized texts and reviews as those of MAHESHWARI (1950) and HESLOP-HARRISON (1972 and 1975). Sexuality, reproduction and alternation of generation in plants were treated thoroughly by several authors in volume 18 of the Encyclopedia of Plant Physiology (RUHLAND, 1967). We shall handle only angiosperms in the first four sections of this chapter. The gymnosperms will be represented by the conifers, which comprise many economically important forest trees. A brief description of the sexual reproduction of the conifers will be given in the last section of this chapter.

3.1.1 The Anther and the Male Gametophyte

3.1.1.1 Differentiation of the Anther

The anther is the particular part of the stamen responsible for the production of microspores; the latter develops further into pollen grains. The anther consists, typically, of four elongated *thecae* (Figs. 3.1 and 3.2). By homology to lower plants, the stamen may be termed *microsporophyll* and the anther—*microsporangium*. In the young anther there is a homogeneous population of meristematic cells surrounded by epidermis. Later, during its ontogeny, the anther attains typically a four-lobed structure. Within these lobes an acheosporium is then formed and a further differentiation into two kinds of tissues takes place. One of them—the *primary parietal* layer—toward the outside, the other—the *primary sporogeneous* tissue—toward the center. The former will form the wall of the anther, which may attain in specific species variable thickness, but in all cases the inner layer of cells surrounding the sporogeneous tissue will form a unique tissue: the *tapetum*. This layer plays an important role in most species, in respect to the nutrition of the microspores. According to recent detailed work reviewed by HESLOP-HARRISON (1972, 1975), it has also an important role during the maturation of the pollen grain and is thus involved in certain incompatibility reactions of the mature pollen.

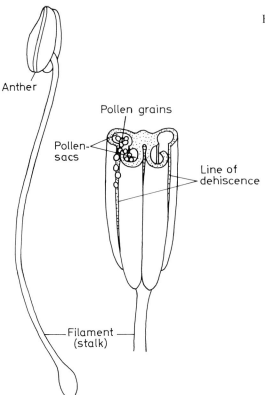

Fig. 3.1. General structure of a stamen

3.1.1.2 The Tapetum

This tissue goes through a programmed sequence of structural and chemical changes in coordination with events in the underlying sporogeneous tissue. The differentiation of the tapetum differs in different plant groups, but in general terms two major types of tapeta can be recognized. In the *parietal* (or secretory) *tapetum*, plasmodesmata previously established among the anther cells are being disconnected in two places: between the tapetum cells and the outer wall cells, and between the tapetum cells and the inner sporogeneous cells. The connections between the cells within the tapetum layer are retained, and at about midmeiotic prophase (see below), these connections are vastly enlarged and cytoplasmic channels are found. Further cuticularization in layers toward the outer wall of the tapetum cells is in some cases established. The inner periclinal wall of the tapetum layer becomes gelatinous and later small spheres of a specific polysaccharide, *sporopollenin*, may be added. Thus, at a certain stage of meiosis in the sporogenic tissue, the whole tapetum can be looked upon as a unit secretory gland with the sporogeneous tissue as its only target. At a later stage, when the tapetum has terminated its nutritional role, it degenerates with its remnants including the sporopollenin spheres investing the pollen grains. This relatively late phase of coating and impregnating the pollen grain with tapetal material is of great significance to the sporophytic type of self-incompatibility.

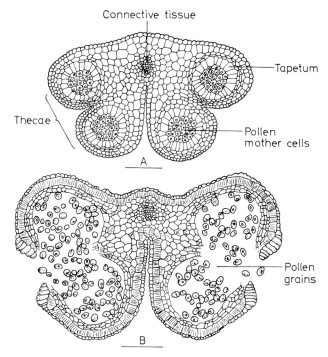

Fig. 3.2 A, B. Cross sections of anthers. (A) at the premeiotic stage; (B) at dehiscence

The *invasive* (or amoeboid) *tapetum* differs principally from the parietal tape-tum in that, in the former a direct contact between the cytoplasmic mass of the tapetum and the sporogeneous cells is established. This type of tapetum is less common but does exist in several monocotyledonous and dicotyledonous families. In species with invasive tapetum, both the walls between the tapetum cells and those between the tapetum and the sporogeneous cells are being comple-tely disintegrated. A continuous cytoplasmic envelope is formed, without an inner wall to separate it from the sporogeneous tissue. When the latter is in a certain phase of the meiotic division (see below), the periplasmodial tapetum mass invades the sporogenic tissue.

3.1.1.3 Development of the Sporogeneous Tissue

The sporogeneous cells undergo a series of changes on their way from the relatively undifferentiated meristematic state through the formation of tetrades and finally pollen grains. We shall review these here only briefly, and emphasize those features which are relevant for understanding sex expression, self-incompati-bility and male sterility. Regarding the alternation of generation, the last phase of the sporophytic generation is attained when the sporogeneous cells become *meiocysts*; these are usually termed *pollen mother cells* (PMC). The differentiation into PMC differs among species. In many species only a fraction of the sporogeneous cells become PMC, with characteristic prominent nuclei and nucleoli; the rest may take the role of nourishing cells (e.g. when the tapetum is only vestigial). When the PMC start the process of meiosis, the transition between the sporophytic and the gametophytic generation ensues.

3.1.1.4 Microsporogenesis

The cytological aspects of this process will be dealt with here rather briefly. The reader is referred to specific reviews (e.g. JOHN and LEWIS, 1973; STERN and HOTTA, 1974) for more details. There are two meiotic divisions, termed *meiosis I* and *meiosis II*. The first is usually divided into five major phases: *prophase, metaphase I, anaphase I, telophase I* and *interphase*. The first of these is the longest and is subdivided into six phases. Meiosis II is divided into three phases: *metaphase II, anaphase II* and *telophase II*. Obviously the process is a continuous one and these phases serve merely as references. The cytologic process differs among plant species but finally, after telophase II, four separate cells, forming the *tetrad*, are established in almost all cases. These are the microspores. The name tetrad stems from the observation that the cells usually occupy the four corners of an imaginary tetrahedron. With the end of meiosis (formation of the microspore) we may speak about the male gametophyte. Its 'sex', as mentioned above, is imposed on it by the character of the sporangium—the anther.

Certain features of the process starting from the diploid ($2n$) PMC and terminating with four haploid (n) microspores, the tetrad, should be noted. These features, although up to now observed in detail in only a few plants, are probably common to angiosperms and characteristic of, if not essential to, the successful development of the gametophyte. First, at a stage of premeiosis, specific in detail for each species, the PMC of each loculus join into a synchronic state. At this phase there are already cytoplasmatic connections between these cells in the form of plasmodesmata. At about midmeiotic prophase these connections start to widen and cytoplasmic channels are produced. The channels develop further and then cover a rather large fraction of each cell's surface—up to 1/5. By this, all the PMC of each loculus produce a unit syncytium which not only goes through meiosis in synchrony but is physically intraconnected, so that when pressed out from the anther it stays in the form of one mass of tissue. The increase of intracellular connections in the PMC parallels the formation of similar connections in the tapetum, as mentioned earlier. Contrary to older claims, there is no evidence that through the cytoplasmatic channels there is an interchange of chromatin *("cytomix")* among the PMC. There is evidence for protoplasmic streaming through these channels and it is not ruled out that cytoplasmic organelles may pass from cell to cell. The cytoplasmic connections are closed at a stage which varies according to species—between *metaphase I* and *telophase II*—so that, at the latest, with the formation of the tetrades the connection between the young microspores is terminated. *Callose*, a polysaccharide specific for the stages leading to the formation of the pollen grain, has probably a major role in securing the final isolation of the microspores.

This callose is first deposited on the inner side of the PMC walls at the stage when the massive cytoplasmatic connections between the PMC are formed. We shall return to the synthesis and degradation of callose in more detail while handling male sterility.

Another important feature is the 'cytoplasmic clearing' of the meiocysts. The ribosome population, quite dense in the young archesporial cells is drastically reduced in number in meiotic prophase. Then, still in prophase, a remarkable simplification of plastids and mitochondria takes place. These cells are thus attaining meiosis I, and in some plants meiosis II, with a 'clear' cytoplasm. HESLOP-HARRISON (1972) discussed this interesting phenomenon and regards it as an essential process in the transition between the diploid sporophytic phase and the haploid gametophyte.

3.1.1.5 From Microspore to Pollen Grain

The young microspore has a centrally located nucleus and dense cytoplasm. It is haploid, having half the number of chromosomes and half the amount of DNA in its nucleus, relative to the somatic cells of the same species. The microspore increases its volume and a vacuole, occupying a considerable portion of its volume, is formed. The nucleus may divide almost immediately upon microspore formation, which is the case in most species, or the division is delayed for days or even months (the long delay occurs in tree species of the colder regions, e. g. *Betula odorata*).

The division of the microspore gives rise to two kinds of cells: the vegetative cell-occupying most of the microspores' volume, and the much smaller generative cell (Fig. 3.3). The latter cell normally takes a position near the cell wall of the microspore but may later move to a more central position. The spatial arrangement of the mitotic spindle in this asymmetric division is not random; there are factors which regulate this arrangement but the discussion of this subject, although of considerable interest, is outside the scope of this book.

At the end of the asymmetric division we are confronted with two very different cells, each surrounded by its own plasma membrane. We should note that in spite of this difference in structure, the nuclei of these cells most probably share the same genetic formation. As exemplified in Fig. 3.3, the vegetative cell has typically an organelle-rich cytoplasm and its nucleus appears to be metabolically active. The generative cell has a small volume of cytoplasm which is poor in organelles and its nucleus is apparently not metabolically active. The two cells develop within one 'pollen-cell' and have a common cell wall system. Some more details on microspore and 'pollen-cell' development will be given in the section dealing with Androgenesis. We shall confine our subsequent discussion to the pollen wall, as this wall and its inclusions have an important role in the process of pollen germination and are of special interest in respect to self-incompatibility.

In gross terms the pollen grain has two main wall-layers. The first-produced, outer layer, is composed of *sporopollenin*, a material characterized by an oxidative

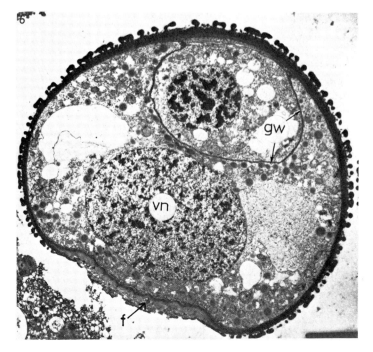

Fig. 3.3. Section through a microspore of *Haemanthus katherina* after first mitotic division.
Note large vegetative cell with vegetative nucleus (vn) situated close to the germination
pore (f) and small generative cell boundered with its plasmalemma (gw). Electron micrograph
(× 4,000) (from SANGER and JACKSON, 1971)

polymer of carotenoids and carotenoid esters. The later-produced, inner layer,
is composed of a *pectocellulosic* material.

The inner layer, the *intine*, is the product of the young gametophyte: its
synthesis starts after the vegetative cell is differentiated. Ultrastructural studies
showed that dictyosomes are located near the cortical layer of the vegetative
cytoplasms. These are probably instrumental in the layering of the intine. However,
the picture seems to be more complicated than the formation of a regular
primary cell wall, as proteinous particles are included in this wall. The intine
is thus not an inert structure, as these proteins include also specific enzymes.
In the intine there usually are one or more germination apertures. The proteins
of the intine are probably involved in the self-incompatibility reaction.

The outer pollen grain layer, the *exine*, shows usually a distinct stratification.
The inner part is termed *nexine*, while the outer—which is typically sculptured—is
termed *sexine*. Most pollen grains have above the nexine a layer of columns
or rods which are mushroom-shaped and termed *bacula*. The 'hats' of these
structures may either be fused to give a roughly closed roof, or unfused. A
structure is thus formed, reminiscent of a hypocaust of the Roman bath. The
space in this hypocaust is closed toward the outside. The outside closure is
not complete, as small pores are revealed by ultrastructural studies (Fig. 3.4).
It is from the outside that tapetal material is impregnated into or deposited

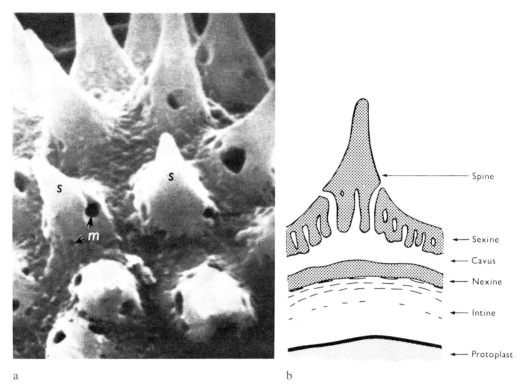

a   b

Fig. 3.4 a and b. Surface of *Cosmos binnatus* pollen. (a) scanning electron micrograph showing spine (s) and microspores (m). (b) diagram of wall stratification, small cavities have been omitted for simplicity (from HOWLETT et al., 1973)

over the pollen wall. This later deposition in most cases includes lipid substances, thus giving the pollen the typical surface layer. It should thus be noted that the pollen wall contains mobile fractions derived in part from the gametophyte, and as this takes place after the cytoplasmic clearing, it is most probably coded by the gametophyte genome; and in part, from the tapetum—therefore, coded by the genome of the sporophyte.

3.1.2 Androgenesis: Production of Haploid Plants by Anther and Pollen Culture

3.1.2.1 Haploid Plants—Occurrence, Induction, and Identification

As pointed out above, the gametophytic and sporophytic generations in higher plants are normally haploid and diploid, respectively. However, haploid sporophytes do exist (e. g. in *Nicotiana, Crepis, Antirrhinum* and *Zea*) and were reported by BLAKESLEE et al. (1922) over fifty years ago. Thus, the haploid condition cannot be regarded as causal for gametophytic differentiation. The potential value of haploid sporophytes for breeding as well as for studying basic questions of plant biology has been mentioned repeatedly since the early years of this century. Reference to the above can be found in the articles and reviews of HARLAND (1955), CHASE (1956), KIMBER and RILEY (1963), MELCHERS and LABIB (1970), COLLINS et al. (1972), and PANDEY (1973).

Haploid sporophytes can originate from either the microspore or one of the haploid cells of the embryo sac. Until 1964 the latter way was the common source of haploid plants. Such plants occurred naturally, usually at very low frequencies, or were induced by facilitating parthenogenesis. Methods were devised to identify haploid plants among progenies composed mostly of diploids. Thus, BURK (1962, 1970) used leaf-color markers to identify tobacco haploid seedlings. STOKES (1963) devised a method to select haploid tobacco plants based on two types of tobacco lines. One line was TMV (tobacco mosaic virus)-sensitive: infected plants could be recognized but were not killed by TMV. The other line was hypersensitive to TMV and killed by infection with this virus. The gene for hypersensitivity (N) is dominant. When a TMV-sensitive plant was pollinated by a TMV-hypersensitive plant, all the diploid progeny produced by normal fertilization were heterozygous and thus killed by TMV infection. Haploid sensitive plants survived and could be rescued, checked for ploidy, and made diploid by colchicine treatment. This method was presented here in some detail because it may, with some amendments, be of general use for the selection of haploid plants. In practice, Stokes' method turned out even in his own hands to be "slow and tedious". DE NETTANCOURT and associates used thermal neutron-irradiated pollen to induce parthenogenesis and leaf markers to select tomato haploids (ECOCHARD et al., 1969). A *Lycopersicon esculentum* line, homozygous for three leaf markers, was pollinated with irradiated *Lycopsicon pimpinellifolium* pollen. Those workers obtained about 20,000 seeds which resulted in 7,400 seedlings, out of which 37 had all three recessive markers but finally only two surviving plants could be verified as haploids. Another method based on interspecific hybridization was developed by KASHA and KAO (1970). It is based on selective elimination of *H. bulbosum* chromosomes from the developing zygote of the cross between this species and *Hordeum vulgare* (barley); thus, haploid barley embryos could be cultured and haploid plants were obtained. These as well as other methods, which can be found in the literature cited above, produced interesting information, but due to inefficiency were generally not very useful in serving as standard aids in plant breeding.

3.1.2.2 The Production of Haploid Callus and Embryoids by Anther and Pollen Culture

The first demonstration of the direct emergence of embryoids from in vitro-cultured anthers resulted unintentionally in an investigation planned for "studying factors which regulate meiosis and cell division". It was reported for *Datura innoxia* by GUHA and MAHESHWARI (1964) of Delhi, who observed that under certain culture conditions, "embryo-like structures projected out of the (cultured) anther from all sides". These investigators inferred, in their first report, that the embryos originated either from the pollen grains or the connective tissue and no mention was made of the ploidy of the embryos; nor were the latter cultured further to give seedlings. Such seedlings were obtained subsequently and they were found to be haploid (GUHA and MAHESHWARI, 1966, 1967). The same investigators were also aware of the potentialities of their discovery but added the remarkable note that "plants raised in this way are bound to be heterogeneous".

The Delhi discovery was apparently forgotten for three years until it was taken up, almost simultaneously, about 4,000 miles east and west respectively of Delhi. Thus, following the above pioneering work, haploids were produced by anther culture in *Nicotiana* (BOURGIN and NITSCH, 1967) and in rice (NIIZEKI and OONO, 1968). While in tobacco as in *Datura* the embryoids seemed to develop directly from the pollen cells, cultured rice anthers produced callus from which, subsequently, plants regenerated. Thereafter, a wealth of reports demonstrated the attainment of haploid callus and plantlets from in vitro-cultured anther and pollen grains. The practice of utilizing anther and pollen grain culture for the attainment of haploid plants (coined by some investigators-androgenesis) became quite fashionable and widespread in respect to both the number of investigators and of plant species in which this method was tried (Fig. 3.5). Albeit, the number of families in which successful embryogenesis was reported is still small. Anther culture was recently reviewed by SUNDERLAND (1973, 1974), who tabulated most of the successful demonstrations of androgenesis up to 1973 and also discussed the culture conditions recommended for the attainment of haploid plants. Table 3.1 is based on SUNDERLAND'S reviews and gives some additional references; it lists 19 genera in which androgenesis was demonstrated. Obviously, this table does not refer to the many unsuccessful attempts made repeatedly by many investigators. Unfortunately, no or only

Fig. 3.5. Androgenesis: emergence of haploid plantlets from an in vitro cultured *Nicotiana tabacum* anther

Table 3.1. Partial list of genera in which an androgenesis was reported[a]

Family	Genus	Reference
Cruciferae	*Arabidopsis*	SUNDERLAND (1974)[b]
	Brassica	SUNDERLAND (1973)[b]; THOMAS and WENZEL (1975b)
Generiaceae	*Saintpaulia*	HUGHES et al. (1975)
Gramineae	*Aegilops*	KIMATA and SAKAMOTO (1971); SUNDERLAND (1974)[b]
	Agropyron	KIMATA and SAKAMOTO (1971)
	Festuca	SUNDERLAND (1973)[b]
	Hordeum	SUNDERLAND (1973, 1974)[b]
	Lolium	SUNDERLAND (1973)[b]
	Oryza	NIIZEKI (1968); WOO and TANG (1972); WOO et al. (1973); SUNDERLAND (1973)[b]
	Secale	THOMAS and WENZEL (1975a); THOMAS et al. (1975)
	Setaria	SUNDERLAND (1974)[b]
	Triticale	SUNDERLAND (1974)[b]
	Triticum	CRAIG (1974); KIMATA and SAKAMOTO (1971); SUNDERLAND (1974)[b]
Liliaceae	*Asparagus*	SUNDERLAND (1973)[b]
Solanaceae	*Atropa*	RASHID and STREET (1974a); SUNDERLAND (1973, 1974)[b]
	Datura	NORREEL (1970); SOPORY and MAHESHWARI (1972); SUNDERLAND (1973, 1974)[b]
	Lycopersicon	SUNDERLAND (1973, 1974)[b]
	Capsicum	SUNDERLAND (1974)[b]
	Nicotiana	BURK (1970); ENGVILD (1974); HARN and KIM (1971); NILSSON-TILLGREN and WETTSTEIN-KNOWLES (1970); RASHID and STREET (1974b); SUNDERLAND (1973, 1974)[b]
	Petunia	SUNDERLAND (1973, 1974)[b]
	Solanum	SUNDERLAND (1973, 1974)[b]; ZENKTELER (1973)

[a] The literature was reviewed up to October 1975.
[b] These are review articles in which reference was made of original papers. As these articles give a thorough coverage of literature on androgenesis, the reader interested in detail is referred to them.

little successful androgenesis was reported up to now in most of the crops of major economic value, such as wheat, corn, soybeans and cotton. Moreover, even in several economic crops where androgenesis was established, the attainment of haploid plants is far more difficult than in *Nicotiana* and *Datura*. Thus, while in some *Nicotiana tabacum* varieties and *Datura* species up to 100% of the cultured anthers resulted in haploid plantlets, only very rarely did anthers of *Gramineae* species (with the exception of *O. sativa*) result in such plants, and even when they did, the plants obtained were in most cases albino. The work of CRAIG (1974) on androgenesis of bread wheat may illustrate the 'state of the art' with such crops. Only one cultivar out of several tested resulted in haploid plants by anther culture. Of this variety, 550 anthers were cultured, after parallel anthers of the same florets were examined for determination of the proper stage. Only one of the anthers resulted in haploid callus which could be regenerated to give nine haploid plants. Recent studies of androgenesis are based on a wealth of experience gained in many laboratories; it is thus anticipated that they will result in expanding the utilization of anther culture in crop plants. Such studies actually resulted recently in success in wheat andro-genesis (PICARD and DE BUYSER, 1975; RESEARCH GROUP 301, 1976).

3.1.2.3 Pathways of Pollen Embryogenesis

It seems now that in most or in all cases in which cultured anthers resulted in embryoids, the latter originated from the haploid gametophytic cells rather than from the diploid tissue of the anther. In cases where callus rather than embryoids result from anther culture, the gametophytic origin is probably also the rule, but this question should be answered in each case in a satisfactory way, e. g. by the use of anthers from plants which are heterozygous for a morpholo-gical marker.

Even in the most successful androgenesis, only a very small number of the microspores are involved in embryogenesis. Other microspores may germinate within the cultured anther, pass through stages which do not lead to androgenesis, or just stay at the same stage as they were at the beginning of culture and ultimately degenerate. Moreover, out of those microspores which start on the right pathway, only very few ultimately produce embryoids.

The early androgenetic cell divisions take place within the intact intine; the latter then bursts and the spherule of emerging cells will either develop directly in an organized way through embryoids to plantlets (common in *Datura* and *Nicotiana* and reported for *Atropa*, *Oryza* and *Petunia*), or these spherules will produce callus and the callus may ultimately differentiate into plantlets (e. g. *Asparagus*, *Brassica*, *Hordeum*, *Lycopersicon* and *Solanum*). One should keep in mind that our information is still incomplete; differences in the above pathways may occur within the same genus and even among varieties of the same species. Culture conditions may affect not only the degree of successful androgenesis but also cause changes in the pathway. Figure 3.6 shows schemati-cally several pathways of androgenesis in *Datura*. From these schemes it should be noted that when the first results of anther culture are callus spherules rather than embryoids, the former may fuse before plantlet differentiation. In hetero-geneous donor plants this fusion may result in chimeras.

The detailed pathways of pollen embryogenesis are being revealed by recent studies (see SUNDERLAND and DUNWELL, 1974; RASHID and STREET, 1974b). These are as yet based on only a very small number of species. The short description given below will thus serve for general orientation; for details, the reader should consult the original publications.

Pollen embryogenesis can be classified into two main pathways (A and B) and an additional, probably less common pathway (C). In these three pathways the diversion from the normal sequences of gametophyte ontogeny does not occur before midphase 2 of the microspore (Fig. 3.7). We may recall that at this stage the microspore is enclosed within the intine. It has a large vacuole which occupies most of the cell's volume; the nucleus is commonly situated near the cell wall and has a prominent nucleolus.

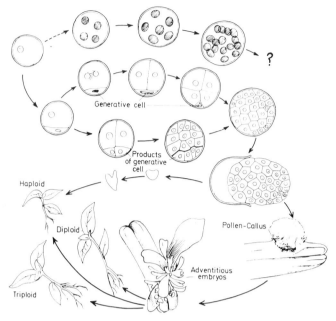

Fig. 3.6. Scheme of possible development of embryoids from *Datura innoxia* microspores following anther culture (from IYER and RAINA, 1972)

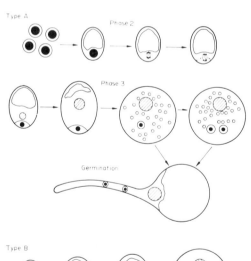

Fig. 3.7. Developmental sequence in angiosperm pollen. *Type A:* common in vivo sequence. *Type B:* probably an anomalous in vivo form which may occur during in vitro culture of anthers (from SUNDERLAND, 1973)

In the *type A pathway* the first mitotic division of the microspore goes to normal completion. This asymmetric division results in two cells: (1) the small generative cell, which is very poor in cytoplasm and most of its volume is occupied by the generative nucleus and is located near the intine; and (2) the vegetative cell, which includes a diffuse nucleus, most of the cytoplasm and the vacuole (Fig. 3.3). The microspore then passes a period of "reorganization" during which there seems to be a clearing of cytoplasmic organelles, reminiscent of the previous clearing in mid-prophase of meiosis. It was even supposed that specific hydrolytic enzymes in vacuole-like vesicles have an active role in the destruction of polysomes, mitochondria and plastids. Thereafter, the vegetative nucleus, which also changed during the reorganization period, passes through a normal mitosis and organelles are reestablished. The generative cell is 'mute' and does not take part in further development. Further normal division will lead to a multicellular structure surrounded by the intine.

In *the type B pathway* the diversion from normal development starts right after formation of the uninuclear microspore. Rather than an asymmetric mitosis, two similar cells are produced by the first mitotic division and no generative cell occurs. Frequently, incomplete cell plate formation and mitotic figures indicated that more than one chromosome complement combined in the same cell. These anomalies may result in partial diploidy or even polyploidy of the cells included in the microspore of the type B pathway. These two pathways are presented diagramatically in Fig. 3.8.

It is obvious that both type A and type B pathways do occur during anther culture. The disagreement concerns only the question as to which of these leads finally into embryoid formation. The argument may be settled if in anthers where *all* microspores have passed mitosis, embryoids will form upon in vitro culture. If this will be affirmed, *type A* pathway will be established at least as one of the two possibilities.

Type C pathway was observed in *D. innoxia* but may occur in additional species. This pathway begins as type A, but the generative nucleus takes part in some embryogenic events, i.e., during some of the mitoses generative and

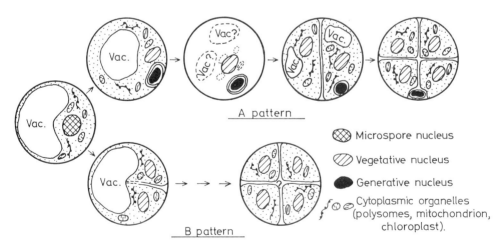

Fig. 3.8. Two possible pathways (*A pattern* and *B pattern*) of androgenesis in microspores of in vitro cultured anthers

vegetative chromosomes probably share the same spindle. The significance of this pathway to androgenesis is not yet clear.

From the viewpoint of the application of androgenesis to plant breeding, which shall be discussed briefly below, the three pathways differ in an important aspect: while type A probably results usually in haploid embryoids, the other two pathways may lead, due to incomplete cell plate formation, to diploids or even to higher ploidy. Nevertheless, one should keep in mind that unless the contents of two or more pollen cells mix during the later phases of androgenesis, the resulting androgenic plants will still be either hemizygous or homozygous.

3.1.2.4 Factors Affecting Androgenesis

Many factors were reported to affect the success of androgenesis. Some of them are confined to specific species or even cultivars. We shall thus not detail them here but rather list them, adding a few remarks in each case. For additional information, the reader is referred to SUNDERLAND (1974).

3.1.2.4.1 Culture Conditions

These include composition of the culture medium and temperature and light conditions during incubation. While some species do not require plant hormones in the medium, others do. On the other hand, auxins tend to favor callus rather than organized development. Among the minerals, the iron level seems to have an important role in the induction of androgenesis. Special techniques, like extracts of anthers, anther precooling, the use of nursing tissue, and the addition of charcoal, were reported to improve the success of androgenesis. In some cases it was found that a shift from one medium, in which the anthers are cultured right after isolation, to a different medium, will facilitate androgenesis.

3.1.2.4.2 Donor Plants

Generally speaking, young and actively flowering plants supply anthers which have a better chance to be induced to androgenesis. Light and temperature conditions may also affect the future behavior of the cultured anther, but this information is based mainly on *Nicotiana*. Some reports claim that treatment with growth regulators (e.g. 2-chlorethyl phosphoric acid in *Oryza*) will increase the anther response.

3.1.2.4.3 Pollen Age

The stage of the microspores in the anthers at the time the latter are removed from the donor plant for in vitro culture, strongly affects the success of androgenesis. In some plants the exact stage can be determined easily by examining one of each flower's anthers: almost all microspores of the same anther are at the same stage. This is not so in other plant species (e. g. *Paeonia*), where some of the microspores lag far behind in their developmental stage. In most plants where successful androgenesis was reported, anthers in which the microspores were at, or just before, first mitosis, gave the best results. On the other hand, haploid protoplasts from pollen tetrads were suggested as the source for haploid plants in *Nicotiana*, and mature pollen cultures of tomato resulted in haploid tissue which could be induced to differentiate into haploid plantlets.

3.1.2.4.4 Anther Stage and Ploidy

The relatively few cases in which the relation between anther stage and the ploidy of the resulting plantlets was followed, seem to indicate that with the advancement of anther stage there is a tendency to polyploidy. Thus, in *D. innoxia*, plantlets from anthers with microspores of up to first mitosis are mostly haploid, whereas when the cultured anthers have binucleated microspores they tend to produce diploid and polyploid plantlets.

3.1.2.5 Application of Androgenesis for Breeding and Genetic Studies

The potential value of androgenesis for breeding and genetic studies has been pointed out repeatedly since its discovery in 1964, but the actual utilization of this technique for such studies has been demonstrated only rarely. An example of the successful application of anther culture for breeding is the work of COLLINS et al. (1974) which dealt with the alkaloid content of tobacco plants for which two dominant genes (A and B) were known. Double recessive lines (aa bb) are very low in alkaloids. COLLINS et al. (1974) used F_1 plants having the Aa Bb genotype, and by anther culture obtained four types of haploid plants: ab, aB, Ab and AB. The ab plants were very low, and the AB plants—very high in alkaloid content. The haploids were made diploid without the use of colchicine, and true breeding diploid lines (aabb, aaBB, AAbb, and AABB) were established.

CARLSON (1973) combined the techniques of androgenesis and cell culture and reported the attainment of tobacco lines resistant to the methionine analogue methionine sulfoximide (MSO). The latter is similar in structure and effect to the toxin produced by the tobacco pathogen *Pseudomonas tabaci* (which causes wild-fire). The tobacco lines obtained by CARLSON which were resistant to MSO also showed milder symptoms after inoculation with *P. tabaci* than plants sensitive to MSO.

MELCHERS and co-workers demonstrated several cases which indicated that hemizygous plants can be obtained from heterozygous tobacco donors by androgenesis (see MELCHERS, 1972). None of these were carried further to the stage of doubling chromosome numbers and testing for homozygosis of the anther culture progenies. Nevertheless, the experiments demonstrated clearly the potentialities of androgenesis for breeding and genetic studies. We shall thus describe one of these experiments in some detail (MELCHERS and LABIB, 1970). It was concerned with two leaf-shape genes: (1) *stalked* vs *winged* petiole (P/p) (Fig. 3.9), and (2) *normal* vs *deformed* blade (H/h). The results obtained by either

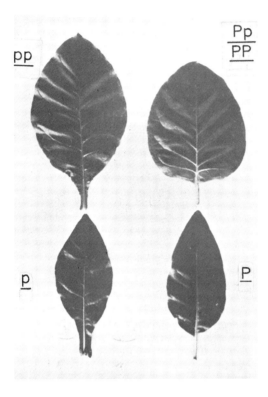

Fig. 3.9. Genes affecting leaf shape in *Nicotiana tabacum* (from MELCHERS and LABIB, 1970)

self-pollination or anther culture of a double-heterozygous plant (PpHh) are summarized in Table 3.2. The attainment of four hemizygous haploid lines of the genotypes PH, Ph, pH and ph by anther culture was thus indicated clearly and the appropriate diploids could undoubtedly be produced.

We may thus adopt the general conclusion that, in plant species where embryoids develop directly from individual pollen cells (e. g. tobacco), the progeny of anther cultures from a multifactorial heterozygote will segregate into hemizygous groups. The number of such groups is expected to be 2^n, n being the number of genes for which the donor is heterozygous. On the other hand, selfing such a donor will result in a progeny with 3^n different genotypes out of a total of 4^n possible combinations. These calculations point to an obvious conclusion that, with the increase of n, the chance of selecting a certain homozygous individual is reduced exponentially. Moreover, in dealing with a character regulated by one or more dominant genes, individuals homozygous for the required gene cannot be identified among the F_2 progeny. Only self-pollination of a great number of individuals with the required phenotype will indicate the homozygous plants. To illustrate the above considerations we shall assume that the

Table 3.2. Progenies obtained by either self-pollination or anther culture of a double heterozygous parental plant. The parental plant was heterozygous to petiole (P/p) and leaf-blade (H/h) controlling genes[a]

A. Progeny of self-pollination

	Stalked petioles Normal blades	Winged petioles Normal blades	Stalked petioles Deformed blades	Winged petioles Deformed blades
Genotypes:	PP HH PP Hh Pp HH Pp Hh	pp HH pp Hh	PP hh Pp hh	pp hh
Phenotypes:	PH	pH	Ph	ph
Expected				
ratios	9	3	3	1
Found	145	20	50	4
Calculated	123.2	41.0	41.0	13.8

B. Progeny of anther cultures

Genotypes:	PH	pH	Ph	ph
Expected				
ratios	1	1	1	1
Found	25	17	3	11
Calculated	14	14	14	14

[a] Data from MELCHERS and LABIB (1970).

breeder is faced with a problem of combining advantageous characters from two parents. For simplicity, we shall also assume that inheritance of these characters is controlled by simple dominant genes and that the two parents differ in respect to only three such genes. The F_1 plants will thus have an Aa Bb Cc genotype. Our breeder is interested in AA BB CC plants. By the conventional breeding system 27 out of 64 (4^3) plants will show the required phenotype, but only one out of 64 will have the AA BB CC genotype. On the other hand, by the anther culture technique one plant out of nine will have the required character and, upon diploidization, it will be homozygous. We do not have to go into further details in order to demonstrate that in cases where hemizygous haploids can be selected by their phenotype, anther culture is of great potential value. The potential advantage of anther culture for plant breeding is also obvious in respect to hybrid seed production based on self-incompatibility. This may be demonstrated in *Brassica*, where S alleles control the ability of pollen to germinate on receptive stigmata, as will be detailed in a later section of this book. For this hybrid seed production method, the breeder is interested in raising parental lines which are homozygous for the S allele (e. g. $S_1 S_1 \ S_2 S_2$ plants). Such plants could be obtained if the $S_1 S_2$ haploid would result from $S_1 S_1 \ S_2 S_2$ plants by androgenesis and diploidized by conventional techniques. Although plantlets were obtained by anther culture of *Brassica* plants, the attainment of haploids by this technique is not yet an established procedure in this genus.

Finally, the recent progress with plant cell cultures (SMITH, 1974) indicates a wide range of practical utilization of haploid cells for solving economically important problems. Thus, one can envision the selection, at the haploid cell culture level, for various characters such as resistance to extreme thermal conditions, to toxins of plant pathogens, and to salinity. The resistant cell lines may then be regenerated into diploid plants having the required character. Haploid plants obtained by anther or pollen culture are an obvious source of haploid cell lines.

3.1.3 The Pistil and the Female Gametophyte

In reference to phylogenetic considerations, the female gametophyte is borne on the *macrosporophyll*. In flowering plants the equivalent of the latter is the *carpel*. The carpel attains in angiosperms a variety of forms and arrangements. Carpels may be solitary or fused, and may form a simple or a compound *pistil*. There may be one or several pistils in each flower. We shall not be concerned here with the almost infinite number of variations of floral organization. They play an important role in plant taxonomy and contribute to the niceties of the plant life surrounding us. For our purpose we shall just refer to the pistil as an organ composed of three functionally distinct components: *ovary*, *style* and *stigma* (Fig. 3.10). One more term should be mentioned—the *gynoecium*. It is the so-called 'female' part of the flower, and may consist of one or more free carpels or of united carpels. Going back to the above terminology, a flower having a gynoecium consisting of a solitary carpel is a flower with a simple pistil. We shall further restrict our discussions to those features and functions in the flower and the female gametophyte which are relevant to the subject of this book.

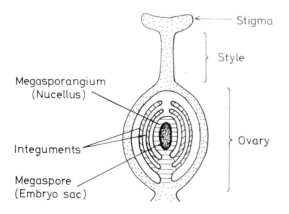

Fig. 3.10. Scheme of an angiosperm pistil

3.1.3.1 The Pistil

The ovary, by definition, bears the ovules. The latter have basically a common organization in angiosperms: attached to the ovary through the *funiculus*, having a cup-shaped structure formed by one, two or rarely three layers of *integuments* which start from the funicular end and almost surround a *nucellus*, leaving a canal opening at the other end, the *micropyle*. The micropyle may eventually close (Fig. 3.11).

The *megaspore mother cell* may develop directly from an archesporial cell which originates in the cell layer located beneath the nucellar epidermis, or the archesporial cell first divides, producing an archesporial tissue; one of the cells of this tissue will then develop into a megaspore mother cell.

The style is the midpart connecting the ovary to the stigma. Three types of styles are distinguished: (1) The *open type*—with a canal lined with glandular epidermis, functioning as transmitting tissue; this type is characteristic for monocotyledons. (2) The *solid type*—having a solid core of transmitting tissue; common in most dicotyledons. (3) The *half-closed type*—having transmitting tissue limited to only one side of the styles' canal, known in members of the Cactaceae (VASIL and JOHRI, 1964). The difference in structure brings about a difference in the path of the pollen tube on its way to the ovary, as will be seen later.

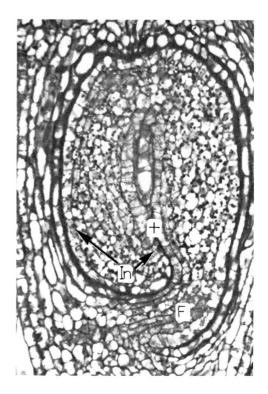

Fig. 3.11. Longitudinal section of an ovule of *Solidago* showing the funiculus (F), the integuments (In) and the micropyle (+). Micrograph (×450) (from O'BRIEN and McCULLY, 1969) Copyright (C) McMillan Pub. Co. Inc. Reproduced by permission

The stigma furnishes the surface which is receptive to the pollen grain. It is glandular and frequently has uni- or multicellular *papillae* ('hairs') (Fig. 3.12). There are 'dry' or 'wet' stigmata. The wet stigmata bear a more or less copious secretion. Such stigmata were found in several plant families (e. g. Solanaceae, Liliaceae, Rosaceae and Onagraceae), in which, as we shall see in handling self-incompatibility, the gametophytic type of self-incompatibility occurs. This secretion forms the germination medium for the pollen grain. It frequently consists of droplets of an emulsion and forms in some cases a 'liquid cuticle' over the papillae. This emulsion contains lipids, phenolic compounds, sugars and polysaccharides. The latter give the secretion its viscosity. Proteins may be a component of this secretion, but information on this is only now accumulating.

Dry stigmata were found in a number of families (e. g. Cruciferae, Compositae, Malvaceae and Gramineae). These are families with sporophytic—or multifactorial gametophytic self-incompatibility systems. The surface of the dry stigma is, at least in some families, not bald but covered with a proteineous layer. This layer has relevance to the self-incompatibility reaction. The proteins are synthesized in the cortical cells of the papillae and diffuse across the cell walls to the outer surface, producing the dry coat.

3.1.3.2 The Female Gametophyte

The megaspore mother cell undergoes usually two meiotic divisions resulting in a 'tetrad' of four haploid nuclei. In rare cases, e. g. *Allium*, there is only

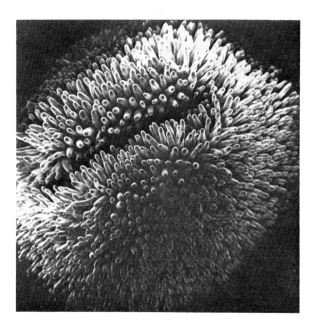

Fig. 3.12. Top view of an unpollinated stigma of *Brassica oleracea* showing dry papillae. Scanning electron micrograph ($\times 65$) (from ROGGEN, 1972)

one meiotic division. Accordingly, four or two *megaspores* are produced. In
some cases more than one megaspore resides within the same cell wall and
they eventually fuse during gametophytic ontogeny. As these megaspores can
have different genetic compositions, this fusion will result in a variety of genetic
consequences, as discussed by HESLOP-HARRISON (1972). As the ovule is the
sporophytic structure which bears the megaspore, it may be termed *megasporan-
gium*. With the formation of the megaspore the sporophytic generation is terminated.

As mentioned above, the meiotic division may result in two or four nuclei;
these may be separated by a cell membrane or the nuclei may reside in the
same cytoplasm. A further variation is the number of these nuclei which take
part in further development: production of the *embryo sac*. The latter variation
is used commonly to classify the female gametophytes of angiosperms into
monosporic, *bisporic* and *tetrasporic*, to mean that in the formation of the embryo
sac, one, two or all four nuclei, respectively, are involved (Fig. 3.13). Our further
discussion will concern only the so-called common monosporic *Polygonum* type.

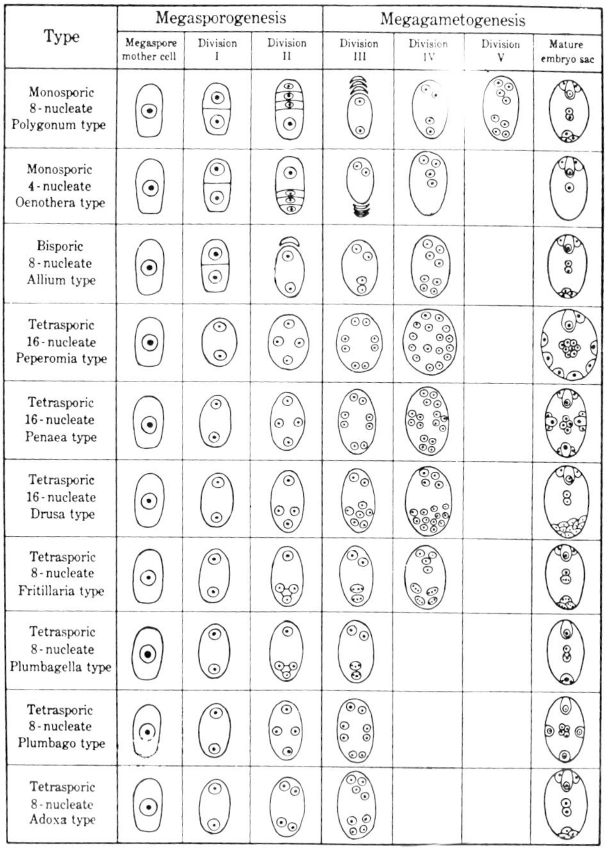

Fig. 3.13. Diagrams of the main types of embryo sacs in angiosperms (from MAHESHWARI,
1950) Copyright (C) McGraw-Hill Book Co. Ltd. Reproduced by permission

In this type the one functional haploid cell undergoes three mitotic divisions to produce an embryo sac with eight nuclei: one egg, two *synergids*, three *antipodals* and two—which may ultimately fuse—*polar cells.*

We shall now go into some details of the embryo sac. The synergid cells attain their position at the micropylar end of the embryo sac. They have a characteristic filiform apparatus and at their *chalasal* end have one or more conspicuous vacuoles. This filiform apparatus, as its name implies, is made up of finger-like projections which extend into the cells. The walls of these projections are rich in pectin and hemicellulose and almost devoid of cellulose. They usually increase the surface of the plasma membrane of the synergids. The two synergid cells have large nucleoli, relatively many mitochondria and ribosomes, and are quite rich in protein. Thus, from ultrastructural appearance they seem to be metabolically active cells.

The egg cell has only few ribosomes and a poor organelle content. Although there seems to be a variability in plant species in respect to the organelle content of the egg cell, this cell is usually vacuolated and has a cell wall partially surrounding it; the wall exists at the micropylar end but is incomplete at the chalasal end.

The polar nuclei in the central cell may either stay separated or fuse before fertilization. The central cell has a plasma membrane and a cell wall which is usually complex but again varies in details among species. It has a large central vacuole, so that the cytoplasm is layered close to the plasma membrane. The cytoplasm is rich in endoplasmic reticulum, ribosomes and mitochondria; in some species the cytoplasm also contains visible plastids and several nucleoli can be discerned within the nuclei. The central cell is usually the largest among the embryo sac cells and from ultrastructural observations it seems to be more active than the egg cell.

The role of the antipods in future development of the embryo sac, i.e., following fertilization, is not clear. In some species these cells undergo additional divisions forming a tissue which probably has a function in nourishing the embryo, taking the role normally played by the endosperm.

3.1.4 Fertilization

3.1.4.1 Contact between Pollen and Stigma

The contact between the pollen grain and the stigmatic surface induces processes which are only now being revealed, and seem to be of great relevance to the germination and penetration of the pollen tube into the style. One of the consequences of this contact is an immediate and continuous leaching of mobile substances from the pollen wall layers. The release process was followed very elegantly by HESLOP-HARRISON, KNOX and co-workers (see HESLOP-HARRISON, 1975, for review) by means of a 'pollen-print' method. In this method the pollen grains are put on an appropriate artificial surface for defined periods and the imprints are developed by methods (e.g. immunofluorescence) indicating the pattern, chemical identity and sequence of the released substances. In studies with species having the sporophytic incompatibility system, it was found that the first to leach, within seconds after contact, are substances coming directly from the outer surface or those released through the micropores of the tectum roof. These are of sporophytic origin and contain mainly glycoproteins and proteins having little or no enzymes activity. It is tempting to speculate that these secretions have a function reminiscent of that of seed lectins, i.e., specific binding affinities to polysaccharides of cell surfaces.

Later, after an interval of a few minutes, substances are released from the intine. These are primarily of gametophytic origin and their release is either through the germination apertures or by diffusion through the exine. We may note that, apart from the role of these leachates in pollen germination, they are of medical interest. Thus, it was found that leachates of *Ambrosia* (ragweed) pollen contain Antigen A, the principal hayfever allergen.

3.1.4.2 Pollen Germination

The pollen germinates usually through one of the germination apertures. The pollen grain probably has a preformed 'machinery' in respect to ribosomes and messenger RNA (mRNA) which are set into action right at the onset of germination—no further mRNA is thus synthesized during this process. The vegetative cell is accordingly packed with ribosome-rich endoplasmic reticulum and usually devoid of nucleoli. With the exception of a few angiosperms (e.g. orchids), the growth of the pollen tube is restricted to a few hours. In orchids this process may take weeks or months. In such plants there is probably a continuous synthesis of ribosomes and a nucleolus is seen within the pollen tube (vegetative) nucleus. Scanning electron microscopy showed that in certain sporophytic-type-incompatibility plants (e.g. *Brassica*), the onset of compatible pollen germination is correlated with the collapse of the papillae (Fig. 3.14). The elongating pollen tube has a typical cell wall synthesizing mechanism at its tip, reminiscent of fungal hyphal tip growth.

In 'open-style' plants the pollen germinates on the surface of the stigma papillae and the tube grows down on the papillae surface to the mucilage-filled canal in the center of the style. This canal has a characteristic 'transfer-type' wall comprising the transmitting tissue. Substances having probably a nutritional role are transferred from this wall to the pollen tube during the growth of the tube.

In 'solid-style' plants the pollen tube penetrates the intercellular spaces of the pistil tissue. The details of this penetration have up to now been revealed in only a limited number of plants. The tube may start its penetration right underneath the cuticle of the papillae and then migrate in an exudate between cuticle and the cell walls. Reaching the style, the tube finds its way to the transmitting tissue and from there grows toward the ovary. In other plants the tube grows first outside the papillar cuticle and only at the base of the cuticle does it penetrate between the stigmatic cells into the transmitting tissue.

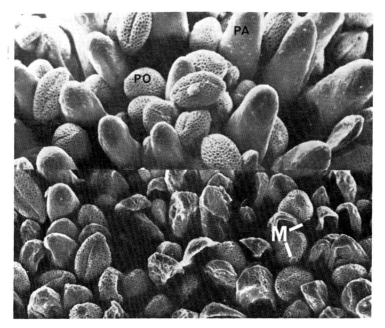

Fig. 3.14. Stigmatic papillae and pollen grains of *Brassica oleracea* var. *gemmifera* (Brussels sprouts). *Top:* 1 h after self pollination, *bottom:* 1 and 1½ h after compatible pollination, showing papillae (PA) and pollen grains (PO) and magnification marker (M). Scanning electron micrograph (× 650) (from ROGGEN, 1972)

3.1.4.3 Pollen Tube Discharge and Double Fertilization

We mentioned above that there may be one or two generative nuclei in the mature pollen grain. In the former case the generative nucleus divides during tube growth. Thus, in any case, at a later stage of pollen tube growth in the style it contains two generative (sperm) nuclei. These nuclei are bordered by a plasma membrane and are poor in cytoplasmic inclusions.

The tube finds its way, by means which are not known, toward the nucellus. From recent studies, mainly those of JENSEN and co-workers (see JENSEN, 1974, for review), it follows that the events from this stage to the completion of the double fertilization are probably similar among angiosperm species (Fig. 3.15). The approach of the pollen tube to the embryo sac is in all or most cases from the micropylar end. The tube then comes in contact with one of the synergid cells and enters into the filiform apparatus. Once inside the synergid cytoplasm, the tube discharges. As a result, part of the tube's cytoplasm, the vegetative nucleus and the two sperm cells, are released into the synergid's cytoplasm. The vegetative nucleus and the nucleus of the synergid cell are pushed aside, resulting most probably in what was termed x-bodies, and the generative cells move—one to the egg cell and one to the central cell. The details of the penetration into these latter target cells are not clear. As a rule, nuclear fusion takes place first in the central cell and then in the egg cell, although this may not be the order of penetration.

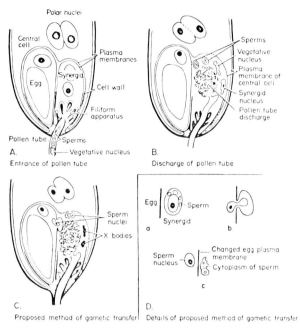

Fig. 3.15. Diagrammatic summary of the events associated with double fertilization in angiosperms (from JENSEN, 1974) Copyright (C) McGraw-Hill Book Co. Ltd. Reproduced by permission

3.1.5 Sexual Reproduction in Conifers

Several economically important evergreen forest trees do not belong to the angiosperms but rather to a group of gymnosperms—the conifers. Among the latter are pines, firs and spruces. The conifers are probably the most successful group of presently existing gymnosperms and are represented in about 50 genera with about 500 species. This group is defined by various authors as subclass, order of family, and thus termed Coniferopsida, Coniferales and Coniferae, respectively.

We shall describe the sexual reproduction of conifers rather briefly, pointing out the features in which they differ from the phylogenetically more advanced angiosperms handled above. The description is based mainly on *Pinus*. The reader is referred for more details and references to texts (e. g. MCLEAN and IVIMEY-COOK, 1958; FOSTER and GIFFORD, 1959).

The conifers are mostly monoecious species. Some species are dioecious. The sexual organs are borne on stroboli (cones) which are characteristically unisexual. In general terms the conifers differ from the angiosperms by having more developed gametophytes. Thus, especially the female gametophyte of conifers is considerably more elaborate than in angiosperms, and the time interval between the arrival of the pollen at the ovule and the formation of the zygote is rather extended in conifers.

The pollen is produced in sporangia on microsporophylls of the male cones. The sporangia are borne on the lower (abaxial) side of the sporophylls. The microspore undergoes three divisions before pollen grain maturation. In the first and second divisions small prothallial cells are produced, pushed aside, and most of the volume of the microspore is occupied by the "antheridal" cell. The latter divides once more to give two functional cells: the generative (antheridal) cell and the vegetative (tube) cell. Thus, the male gametophyte at the stage of pollen maturity consists of four cells. In many species (pines) the pollen grain has a morphology which facilitates its dispersal by wind.

The female gametophytes develop on female stroboli, which are considerably larger than the male ones. The female strobilus consists of spirally arranged bracts, each bearing a scale on which the ovules develop. The megaspore mother cell develops within the nucellus of the ovule in a way basically similar to that described for angiosperms. The meiotic division results in two to four macrospores; only one of which develops further. This macrospore, the beginning of the gametophytic generation, divides many times producing a female prothallium. On the micropylar end of the prothallium, archegonia are produced. Each of the latter consists of several cells; one of which is the enlarged egg cell.

Pollination starts with the approach of the pollen grain to the micropylar end of the ovule, where it is attached by an ovular secretion to the nucellus. At this stage the female gametophyte is still rather immature. The germination of the pollen grain and the approach of the tube to the female gametophyte is a rather long process, e. g. several months. Contrary to the tube of angiosperms, the tube of conifers may ramify during growth. The generative cell divides and one of the resulting cells (sperm, or body cell) divides once more to give two male gametes. At this stage the male prothallial cells have degenerated and the male gametophyte consists of four cells. All four penetrate the egg cell but only one of the two gamete nuclei takes part in the formation of the zygote.

3.2 Control and Modification of Sex

After having reviewed in Chapter 1 the various forms of sex expression as well as the structural and developmental aspects of reproduction in those groups of higher plants to which most economic crops belong (i. e., angiosperms and conifers), we can now review the genetic and environmental factors which regulate sex expression in these plants. We shall first confine ourselves to understanding the principles of the genetic control of sex determination and treat, rather briefly, the modifications of sex by nongenetic factors. Thereafter, we shall handle in more detail a number of representative economic crops, where sex expression should be considered in respect to breeding and crop productivity.

3.2.1 The Genetic Control of Sex Determination

Sex expression in plants is in many cases strongly affected by environmental conditions as well as by applied chemical agents. Several examples of such effects will be given in subsequent sections. The modification of sex expression by high temperatures was reported for cucumbers by KNIGHT (1819) and was probably known in horticultural practice many years before this. The genetic basis of sex determination in plants was indicated experimentally shortly after the rediscovery of the Mendelian laws of inheritance, and by one of the rediscoverers of these laws—CARL CORRENS. Briefly, CORRENS (1906) based his conclusions on three main test crosses. In one he crossed female *Bryonia dioica* plants with plants of the monoecious species *Bryonia alba* and obtained 11 plants, all very similar, being either pure females or females with a few early and nonfunctional staminate flowers. In another test CORRENS crossed female *B. dioica* plants with male *B. dioica* plants and obtained 41 flowering plants of which 21 were males and 20 were females. In a third test, pistillate inflorescences of *B. alba* were crossed with male *B. dioica* plants and 76 flowering plants were obtained, exactly half of which were males and half were females. Incidentally, in the plants resulting from the first-mentioned cross which did produce staminate flowers, these flowers were produced very early in the plants' ontogeny, on the main shoot, and never occurred in the subsequent year. Correns attributed this to unusual soil conditions, but as we shall see later this is a rather common phenomenon in monoecious species.

On the other hand, SCHAFFNER (see SCHAFFNER, 1935, for references) reported many cases of environmental modifications of sex expression and from these and theoretical considerations, concluded that there is no genetic basis for sex determination.

Such contradictory conclusions were not restricted to sex expression, but stemmed from a lack of full understanding of the interaction of genetic information with external and internal stimuli. Our present knowledge of how genetic information is expressed is such that cases of environmental modifications of form and function will by no means lead us to negate the role of inheritance in the final expression of the organism. We know that even the synthesis of a direct gene product, such as ribosomal RNA, can be strongly affected quantitatively by extracellular factors. An extrapolation to the far more complicated phenomenon of sex expression and its modification by environment is thus reasonable. ALLEN (1932) pointed out that those who negated the involvement of gene effects in sex expression neglected two facts: that the hereditary constitution provides only potentialities and that these potentialities, according to the prevailing conditions, may or may not be manifested.

Keeping in mind that sex expression in plants is a result of interaction between hereditary and nonhereditary factors, we shall now handle cases where the genetic determination of sex is evident.

We shall be concerned in this section mainly with the genetics of dioecism. Other sexual forms will be handled in subsequent sections in which the sexual regulation of specific economic crops will be discussed.

3.2.1.1 General Considerations Concerning the Genetic Control of Dioecism

We have seen previously that stamens and carpels not only differ vastly in their morphology, but also that in each of them a rather different and elaborate process leads to either the production of the male gametophyte—the pollen grain, or of the female gametophyte—the embryo sac. It is, therefore, reasonable to assume that for the morphogenesis of stamens and carpels a great number of genes is required. In hermaphrodite plants we can thus reason that all the genes for both types of flower organs *exist* and *operate*. However, while all these genes probably exist in every cell of the plant, they obviously have no formative manifestation in the vegetative part of the hermaphrodite plant. Even within the flower itself, the genes responsible for stamen differentiation, say, *A, B, C,* are not expressed in the rest of the floral members, and vice versa, the genes responsible for carpel differentiation, say, *D, E, F,* are restricted in their morphogenetic expression to the carpel.

We shall take this reasoning further and argue that even in the stamen itself, *A,* or any other individual stamen-gene, is functional only at a specific time and site. We thus come to the obvious conclusion, already mentioned above, which is fundamental to all processes of differentiation, that it is the *expression* of the genes and not the mere presence of genetic information which results in a specific type of morphogenesis. The lack of a form or a function can thus stem from either the suppression of expression of the genes concerned, or from their absence.

We shall turn now to plant species, or lines, which show dioecism, i.e., they consist of female plants having only carpellate (or pistillate) flowers, and of male plants having only staminate flowers. In these species or lines, by definition, crosses between female and male plants result again in female and male plant segregants. Based on the foregoing considerations there are two main possibilities which lead to this sexual dimorphism: (1) the genes for stamens and carpels exist in both male and female plants, but only the stamen-genes are expressed in the male plants and only the carpel-genes are expressed in the female plants; (2) either carpel-genes or stamen-genes are excluded from the respective plants. By integrating the two possibilities we can visualize three models. In the first, stamen-genes are excluded from female plants and carpel-genes are excluded from male plants. In the second model, stamen-genes are excluded from female plants but both stamen-genes and carpel-genes are present in the male plants, the latter not being expressed. In the third model, both stamen-genes and carpel-genes are present in female plants but carpel-genes are excluded from male plants; the stamen-genes are not expressed in the female plants.

Let us consider the first possibility, according to which both stamen-genes and carpel-genes are present in female and male plants and sex dimorphism is determined by the expression of these genes. LEWIS and JOHN (1968) schematized this model in the following way:

$$Mm \begin{cases} \text{AA BB CC Active} \\ \text{DD EE FF Inactive} \end{cases} \rightarrow male$$

$$mm \begin{cases} \text{AA BB CC Inactive} \\ \text{DD EE FF Active} \end{cases} \rightarrow female$$

The genes *A*, *B* and *C* are here the stamen-genes and *D*, *E* and *F* are the carpel-genes. The allele pair *M/m* has the key role: mm will allow only the expression of the carpel-genes and Mm will allow only the expression of the stamen-genes. Both the carpel-genes and the stamen-genes are always in the homozygous state.

Thus, an unlimited number of genes may be involved in the differentiation of stamens and carpels but sex determination is controlled by just one locus: M/m, acting as a trigger.

3.2.1.2 Artificial Dioecism in *Zea mays*

Zea mays (maize) is normally a monoecious species. The pistillate flowers[1] are borne on axillary inflorescences which produce the modified spikes, the 'ears'. The pistillate flowers have very long stigmata commonly termed *silk*, giving the ears a "bearded" appearance. The staminate flowers are borne in a terminal inflorescence—the *tassel*.

A number of mutants were found in maize which either suppress the pistillate inflorescence (the plant becomes *silkless*) or affect the tassel. The tassel may be affected such that it produces pistillate flowers rather than staminate ones. The terminal inflorescence is thus converted into *tassel seed* (Fig. 3.16). JONES (1934) utilized such mutations to convert maize into dioecious populations. He used two genes: (1) sk (*silkless*, on chromosome 2) and (2) ts_2 (*tassel seed$_2$*, on chromosome 1). Obviously a plant which is both *silkless* and *tassel seed* is a functional female plant, and a plant which is *silkless* but with a normal (staminate) tassel is male. Thus both males and females in such a population are *silkless*. The *silkless* phenotype is expressed only in a homozygous recessive genotype sk sk, and likewise the *tassel seed* phenotype is homozygous: $ts_2 ts_2$. A plant with the genotype sk sk $ts_2 ts_2$ is thus female. The male plants can have either of the two genotypes: sk sk $Ts_2 Ts_2$ or sk sk $Ts_2 ts_2$. JONES chose sk sk $Ts_2 ts_2$ males and pollinated the sk sk $ts_2 ts_2$ females with them. Such a cross will always result in only the two parental types in their progeny (Fig. 3.17). We may now refer to our model, schematized above, and regard the Ts_2/ts_2 allele pair as the trigger (M/m). The two sexes—male and female plants—have the *genotype* sk sk in common as well as the many other genes involved in the differentiation of stamens and carpels. This is an artificial dioecious population where the males are heterozygous and the females homozygous for the sex-controlling gene.

As all the genes but the trigger one are common to both sexes and always homozygous, no gene redistribution (with the exception of Ts_2) by crossing-over is possible. They may either all be located on one pair of chromosomes or distributed in several chromosomes. In this model the male plants always produce two kinds of gametes with either Ts_2 or ts_2, while the female plants produce only one kind of gametes with ts_2.

---

[1] The flowers of Gramineae are termed *florets*: these are grouped in *spikelets* and the latter are borne on *spikes*—the typical inflorescences of this family.

Fig. 3.16. The tassel-seed (ts_2) character in maize *(Zea mays)* (from NICKERSSON and DALE, 1955)

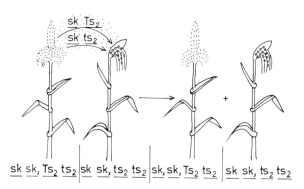

sk sk, Ts_2 ts_2 | sk sk, ts_2 ts_2 | sk, sk, Ts_2 ts_2 | sk sk, ts_2 ts_2

Fig. 3.17. Scheme of artificial dioecy consisting of two maize mutants: silkless (sk sk, Ts_2 ts_2) and tassel-seed (sk sk, ts_2 ts_2)

Dioecism in maize was also achieved by a slightly different gene combination (EMERSON, 1932). The gene ba (causing *barren stalk*, on chromosome 3) is recessive and in ba ba genotypes only the tassel is produced. Thus ba ba, ts_2 ts_2 are female plants and ba ba, Ts_2 ts_2 are male plants. As above, the male is the heterogametic one. Still another tassel-seed gene can be utilized—Ts_3; but when this gene is used rather than ts_2, the female is heterogametic, because Ts_3 is dominant and thus $ba\,ba$, $Ts_3\,ts_3$ plants are female, while $ba\,ba$, $ts_3\,ts_3$ plants have no tassel seeds and are male plants. From this we see that males may be either heterogametic or homogametic according to the specific genes involved.

A word of caution shall be added here. Although these *Z. mays* systems are widely and frequently cited as examples of artificial dioecism, a close examination shows that plants with $ts_2\,ts_2$ as well as with Ts_3 genotypes are not always pure females. The morphology of these mutants was investigated by NICKERSON and DALE (1955), who compared their observations with those of EMERSON et al. (1935). In tassels of $ts_2\,ts_2$ plants, NICKERSON and DALE (1955) observed hermaphrodite spikelets in addition to pistillate ones, while EMERSON et al. (1935) observed staminate spikelets in addition to pistillate ones. Moreover, even without $sk\,sk$, the $ts_2\,ts_2$ plants had rather undeveloped ears. In tassels of Ts_3 plants, fertile staminate spikelets were usually found (Emerson et al., 1935; NICKERSON and DALE, 1955). Thus the system is obviously not a 'pure' dioecious one. Nevertheless, as the plants varied in respect to the production of staminate spikelets in the tassels, the elimination of staminate spikelets by selective breeding thus seems very probable.

Note that in the artificial dioecious systems described above, the ts genes have actually a unique effect: they not only suppress the staminate flowers on the tassel but also induce there the formation of pistillate flowers. By this we do not mean that two separate activities are involved; we can visualize a situation where the mere suppression of the stamen initials in the floret, the ontogenetically later-appearing carpel initials, can differentiate to their full functional state. Nevertheless, to enable such a situation we have to assume that in *Z. mays* there is a gene (or genes) for the *trigger* mechanism which will allow in any individual floret the differentiation of either stamens or carpels only. We shall return to this presumed *trigger mechanism* in our subsequent discussions.

3.2.1.3 Artificial Dioecism Caused by Suppressive Genes—Linkage between Genes as a Prerequisite for Sex Dimorphism

The models of artificial dioecism in maize described above were achieved in a 'simple' system which did not require any linkage between genes. This was possible because of the natural separation in this plant between the staminate and pistillate inflorescences and the particular effect of the ts genes. A different system exists in *Rubus idaeus* (see LEWIS and JOHN, 1968). In this normally hermaphrodite species, two genes affecting sex determination were found, one suppressing stamen differentiation (m), the other suppressing carpel differentiation (f). Both genes are recessive and act independently of each other. Thus, the genotypes FF mm and Ff mm result in female plants while ff MM and ff Mm result in male plants. As would be expected from the independent effects of the m and f genes, the double recessive has no functional flowers as both stamens and carpels are suppressed. Therefore, in this system, a dioecious line could not be constructed. Consider the two possibilities of crosses between 'female' and 'male' plants:

1. FF mm × ff Mm (heterogametic male plants)
2. Ff mm × ff MM (heterogametic female plants)

In both cases we expect hermaphrodite (Ff Mm) plants among the progeny.

However, in this system we could visualize the establishment of dioecism under two additional conditions: (1) that one of the two suppressing genes is dominant; and (2) that there is a complete linkage between the two genes. We shall consider it using the gene symbols of *Rubus*. We may first suppose that the female suppressor is dominant: thus, FF MM, Ff Mm, FF Mm will all be 'male' plants, while ff mm will be 'female'. If we now cross a ff mm female with a Ff Mm male we shall obtain four different genotypes, because the 'male' produces four types of gametes, FM, Fm, fM and fm, while the 'female' produces only one type of gamete: fm. Under such conditions a cross between female and male plants will again result not only in 'female' and 'male' plants, but if we add the other restriction, complete linkage of F with M (FM) and f with m (fm), to be symbolized as $\frac{FM}{fm}$, the 'male' will produce only two kinds of gametes: FM and fm. A cross between the double homozygous recessive female (ff mm) and the aforementioned male will then result in only the two parental types; a dioecious system will thus be established.

The above-mentioned model may seem complicated but is a rather common basis for sex dimorphism in both animals and plants. Moreover, it leads from the *genetic* sex determination to the probably more advanced and stable *chromosomal* one. The best way to prevent crossover between the two opposite suppressive genes would be the establishment of a mechanical barrier on the chromosomal level. Such a mechanism should prevent, during the meiotic prophase, pairing and chiasmata formation, among those chromosomal segments containing the linked genes. Sex chromosomes which are heteromorphic for at least the segment in question will furnish such a mechanism. This rationale was already formulated by DARLINGTON (1932).

3.2.1.4 Chromosomal Control of Sex Determination

We have indicated above that sexual dimorphism based on stamen- and carpel-suppressing genes would operate most efficiently when these genes are linked and that such linkage is secured by a mechanism preventing crossing over in the chromosomal segment containing the suppressing genes, leading to the establishment of *heteromorphic sex chromosomes*. From this we may deduce that in many cases sex chromosomes developed evolutionarily at a later stage, in species in which dioecism already existed. We should, therefore, be faced with many cases of dioecious species, in which cytological heteromorphism of sex chromosomes is slight or can be recognized only at specific meiotic phases. Moreover, in the very same dioecious species the degree of sex-chromosome dimorphism may vary in extent, resulting in contradictory reports by several authors handling different specimens of the same species.

This rationale is actually manifested and a great number of examples were reviewed by WESTERGAARD (1958) and by LEWIS and JOHN (1968). The former listed cases in which convincing evidence for chromosomal dimorphism was available. In this list there are only 13 species of three families. WESTERGAARD'S list for doubtful cases of chromosomal dimorphism is much more extensive, containing about 70 species of 24 families. An up-to-date revision may change the ratio of species between the two lists. Careful cytological observations and new staining techniques (e.g. use of Giemsa staining or of quinocrine-mustard staining, and evaluation by fluorescence microscopy as applied by KAMPMEIJER, 1972) may move some species from the second to the first list.

Intermediary cases of sex chromosome heteromorphism are known. In some dioecious species, such as *Carica papaya*, no chromosomal heteromorphism was observed, but one chromosome pair showed a delayed separation in anaphase. There remains, of course, the question whether or not heteromorphic chromosomes, or pairs delayed in anaphase separation, are indeed the sex chromosomes. We shall refer to this question below. Still, delayed anaphase separation is rather common in dioecious animal systems and is therefore considered to be an incipient X/Y system (LEWIS and JOHN, 1968). However, even taking possible future revisions into account, the majority of dioecious plants lack sex chromosome heteromorphism. This should not surprise us, because even in animals, in which dioecism was relatively a much earlier evolutionary development than in flowering plants, whole classes have no cytologically recognizable sex chromosome heteromorphism and in others (e.g. birds) chromosome dimorphism was established only recently (see MITTWOCH, 1973).

We shall now consider the simplest kind of chromosomal dimorphism, the X/Y system. In this system, when the male plant is heterogametic, XX plants are female and XY plants are male. In previous considerations we deduced that at least a segment of the X chromosome should not pair with its heteromorphic partner, the Y chromosome. On the other hand, it should be assured that during meiosis the two chromosomes will segregate in an orderly way. This will be achieved only when X and Y also contain a homologous segment, thus enabling pairing in the meiotic prophase and keeping this pairing until the proper segregation into the gametes is secured. The sex chromosome should thus be constructed of at least two kinds of segments: one in which pairing is prevented, and one in which pairing is assured.

From the above simple X/Y system we can extrapolate to a great number of more complicated systems. There may be more than one Y chromosome or more than one X chromosome, as well as other variations. Many complicated cases of sex chromosome heteromorphism were actually revealed in animals (see LEWIS and JOHN, 1968), but only a few are known in plants. Some of the latter will be mentioned when describing sex determination in specific plants.

A special case would be a system of unpaired sex chromosomes, the X/0 system. Its existence was claimed in only three plant species of the genus *Dioscorea*, but the validity of this claim is now in doubt (see MARTIN, 1966; LEWIS and JOHN, 1968). From cytological considerations, an X/0 system requires a mechanism ensuring the movement of an unpaired chromosome in the meiotic division. On the genetic level, the X chromosome should contain no information required for the haploid generation. The latter is worth notice in angiosperms and gymnosperms where this generation, the gametophyte, goes through two or more mitotic divisions, respectively, and, as detailed previously, has an elaborate differentiation. It is, therefore, not surprising that the X/0 system is quite prevalent in animals but probably missing from plants. Even in animals, in certain cases when X/0 individuals were experimentally or accidentally produced in species with a normal X/Y system, these individuals expressed anomalities. Thus, X/0 *Drosophila* flies are male but their sperm is not functional, and in man, X/0 individuals are female with degenerate gonads, and in most cases have other anomalities including a mental deficiency termed TURNER'S syndrome.

3.2.1.5 Main Methods for the Study of the Genetics and Cytology of Sex Determination

WESTERGAARD (1958) considered that three consecutive steps should be followed in the analysis of sex expression:
    1. Establish which of the sexes is heterogametic;
    2. Localize the sex-determining gene complex(es) on the sex chromosomes and understand their interaction with the autosomal sex genes;
    3. Break down the gene complexes and explain the results in terms of individual gene action.

These suggestions were based on WESTERGAARD'S thorough studies of the determination of sex expression in *Melandrium*, which we shall consider below. Still, with appropriate adaptations, WESTERGAARD'S suggestions are applicable to other heterosexual systems, including those not strictly dioecious and those in which no sex chromosomes are cytologically identifiable.

There are several ways to establish which of the sexes is represented by heterogamous plants. CORRENS (1928) suggested, and actually employed, four ways, which were also utilized by other investigators: (1) cytological methods— search for chromosomal heteromorphism; (2) identification and genetic analysis of sex-linked characteristics; (3) competitive pollination—pollination with different quantities of pollen grain or with aged pollen, and search for causally related changes in the ratios between the resulting progenies; (4) reciprocal crossings between dioecious species, which may result in either homomorphic or heteromorphic progenies.

To CORRENS' suggested methods WESTERGAARD (1958) added others, useful not only for the detection of heterogametic plants but also in a general way for the investigation of sex determination: (1) selfing or sibbing of occasional bisexual individuals in dioecious species; and (2) crossings between experimentally produced polyploids and diploids and obtaining series of plants with different numbers of autosomes and sex chromosomes.

The first of these two methods has a drawback: the progeny of plants which are not 'pure' staminate or pistillate is usually composed of plants in which the classification of sex expression is rather difficult.

Two additional methods can be suggested, in the light of recent investigations: (1) selfing of normally 'pure' pistillate and staminate plants can be undertaken in certain species after appropriate chemical treatment inducing in the above plants the development of either staminate or pistillate flowers, respectively; and (2) obtaining haploid plants by anther culture. Some of these methods will be illustrated in the next section and others while dealing with the sex expression of specific species.

3.2.1.6 Genetic Regulation of Sex: Representative Examples

Two main systems will be discussed. Unfortunately, because of lack of suitable examples, our test cases will not be drawn from economic crops but rather from wild species.

3.2.1.6.1 The Active Y Chromosome System

This system will be represented mainly by *Melandrium* (Caryophillaceae). This genus was utilized in the early years of the rediscovery of the Mendelian laws by pioneers of genetics, e. g. DE VRIES, BATESON and CORRENS. The experiments of CORRENS are of relevance to us because he focused his attention on sex expression. CORRENS employed various experimental approaches to reveal the inheritance of sex expression (see CORRENS, 1928). Two of these were technically simple and involved the pollen grains. The rationale was that if the male is heterogametic, its pollen should consist of two types. These two types may differ in their ability to reach the embryo sac and bring about fertilization. One method was 'pollen competition'. CORRENS argues that if very few grains are used there will be no competition and about equal number of male and female plants will result; but if many pollen grains are used, the pollen which is more efficient will have a better chance to fertilize the embryo sac and the ratio of male to female plants among the progeny will not be equal. This was actually the case: with increasing amounts of pollen the ratio of female/male increased. By this he decided that *Melandrium* male plants are heterogametic, and the females are homogametic. He then used 'pollen ageing' with the same basic argument, i. e., if one kind of pollen grain is more resistant, ageing should change the ratio of females to males in the progeny. CORRENS' data substantiated his reasoning. Moreover, CORRENS made similar experiments with *Fragaria* and found no changes in the ratios of the females to males in the progeny, concluding that in this genus the females are heterogametic.

Although CORRENS' pollen-ageing experiment could not be confirmed (VAN NIGTEVECHT, 1966b), both of CORRENS' conclusions were subsequently verified. For the heterogamy of male *Melandrium* plants BLACKBURN (1923) and WINGE (1923) furnished cytological evidence. The former also established a size difference: the Y-chromosome is longer than the X-chromosome. These results were substantiated by several authors, so that the chromosomal constitution of *Melandrium* (both *Melandrium album* and *Melandrium dioecum*) was established as $22 + XX$ for female and $22 + XY$ for male plants.

Several sex-linked genes were found in *Melandrium* by a number of authors (see VAN NIGTEVECHT, 1966a), but they contributed little to the further genetic studies of sex expression in this genus. On the other hand, information on construction of polyploids, obtained quite independently in three laboratories in the United States (WARMKE and BLAKESLEE, 1939), Sweden (WESTERGAARD, 1940) and Japan (ONO, 1939) was the key for the studies which produced one of the best understood systems of genetic and chromosomal control of sex expression.

Before going into the details of these studies, some words of caution are in order. The first concerns nomenclature. The genus *Melandrium* is termed in some studies (e.g. SHULL, 1910) as *Lychnis*, and some authors included it in *Silene*. More confusing, the species termed by WESTERGAARD as *M. album*, was included by WARMKE and BLAKESLEE (1940) and WARMKE (1946) in *M. dioica* [in conformation with giving *album* a subspecies rank (*M. dioecum* ssp. alba) by LÖVE (1944)]. Moreover, the *album* spp. used by the American investigators (e.g. WARMKE and BLAKESLEE) was probably not pure and included genes from the true *M. dioecum*. This last assumption, put forward by VAN NIGTEVECHT (1966a), could be the source of somewhat different results obtained in different laboratories with presumably the same species. Furthermore, the flowering pattern and especially the tendency to divert from pure dicliny in the two 'species' *M. album* and *M. dioecum*, are different. Due to this confusion we shall neglect the species designation in our following discussion (as also done by WESTERGAARD, 1958), but keep in mind that the experiments were actually undertaken with different plant material. The final word of caution regards the definition of the sex types. *Melandrium* in general has a tendency to show a certain degree of hermaphroditism in its flowers, e.g. in male plants staminate flowers may have pistil initials in various degrees of development. In some cases, these flowers are female fertile and were used to self-pollinate 'male' plants. This resulted in progenies which were even less unisexual, causing great difficulties in the classification into sex types. The tabulation of results is, therefore, in some cases a poor representation of the results. We may now turn to the findings made by several workers in respect to the chromosomal control of sex expression in *Melandrium*. The detailed references to their studies may be found in WESTERGAARD (1958), VAN NIGTEVECHT (1966a, b), and LEWIS and JOHN (1968).

It was found that *Melandrium* is readily amenable to colchicine treatment and tetraploids with the chromosomal constitutions $44 + XXXX$ and $44 + XXYY$, being female and male, were obtained. These served in various types of crosses from which individuals with a specific number of autosomes and sex chromosomes could be identified with respect to both their sex expression and their cytology. One should obviously keep in mind that although two plants may have the same overall chromosomal constitution, e.g. $3A + XXY$, their autosomal constitution can, and probably does, in most cases, differ. There was no way to identify all autosomes either cytologically or by genetic means (Fig. 3.18).

Considering chromosomal constitution and sex expression of individual *Melandrium* plants, the conclusion was drawn that in all cases where the level

Fig. 3.18. Tetraploid male and
female flowers of *Melandrium
album* and their karyotypes (after
Warmke, 1946)

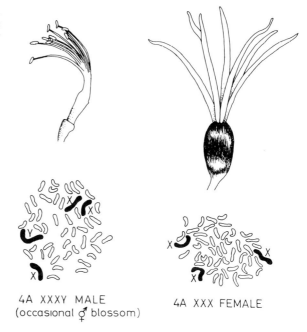

4A XXXY MALE   4A XXX FEMALE
(occasional ♂ blossom)

of ploidity was 2, 3 or 4 and the number of X chromosomes between 2 and
5, but without Y chromosomes, plants were female. This was an obvious diversion
from results with polyploid *Drosophila* flies (Bridges, 1939), as can be seen
from Table 3.3.

Table 3.3. The relation between chromosome constitution and sex in *Melandrium* and
Drosophila

Chromosome constitution	X/A Ratio	*Melandrium*	*Drosophila*
1. 2A + XX	1.00	♀·♀	♀
2. 2A + XXX	1.50	♀·♀	♀
3. 3A + X	0.33	—	♂
4. 3A + XX	0.67	♀·♀	♀
5. 3A + XXX	1.00	♀·♀	♀
6. 4A + XX	0.50	♀·♀	♂
7. 4A + XXX	0.75	♀·♀	☿
8. 4A + XXXX	1.00	♀·♀	♀
9. 4A + XXXXX	1.25	♀·♀	—

From data of Westergaard (1958).

    In the eight combinations of chromosomal constitutions of *Melandrium* the
X/A ratios varied from 0.50 to 1.50, but this had no effect on sex expression.
On the other hand, sex in *Drosophila* flies is dependent on the X/A ratio:
ratios of 1.00 or higher 'cause' female sex expression and ratios of 0.50 or
lower 'cause' male sex expression. This latter sex-determining system, designated
'X-autosome balance', thus differs substantially from the *Melandrium* system.

A series of polyploid *Melandrium* with different numbers of Y-chromosomes was also produced, e. g. 2A + XY, 2A + XXY, 2A + XYY, 3A + XY, 3A + XXY, etc. Here the picture was not so clear as in plants without a Y-chromosome. Basically, all plants having at least one Y-chromosome showed at least some tendency to maleness: none of them was female. However, in details, the results of WARMKE differed from those of WESTERGAARD. This was expressed mainly in plants with only one Y-chromosome and several X-chromosomes as well as several sets of autosomes (e. g. 4A + XXXXY). In WARMKE'S material such plants were almost hermaphrodite, while in WESTERGAARD'S material they had a clear male tendency. Furthermore, when the information is summed up, as in Table 3.4, we can see that the Y-chromosome is not the only factor of

Table 3.4. The effect of different Y- and X-chromosome numbers on sex expression of *Melandrium*. Results are expressed in percent male plants. Total number of plants per chromosomal constitution is given in parentheses[a]

Number of Y-chromosomes	Number of X-chromosomes			
	1	2	3	4
1	100 (15)	89.3 (28)	35.1 (205)	0 (8)
2	—	88.1 (42)	58.3 (12)	0 (1)
3	—	100 (?)	—	—

[a] Data of WESTERGAARD (1958).

sex expression: an increase in X-chromosomes brings about a lower percentage of male plants, while an increase in Y-chromosomes seems to cause an increase in percentage of male plants.

In certain chromosomal constitutions the effects of both X-chromosomes and autosomes could be revealed. We may recall here that although the autosomes are regarded as groups of 'A', an extensive exchange of individual chromosomes is not excluded—meaning that two plants having, for example, the same overall designation of 3A + XXXY, may differ vastly in their autosomal composition. Knowing now that autosomes do affect sex determination, it is no wonder that such two plants can be genetically, as well as phenotypically, rather different. The effect of autosomes on sex expression is also evident from selfing of XY individuals, which had some hermaphrodite flowers. In such plants selection can be done resulting, after some generations, in hermaphrodite plants (VAN NIGLEVECHT, 1966b). It was ruled out that in these plants the sex chromosomes were changed, because out-crossings to normal female plants resulted in progeny which segregated in the usual way to female and male plants. Such hermaphrodites differed from others of independent origins. In the latter, indications were found that changes in sex chromosomes did occur. This leads us to another aspect of *Melandrium* sex determination: the genetic composition of the sex chromosomes.

As stated above, normal Y-chromosomes are longer than X-chromosomes; moreover, the former are not metacentric. Thus, fragmented Y-chromosomes

(e. g. Y^1) do have a length similar to X-chromosomes but XX have two chiasmata in metaphase while XY^1 have only one; the fragmented Y-chromosome can thus be identified. Plants with a fragmented Y were isolated in various ways but mainly among progenies of triploids (WESTERGAARD, 1946).

Figure 3.19 shows three types of fragmented Y-chromosomes. The XY^1 plants were found to be true hermaphrodites and out-crossing to normal XX-female plants results in female and hermaphrodite plants. It was further found that

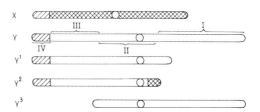

Fig. 3.19. Scheme of normal x- and y-chromosomes and fragmented y-chromosomes of *Melandrium* as revealed by cytogenetic studies

when segment I was missing from Y in such XY^1 plants, normal pistil development occurred in their flowers. WESTERGAARD concluded that this segment has a *female-suppressing* effect, because it did not interfere with the male expression. However, as pointed out previously, a *lack of trigger mechanism* is also compatible with these results. In the present case the normal function of this trigger would be to prevent further differentiation of pistil initials, once anther initials attained a certain stage. The point is stressed here because it involves basic concepts of floral differentiation. As we shall see in our subsequent handling, the 'trigger mechanism' is a rather attractive concept. It is useful to interpret many other cases of a single gene difference between plants having only diclinous (unisexual) flowers and those having (also or only) hermaphrodite flowers. The requirement for a trigger mechanism is rather simple: a mutual, or unidirectional inhibition of stamen and carpel initials. We shall detail the discussion of this mechanism while dealing with the sex expression in *Cucumis*. The "trigger" concept (GALUN, 1961) does not contain a mechanistic explanation. It is merely used to describe a situation where, within the same floral bud, either the staminate or the pistillate initials, but not both, can attain further development.

A second type of aberrant Y-chromosome, designated Y^3, lacks segments III and IV; it thus has no pairing segment. Plants with XY^3 are male, but sterile. Meiosis proceeds normally but the gametophytes degenerate and no viable pollen is produced. Because in XY^3 no carpels were produced, WESTERGAARD considered that the segments III–IV are not involved in female suppression but in the completion of anther differentiation.

By attributing a female-suppressive effect to the distal end of the Y-chromosome (I), and a late-stage anther differentiation control to the proximal (III–IV) segments of Y, WESTERGAARD deduced that the central segment (II) must include genes which control initiation of anther formation. WESTERGAARD'S conclusion, though compatible with his results, is obviously not the only one. Unfortunately, not many physiological investigations on the sex expression of *Melandrium* were made. The information which is available does not favor WESTERGAARD'S inter-

pretations. The most clear evidence against an absolute Y-chromosome requirement for anther initiation comes from a phenomenon known for many years
(STRASBURGER, 1900; detailed description by BAKER, 1947); when female (XX)
Melandrium plants are infected by the smut *Ustilago violaceae*, the staminate
initials do differentiate into anthers and hermaphrodite flowers are produced;
the anthers, however, are filled with fungal spores rather than with pollen grains.
The differentiation of anthers in flowers of genetic XX, female plants was also
reported (LÖVE and LÖVE, 1945) to occur after application of animal steroid
male hormones (e. g. testosterone), but this observation probably requires confirmation.

Several genes located on the X- and Y-chromosomes are known in *Melandrium*.
Such sex-linked genes, as well as markers for specific autosomes, could be rather
useful in genetic experiments in conjunction with polyploids and fragmented
Y-chromosomes to obtain a better understanding of the location of genes controlling sex regulation. No data resulting from such studies are available. It would
be fair to argue that *Melandrium* deserves an experimental, rather than an
interpretational approach to the physiology of sex regulation. Future experimental
work should utilize the wealth of genetic sex types. The regulation of sex expression
in *Melandrium* may then be investigated by means of thermophotoperiodic studies,
by application of growth regulators and by floral bud culture. The last of these,
used several years ago for the study of sex expression in *Cucumis* (GALUN
et al., 1962, 1963; PORATH and GALUN, 1967; GALUN and PORATH, 1970),
has yielded interesting results recently in the study of the sex regulation of
Cleome (JONG and BRUINSMA, 1974a).

Several other plant systems in which the Y-chromosome has a decisive role
in sex determination were reported (WESTERGAARD, 1958; LEWIS and JOHN,
1968). The information on these is less detailed than on *Melandrium* and, therefore,
will not help us to understand this system better. We may, however, refer to
one of these—the sex determination in species of *Rumex* subgenus *acetosella*.
This subgenus consists of four species with 7 chromosomes as the basic number
(x), the 2n chromosomes numbering thus 14, 28, 42 and 56, respectively. They
are all dioecious and have basically the *Melandrium* type of Y-chromosome
sex determination. SINGH and SMITH (1971) reported on a hexaploid $(6x=42)$
R. acetosella with homomorphic "sex chromosomes" designated S^F and S^m.
The S^m chromosome plays a decisive role in sex determination, probably by
inhibiting, according to these authors, the female potency located in the S^F
chromosome. The autosomes were not assigned a major role in sex regulation.
Sex was therefore assumed to be controlled by a few major- or several minor genes
in the two homomorphic sex chromosomes. Thus, we are faced here with a
unique situation: a hexaploid with only one pair of sex chromosomes, and
these being homomorphic.

3.2.1.6.2 The X-Autosomal Balance System

This is the classical "Drosophila System" studied by BRIDGES (1939). It is represented in plants by species of the *Rumex* belonging to the subgenus *acetosa*.

This subgenus includes species with a wide range of sex expressions, e. g. hermaphrodite, monoecious, and dioecious. The chromosome number is also quite variable among these species (LÖVE and SARKAR, 1956) and even within the same species. Moreover, this subgenus includes also a species, *Rumex paucifolius*, having the *Melandrium* type of active Y-chromosome sex regulation (SMITH, 1968). Therefore, *Rumex* subgenus *acetosa* is well suited for investigators interested in the evolution of sex determination in plants. The dioecious species of subgenus *acetosa*, with the exception of *R. paucifolius* ($2n = 14$ or 28), have an overall chromosome constitution of $2A + XX$ in female and $2A + XY, Y_2$ in male plants. Detailed analyses of the role of sex chromosome in sex determination in *R. acetosa* were performed first by ONO and subsequently by several other investigators (see SINGH and SMITH, 1971, for references). The results of these studies, in which polyploids and aneuploids were produced and utilized, revealed a regulation which functions through the balance between the number of X-chromosomes and the number of genomal sets (A) of autosomes. Thus, in typical *R. acetosa*, the female plants are $2A + XX$ and the male plants are $2A + XY, Y_2$, representing ratios of 1.00 and 0.50, respectively. In the experiments with polyploids it was found that plants with ratios of 0.50 and lower, e. g. $3A + XY, Y_2$ were male plants and plants with ratios of 1.00 and higher, e. g. $3A + XXXX$, were female. Plants with ratios between 0.50 and 1.00, e. g. $3A + XX$, were intermediate in their sex expression. The conclusion was reached that female-promoting genes are located on the X-chromosome, and male-promoting genes on the autosomes. A detailed analysis by YAMAMOTO (1938) even assigned sex genes to specific autosomes; chromosomes a_1, a_4 and a_6 were found to promote maleness. Moreover, YAMAMOTO also found that chromosomes a_2 and a_5 promoted femaleness. Thus, female promotion is not limited to the X-chromosome. The autosomes therefore seem to contain both male- and female-promoting genes, but the net effect is male promotion. The final sex expression is thus a result of balance between the autosomes and X-chromosomes. This final conclusion was verified more recently by ZUK (1963), who produced a series of polyploids using two other species of the subgenus *acetosa*, as shown in Table 3.5.

By analyzing ZUK's results we may observe that although they conform to the X-chromosome/autosome balance hypothesis, actually there were no male plants without a Y-chromosome. Thus, the Y-chromosome may have a certain sex-regulating effect. On the other hand, in the active Y-chromosome system of *Melandrium*, both the autosome and the X-chromosome also affected sex expression. We may thus conclude that there is no sharp distinction between the two systems, nor should there be one on theoretical grounds. This is not surprising, considering *R. paucifolius*, which systematically belongs to the subgenus *acetosa* (where the X-chromosome/autosome balance prevails) but has an active Y-chromosome system for sex regulation.

3.2.1.7 Genetics of Sex Determination in Some Economic Crops

The genetic basis of sex determination of nine economic crops is summarized in Table 3.6. In this summary only the main facts concerning the major genes

Table 3.5. Karyotype, sex and sex index of *Rumex thyrsifolia* and *Rumex arifolia* and their polyploid hybrids[a]

Plant	Karyotype	Sex expression	Sex index
Rumex thyrsifolia	2A + XX	♀·♀	1.00
Rumex thyrsifolia	2A + XYY	♂·♂	0.50
Rumex arifolia	2A + XX	♀·♀	1.00
Rumex arifolia	2A + XYY	♂·♂	0.50
Polyploid hybrid	3A + XXX	♀·♀	1.00
Polyploid hybrid	3A + XXY	☿·☿	0.67
Polyploid hybrid	3A + XXX	♂·♂ → ♂·☿	0.67
Polyploid hybrid	3A + XXYY	☿·☿	0.67
Polyploid hybrid	3A + XXYYY	♂·♂ → ♂·☿	0.67
Polyploid hybrid	3A + XYYY	♂·♂	0.33
Polyploid hybrid	3A + XX	☿·☿	0.67
Polyploid hybrid	4A + XXXX	♀·♀	1.00
Polyploid hybrid	4A + XXYYYY	♂·♂	0.50
Polyploid hybrid	4A + XXXYYY	☿·☿	0.75
Polyploid hybrid	4A + XXXYY	♀·♀ → ☿·♀	0.75

[a] From ZUK (1963), with some omissions and change of index to conform with the Drosophila indexing.

affecting sex are mentioned and those interested in details are referred to a choice of publications. It should be noted that in some cases there are disagreements among authors; therefore, for a real acquaintance with the inheritance of sex the references should be consulted. Table 3.6 also gives short remarks on the utilization of sex expression for breeding. In addition to the references cited in the above-mentioned table, the reader is referred to the following reviews: ALLEN (1940), WESTERGAARD (1958) and LEWIS and JOHN (1968). Three of the crops mentioned in Table 3.6 (cucumber, hemp and maize) are described in Section 3.2.3 below.

3.2.2 Modifications of Sex Expression

3.2.2.1 Introduction

There exists a wealth of information on modifications of sex expression in spermatophytes. This information was reviewed and discussed thoroughly several times during the last half century (e. g. CORRENS, 1928; LOEHWING, 1938; HESLOP-HARRISON, 1957, 1972; NAPP-ZINN, 1967). We shall thus refrain from repeating the information here. We shall rather try to analyze the nongenetic factors involved in modifications of sex expression, giving a few examples for illustration.

Sex modification has a clear and unambiguous meaning in three types of plants: hermaphrodite, androecious and gynoecious. Thus, any diversion from the normal development of functional stamens and pistils in each flower of a plant known to be hermaphrodite while unaffected by modifying factors is

Table 3.6. Genetics of sex in some economic crops

Crop	Chromosome number	Main forms of sex expression	Heteromorphic sex chromosomes	Main genes affecting sex expression	Remarks	Main references
Asparagus *Asparagus officinalis*	$2n = 20$	Old conservative varieties are dioecious; in male plants, hermaphrodite flowers may appear. Some new varieties are genetic males or andromonoecious. Male plants are favored in agricultural practice.	Heteromorphic sex chromosomes not firmly established, but one pair divides precociously in meiosis.	Basically determined by sex chromosomes interacting with autosomal genes; thus XX = ♀♀ and XY = ♂♂ and a and G—autosomal, causing ♂ flowers in male plants. Therefore: XYaaGG = strongly ♂♀ and YYaaGG = ♂♀	Andromonoecious (XY) lines can be propagated by selfing, producing commercial XY hybrids. Anther culture was suggested to produce male lines.	FRANKEN (1970) HONDELMANN and WILBERG (1973) MARKS (1973) RICK and HANNA (1943) WRICK (1968, 1973)
Castor *Ricinus communis*	$2n = 20$	Normally ♂♀. Racemes bear ♂ and ♀ flowers on the lower and upper part, respectively, but genetic females are known. Location of racemes, ratio of ♂ to ♀ flowers in racemes and appearance of late ♂ flowers in racemes are affected by genetic and nongenetic factors. SD and high temperature promote ♂.	None	f is a recessive gene causing femaleness (male flowers may occur under certain conditions); this gene is rather stable ontogenetically and only slightly affected by environment. Polygenes affect sex ratio in racemes. An unusual genetic factor (not stable?) causes femaleness and is affected by modifiers regulating time and frequency of reversion to monoecism; this factor is also strongly affected by environment.	By utilizing the different genes affecting sex, hybrid seeds are commercially in use.	ATSMON et al. (1962) CLASSEN and HOFFMAN (1950) JAKOB and ATSMON (1965) SHIFRISS (1956, 1960) WEISS (1971)

Table 3.6. Continued

Crop	Chromosome number	Main forms of sex expression	Heteromorphic sex chromosomes	Main genes affecting sex expression	Remarks	Main references
Cucumber *Cucumis sativus*	$2n = 14$	Commonly ♂,♀ but some ♂,♀ varieties exist; ♀,♀, ♂♀ and ♂,♂ were produced during breeding. All forms except to ♀,♀ are strongly affected by environmental conditions.	None	The allele pair st⁺/st controls the flowering pattern with st causing shortening of the male and mixed (male and female) phase; additional alleles may exist; unrelated modifying genes also control sex expression. The allele pair M/m controls bisexuality of individual flowers, M causing unisexuality is dominant.	Commercial F_1 hybrid seed production is based on genetic females as seed parents; the use of hermaphrodites as pollinators in such plants is being developed.	ATSMON et al. (1962) GALUN (1961, 1973) KUBICKI (1965, 1969) PIKE and PETERSON (1969) SHIFRISS (1961)
Hemp *Cannabis sativa*	$2n = 20$	Normally dioecious, and heteromorphic—♂♂ and ♀♀ plants have a loose and a compact inflorescence, respectively; other sex types (e.g. ♂♀) were established by breeding.	Heteromorphic sex chromosomes of the X/Y type, both larger than autosomes and Y larger than X.	XX are ♀,♀ and XY are ♂,♀. xm causes appearance of ♂ flowers in female inflorescences, thus: xm/xm = ♂,♂ with female-type inflorescence.	In order to cause uniformity, monoecious plants were bred.	HOFFMANN (1941, 1952) KÖHLER (1964a) YAMADA (1943)
Hop *Humulus lupulus* *Humulus japonicus*	$2n = 20$ $2n = 16$ and 17	Normally dioecious, but monoecious individuals are known; ♂♂ are strongly affected by environmental conditions—thus bisexual flowers are produced.	*H. lupulus* = X:Y (Y smaller than X). *H. japonicus* ($2n = 16$)- ♂♂ = XY, Y₂ ♀♀ = XX.	Probably chromosomal sex determination, but no detailed study available; plants which are not unisexual are more common in polyploid plants.	Only female plants are of commercial value; they may be propagated vegetatively.	HAUNOLD (1972) JACOBSEN (1957)

Species	$2n$	Sex expression		Genetics	Notes	References
Maize *Zea mays*	$2n = 20$	Monoecious with separated male (tassel) and female (ear) inflorescences. Female flowers may appear on tassels under some environmental conditions.	None	Several genes affect tassel, causing information of female flowers (e. g. t_s, ts_3); other genes prevent ear formation (e. g. ba, sk).	Commercial varieties are pure monoecious.	BONNETT (1948) JONES (1939) NICKERSON and DALE (1955) RICKEY and SPRAGUE (1932)
Melon (cantaloupe and muskmelon) *Cucumis melo*	$2n = 24$	Mostly andromonoecious, some varieties monoecious; hermaphrodite and gynomonoecious types are known.	None	Genetic control of sex basically similar to cucumber, but different in designation and dominance relationship. The allele pair G/g controls length of male phase (gg: no male phase and M/m controls unisexuality (dominant) versus bisexuality of flowers; several additional genes modify flower morphology and pattern of flowering.	Attempts to utilize sex expression for F_1 hybrid seed production are in progress.	KUBICKI (1969b) POOLE and GVIMBALL (1934) ROSA (1928)

Table 3.6. Continued

Crop	Chromosome number	Main forms of sex expression	Heteromorphic sex chromosomes	Main genes affecting sex expression	Remarks	Main references
Papaya *Carica papaya*	$2n = 18$	Mostly dioecious, but hermaphrodites also exist; female plants are stable in various environments and rarely produce male flowers after selection; male plants are affected by environment and selection and may produce bisexual flowers.	None	Females are homogametic and males and hermaphrodites are heterogametic having the formulae: m/m, M_1/m and M_2/m respectively; M_1/M_2, M_1/M_1 and M_2/M_2 plants do not exist, as M probably represent lethality (deletion in one of the sex chromosomes?); probably female and male promoting genes are on the sex chromosomes and autosomes, respectively.	The interpretation of results are different in HOFMEYR'S and STOREY'S publications, but both authors agree on the overall genetic basis of sex determination.	HOFMEYR (1938, 1953, 1967) STOREY (1953, 1967)
Spinach *Spinacia oleracea*	$2n = 12$	Originally dioecious, but during breeding work monoecious types were developed.	None in some stocks, but in others chromosome No. 1 is heteromorphic.	Basically sex is controlled by a series of alleles — Y, X^m and X, thus $XX = ♀♀$, X^mX and $X^mX^m = ♂♀$, XY and $X^mY = ♂♂$. In addition, genetic factors exist causing changes which enable a ♂♀ phenotype in XY plants and even YY plants. Such factors probably do not exist in all breeding material, and are therefore refuted by some authors.	Many new varieties are monoecious, but occasional male plants occur in them; attempts are made to produce F_1 (mainly triploid) hybrid seeds. Strongly pistillate ♂♀ diploids or tetraploids serve as seed parents. Triploids are especially vigorous.	AKKOÇ (1965) DRESSLER (1958, 1973) JANICK and IIZUKA (1962) ROSA (1925) THOMPSON (1955)

obviously a sex modification. This definition also holds true for plants in which we have a clear previous indication that they should develop unaffectedly into either pure androecious or pure gynoecious plants. When such plants, as a result of exposure to modifying factors, show development of pistils and stamens respectively, this is a manifestation of sex modification. In other sex types listed in Chapter 1, namely, monoecious, andromonoecious, gynomonoecious and tri-monoecious, sex modification has an ambiguous meaning. This is not merely a semantic problem, but is rather basic to a clear understanding of the control of sex expression and has also an important bearing from the practical point of view.

We shall be concerned first with monoecious plants. Our reasoning for other sex types, listed above, will become obvious by simple extrapolation. In monoe-cious plants there are only two types of flowers—male (staminate) and female (pistillate) (Fig. 1.5). Referring to Chapter 1 of this book, let us assume that all flowers are potentially bisexual and that only stamens or the pistil may achieve actual differentiation in monoecious plants. We are now ready to consider sex modifications from two completely different points of view. First, we may look at the plant as a 'whole', and at certain times determine the number of male and female flowers and from this count derive a ratio of male-to-female flowers. This ratio will serve as a parameter of "sex tendency". A slight variation of this approach will be to count continuously all male and female flowers until a certain day. When investigating factors which modify sex expression, the "sex-tendency" of plants exposed to such factors will be compared with the "sex-tendency" of control plants.

On the other hand, we may focus our attention on individual flower buds located at specific sites on the plant, and ask whether or not exposure to certain factors will modify the sex of these flowers.

The latter approach has advantages in studies intended to improve our understanding of the mode of sex regulation, as will be detailed elsewhere. However, if we are only asking whether or not a certain factor (e.g. high tempera-ture) affects sex expression, both approaches are applicable, provided one assump-tion is validated: throughout the flowering pattern of the plant there is a completely random distribution of male and female flowers. Such a random distribution was never actually shown for any monoecious plant. On the contrary, many monoecious plants (e.g. maize) bear male and female flowers on different parts of the plant. In other monoecious plants the ratio of male-to-female flowers changes gradually during ontogeny in a direction which is typical for the plant in question. The change may take place by a gradual increase in the numbers of female flowers in the course of maturation, as is the case in many cucurbits (see NITSCH et al., 1952); or the change occurs within an indeterminate inflores-cence, as in banana (*Musa* spp.) and castor *(Ricinus communis)*. In the banana plant the change is from female to male flowers, while the opposite occurs normally in castor racemes. It can, therefore, be concluded that male and female flowers are not randomly distributed in monoecious plants and that in these plants there exist spatial differences in the ratios of male to female flowers as well as gradual changes in this ratio during ontogeny. Further examples will be given in our future handling of individual species. At this stage we should note that the nonrandom distribution of flowers may lead to erroneous deductions if sex expression is evaluated by "sex-tendency". Reference to Fig. 3.20 will make our point clear. In this figure we presented a scheme of a monoecious plant. Its flowering pattern is divided into three zones (I, II and III). If "sex-tendency" is determined when the plant is flowering only in zone I, the plant will be regarded as pure "male": 9 male flowers/no female flowers. Obviously when "sex-tendency" is evaluated from both zones I and II, a ratio of 17 male to 7 female flowers will be obtained; if only zone III is used for "sex-tendency" determination, the same plant will be regarded as pure female. From this schematic example we can deduce that any factor—even one not involved at all with sex regulation of the floral bud—which will cause the plant to stop flowering (e.g. at zone I or II), or, on the contrary, cause the plant to attain quick maturation and thus facilitate flowering (e.g. early flowering in zone III), will cause an *apparent* change in sex expression, when indeed sex expression is evaluated by counting male and female flowers. Factors such as temperature, light, nutrition and pruning obviously affect maturation, habitat, and time of attainment of the flowering stage in plants. Thus, changes in the ratio of male-to-female flowers caused by these factors cannot be accepted as unequivocal indications of sex modification. The unfortunate fact is that more often than not, claims as to the effects of sex-modifying factors were based in the literature on such changes in "sex tendency".

Another difficulty in deriving correct conclusions on the effects of environmental factors on sex expression in plants results from experimental procedures. In very few reported experiments could the true effect of a single factor be determined, because several of them were changed together. Thus, total light energy and duration of illumination were rarely if ever separated. Light and temperature effects can be tested only under controlled conditions and these exist only in 'ideal' phytotrons. Even in phytotrons the control of the temperature experienced by the tested plants, is questionable. Moreover, even for apparently "simple" factors such as mineral nutrition, most of the available literature gives ill-defined experimental conditions. With these reservations in mind, we can deal with the nongenetic modifying factors of sex expression.

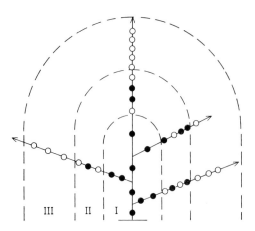

Fig. 3.20. Changes in the ratios of male (●) to female (○) flowers ("sex tendency") during the ontogeny of a monoecious plant

3.2.2.2 Mineral Nutrition and Edaphic Factors

One of the earliest reports based on experimental work on sex modification caused by edaphic factors, was concerned with the effect of soil type on sex in hemp. MOLLIARD (1897) reported that 'poor' soil favors the production of female flowers in this plant: he observed a ratio of 425 female to 100 male plants in poor soil, as opposed to a ratio of close to 1:1 of such plants in rich soil. This conclusion seems now, with the information of many experiments on hemp at hand, rather doubtful, if not wrong. The plants of MOLLIARD growing in 'poor soil' were also cultured under a short-day and low temperature regime; moreover, no chemical definitions of the 'poor' and 'rich' soils were furnished. This deficient definition of experimental conditions and lack of separating the modifying factors are quite typical of the many studies of the effect of edaphic factors on sex in plants. Thus, in spite of many publications in this field, which were summarized in the reviews cited above, no clear-cut conclusions can be drawn from such studies.

Another example will illustrate some points noted in Section 3.2.2.1. *Arisaema japonica (Araceae)* was regarded in old texts as a dioecious species, due to the fact that in natural habitats both female and male plants were observed. Detailed investigation by MAEKAWA (1924) showed that this species is actually monoecious. *A. japonica* is a perennial having a corm which each year forms two foliar leaves and a scape with a terminal spadix on a rachis. The spadix produces unisexual flowers; it develops from the axil of the upper leaf while the axil of the lower leaf holds the bud of the next year's shoot. MAEKAWA found that the same plant which was devoid of flowers during the first one or two years, produced only male flowers in subsequent years on its spadix, while in later years this plant produced only female flowers. Moreover, MAEKAWA found that culture conditions affect sex expression in *A. japonica*. Plants grown under conditions which result in a larger corm tended to become 'female' earlier. Thus, nearly all the plants in a field with corms of over 25 g were 'female', while those with corms of between 4 and 21 g were 'male-plants'. Replanting 'female-plants' in sand and under low humidity conditions, or removal of the foliar leaves, resulted in production of male spadixes by some of the transplanted plants. From MAEKAWA'S investigations we can conclude that the differentiation of the floral buds was affected by nutritional (or, indirectly, hormonal) conditions prevailing during the preceding season. Still, the nutritional conditions were not defined and under the same field conditions the plants showed a range of responses. Thus, a real causal relationship between nutrition and morphogenetic regulation cannot be drawn from the above mentioned experiments. *A. japonica* was not the only species which was erroneously defined as dioecious; a similar case is *Myrca gale* (DAVEY and GIBSON, 1917).

Begonia semperflorens will serve as another example of an apparent effect of nutrition on sex expression. This species is monoecious and METZGE (1938) reported that in rich soil the ratio of female to male flowers was 1 to 2.8, while in poor soil the ratio changed to 1 to 4.5. Such results would indicate a simple edaphic effect on sex expression. However, a closer look will reveal some uncertainties. First, the total number of flowers is drastically reduced in poor soil. Then, we should look at the flowering pattern of *B. semperflorens*, as presented schematically in Fig. 3.21. The inflorescence of *B. semperflorens* is a dichasium passing into a cincinnus; the terminal flowers are female. From this scheme it is rather clear that prevention of the inflorescence from reaching the terminal female phase will cause a shift to maleness. The flowering pattern of this plant is thus reminiscent of the theoretical scheme presented in Fig. 3.20, and the rationale is obvious.

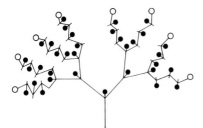

Fig. 3.21. Scheme of an inflorescence of *Begonia semperflorens* showing the pattern of male (●) and female (○) flower distribution

Turning to the individual floral bud we may ask whether or not there is a promotion or inhibition of either staminate or pistillate differentiation imposed by nutritional conditions. A study which came close to answering these questions was reported by DE JONG et al. (1974), who cultured isolated floral buds of *Cleome iberidella* in vitro under variable conditions. This species has male and hermaphrodite flowers (i. e., it is an andromonoecious species). It was found that the addition of various organic nitrogen sources (e. g. glutamic acid, urea) shortened the pistil's length, and so did the addition of NO_3 and succinate to the normal basal medium. Nevertheless, these authors decided that neither nitrogen nor carbohydrate nutrition had a direct effect on floral differentiation and rather favored the possible involvement of hormones in floral differentiation. Unfortunately, the *Cleome* floral buds were put into culture *after* sex determination; thus, all variations in relative pistil length are not relevant to the basic question regarding the factors which are decisive at the early, potentially bisexual, stage of the floral bud.

The few examples presented above briefly illustrate the general lack of precise information on the effect of nutritional and edaphic factors on sex expression in plants. Still, as noted above, the quantity of information available on the effect of these factors on sex is rather extensive and from it most reviewers of this subject concluded that, in general terms, 'rich' soil and a high nitrogen supply tend to promote femaleness.

3.2.2.3 Light

As opposed to edaphic and nutritional factors, light is a rather defined entity and thus easily applicable to experimental manipulations. Most of the information at hand concerns the effects of various daily durations of illumination on sex expression and floral differentiation, while little or nothing is known on the effects of light intensity and quality on these morphogenetic processes.

From the bulk of the information on this subject it may be generalized that, with the exception of some Long-Day (LD) plants, shortening the duration of daily illumination and/or reducing light intensity, shift sex expression toward femaleness. Or, in an even more general way, light conditions which promote flowering also promote female sex expression. This latter overall conclusion was arrived at by several authors who reviewed this subject (e. g. HESLOP-HARRISON, 1957; NAPP-ZINN, 1967); nevertheless, this conclusion was challenged. HEIDE (1969) reported results with *Begonia* which apparently showed that promotion of flowering and of female sex expression are not necessarily connected. A

Table 3.7. The effect of temperature and day length on sexuality in *Begonia cheimantha*. Flower number refers to seven plants in each treatment

Temperature (°C)	Day length (h)	Number of flowers male	female	Percent female flowers
18	8	1,512	8	0.5
21	8	862	7	0.8
18	24	889	148	14.3
21	24	372	131	26.0

Data from HEIDE (1969).

summary of her results is presented in Table 3.7. This plant is monoecious (see Fig. 3.21), and flowering is promoted by low temperature and short days. Daily light durations and temperature interact within a certain range; flowering occurs even in continuous light, provided the temperature is kept below 24 °C. Thus, *Begonia* cannot be classified as a 'true' Short-Day (SD) plant. The data of Table 3.7 show that under both 18 °C and 21 °C, lengthening daily duration of illumination caused an increase, rather than the expected decrease, in femaleness. Still, we should note a reservation: from HEIDE'S results it is not clear whether or not an 8-hour daily light period is optimal for flowering; it may be too short, i.e., suboptimal. If the latter possibility is correct, then one cannot accept HEIDE'S claim because several cases are known where a very low light intensity or a very short daily illumination reverses the effect of normal SD conditions. We shall see below that this was found in cucumber, where there is good evidence that SD promotes femaleness.

Xanthium pennsylvanicum is a monoecious species, which served in numerous experiments as a model SD plant. NIEDLE (1938) reported that LD treatments of plants which were previously induced to flower (by exposure to SD), caused an obvious shift toward maleness, in conformation to the generalization stated above. Supporting evidence along this line came from NAYLOR (1941). However, more recent results (VON WITSCH, 1961) with other *Xanthium* species which are Day-Neutral in respect to photoperiodism, indicated that lengthening the daily illumination shifted sex expression toward maleness; there is thus no evidence in *Xanthium* for a fixed connection between floral retardation and sex expression.

Let us at this stage reconsider the subject of sex modification by looking at it as a process which has to be determined in each individual floral bud, or as a result of a process which is determined at the whole-plant level. The early investigators, at the beginning of this century, looked at male and female plants just as zoologists looked at male and female animals, i. e., they assumed that the determination of sex occurs only once in the life span of the plant. Working with dioecious plants they therefore estimated the effects of sex modifiers by changes induced by them in the ratio of male to female plants in a segregating population. This is exemplified by the experiments of TOURNOIS, which are worth special attention. In studies with *Humulus japonicus* (Japanese hop), TOURNOIS (1911) was the first to demonstrate photoperiodism: flowering was induced by keeping the plants for 18 h daily in black boxes, while reversion to vegetative growth occurred when the plants were exposed to LD. SD conditions also affected sex expression: rather than the usual 1:1 ratio of male-to-female plants, TOURNOIS observed under SD conditions a ratio of 30 to 48. However, as already indicated by TOURNOIS, even the 30 male plants had some stamens with carpelar structures. Thus, sex expression could not be regarded as determined explicitly on the whole-plant level. TOURNOIS also reported a change in the ratio of male to female plants caused by light regime in another dioecious species: *Cannabis sativa* (hemp). These as well as quite a number of similar cases were reviewed by several authors, as indicated above.

The many observations which indicated vast effects of environmental conditions on sex expression convinced SCHAFFNER (see SCHAFFNER, 1935, for review of his publications) that genetics has nothing to do with sex determination, either in plants or in animals. This view, now obviously known to be wrong, was held by SCHAFFNER and some of his contemporaries for several years. Experimental evidence against SCHAFFNER'S assumptions is overwhelming. We shall mention here one study of BORTHWICK and SALLY (1954) on hemp. In conformation with previous reports, these authors showed that when daily illumination exceeded 17 h, flowering was inhibited in hemp; in photoperiods of between 10 and 19 h, the ratio of male to female plants gradually increased. However, BORTHWICK and SALLY found that individual genetic lines differed significantly in their response to the light regime. Another observation of this study was that when plants were pruned to obtain a Y shape, and each of the two main branches was exposed to a different photoperiod, there was no transfer of effect between these branches. A final note about this study: photoperiods causing incipal flowering induced only male flower differentiation; moreover, these flowers were initiated in leaf axils which were quite far removed from the shoot tip; and this is unusual for male flowers of hemp, which under normal inductive photoperiods are borne very close to the shoot tip.

A very sound approach to the study of the effect of photoperiods on floral differentiation was taken by BATCH and MORGAN (1974), who studied spike and floret development in barley. They used plant material which reacted strongly to day length: LD promoted flowering and SD suppressed it. BATCH and MORGAN exposed barley plants to LD and found that under the conditions of their experiment, spike differentiation went through a sequence of developmental stages. Groups of plants were thus removed from LD at specific stages, and transferred to SD conditions. This experiment is summarized in Table 3.8. It was found that the shift to suppressive photoperiods at any time after 21 LD

Table 3.8. Effects of photoperiod on fertility of barley. Plants were exposed to photoinductive LD and after different periods were shifted to SD. The developmental stages of the spikes of the main shoot, at the time of shift to SD, were determined by dissection

| | Days in LD before shift to SD | | | | |
| | 21 | 35 | 42 | 56 | |
Develop. stage of spike (main shoot) at transfer to SD	—	"Double-ridge"	Stamen initials	Awn initials	Final spike differentiation
No. of spikes/plant	3.9	3.7	3.4	3.9	3.6
No. of grains/plant	63	60	60	52	27
No. of florets/plant	98	96	92	101	92
Percent fertile florets/plant	65	63	64	51	30
Percent fertile florets/main shoot	90	76	70	15	67
Percent fertile florets/shoot of 3rd leaf tiller	55	61	77	68	6

Data from BATCH and MORGAN (1974).

did not reduce the number of spikes per plant. The number of florets per plant was also not affected. A late transfer to SD did reduce the number of grains per plant. The fertility of florets was strongly affected by the transfer to SD; moreover, in those plants transferred to SD after 42 days under LD, the florets of the main shoot were almost completely infertile. Further examination showed that actually only the pollen of these florets had degenerated, whereas ovary development was not affected. The "awn initials" stage coincided with the beginning of meiosis in the anthers; thus, it seems that this stage is most sensitive to the photoperiodic regime. The fertility of florets of the spike of the third leaf tiller was strongly reduced by transfer to SD at 56 days after the beginning of LD treatment. This 14-day delay relative to the main shoot was an obvious result of the later stage of floret differentiation in the spikes of the tillers. This approach of relating the effect of an environmental factor to a specific sensitive stage and tissue within the differentiating floral bud is instrumental for a real understanding of the morphogenesis of sexual organs (HESLOP-HARRISON, 1972). We may recall that according to HESLOP-HARRISON's scheme (Fig. 1.2), once initials of a certain floral member are "committed", they will continue their differentiation in a predetermined way. Thus, stamen initials will either stop their development or develop into normal stamens. This is the normal trend in most plants and under usual conditions, but there are exceptions, many of which were reported by CORRENS (1928). In these exceptions floral

Fig. 3.22. Stigmatized anther of a *Nicotiana tabacum* mutant regenerated from an X-ray-radiated mesophyll protoplast cultured in vitro

members change their normal development during ontogeny, resulting in 'sex-reversions'. This may result in anthers with carpels or stigmata-like structures. Certain light conditions seem to evoke such modifications, as mentioned above for hemp and as known in *Carica papaya* and several other dioecious species. Such modifications are not restricted to dioecious species, nor are they caused by environmental conditions only. 'Sex-reversions' within a floral member are known even in hermaphrodite plants, and some of these are controlled genetically. An example is presented in Fig. 3.22.

3.2.2.4 Temperature

The effect of temperature on sex expression of plants has been recorded for a number of dioecious and monoecious species, among them several important crops. Although few of these reports were based on very carefully executed experiments, they led to the conclusion that within external conditions allowing normal growth and development, high and low temperatures promote maleness and femaleness, respectively (see for references NAPP-ZINN, 1967; HESLOP-HARRISON, 1972).

In some dioecious species (hemp), a shift from higher to lower temperature usually caused not only an increase in the ratio of female to male plants, but also the appearance of female flowers on male plants. It thus seems that the change in sex regulation can take place at various stages in the ontogeny of the plant. A very early temperature shift will revert 'genetic male' plants into female phenotypes, a later shift will cause a change in sex expression in individual floral buds resulting in monoecious plants, and a very late shift may induce the occurrence of bisexual flowers. Bisexual flowers were actually recorded in normally dioecious species under certain thermal conditions.

Temperature-induced changes in sex expression in 'genetic females' are clearly observable in perennial dioecious plants. Thus, LANGE (1961) reported femalization of male papaya trees caused by low temperature.

Although in general terms trees seem to conform with annual crops in showing stronger female tendency at lower temperatures, the factual data are scarce. A study was made by BADR and HARTMANN (1971) in which controlled temperature conditions were applied to olive trees. The olive is usually an andromonoecious plant, but when exposed to an insufficient cold floral induction period, male flowers (with rudimentary pistils) predominate. The temperature regime during the inductive periods seems to affect sex expression quite considerably (Table 3.9).

Table 3.9. The effect of temperature regime on sex expression of 'Ascolano' olives

Treatment No.	Environmental conditions	Flower buds produced (%)	Perfect flowers produced (%)
1	Outdoors: Dec. 13–Feb. 10	23	68.5
2	7 °C for 20 h, 26 °C for 4 h: Dec. 13– Feb. 4	62	21.3
3	12.5 °C for 20 h, 21 °C for 4 h: Dec. 13–Feb. 4	63	3.5

Data from BADR and HARTMANN (1971); number of trees per treatments 1, 2 and 3 was 5, 10 and 10, respectively; all trees were grown outdoors until Dec. 12, and returned to the greenhouse after treatment; outdoor winter temperatures were approximately 0–5° and 10–15 °C during night and day, respectively; the greenhouse was kept above 16 °C.

In monoecious plants, changes in sex expression caused by shifts to different thermal conditions are easily recordable. In some of these (cucumber and maize) the effect of temperature on sex expression was examined under defined conditions, and it was clear that under low temperature regimes female flowers occurred at sites normally occupied by male flowers in plants kept under higher temperature regimes. Examples of such experiments will be presented below, along with a detailed discussion of the sex expression of these plants.

In crops belonging to hermaphrodite species, by definition, there is no 'trigger mechanism' in their floral buds; thus, both stamens and pistils attain their final morphological development. In such species, changes in sex expression are manifested by the relative final size and development in the reproductive members of the flowers. Unfortunately, we have no reliable information on the effect of temperature on the sex expression of these species, as separated from other environmental conditions. The observations on temperature-induced sex modifications in hermaphrodite species were based almost exclusively on plants grown in a complex of environmental conditions which did not allow strict separation of the effect of temperature. Thus, HOWLETT (1939) reported a vast variation in the relative size of pistil and stamens of tomato varieties grown under different environmental conditions. He found that it was due mainly to an increase in pistil size of plants grown in the winter months, that an increase occurred in the pistil/stamen ratio. HOWLETT attributed this change to the prevailing short day, giving little or no reference to temperature. Based on observations by SMITH (1932) and later workers (e.g. OSBORNE and WENT, 1953), we tend to conclude that low temperatures, by themselves, promote pistil development and suppress functional pollen production, rendering winter-grown tomato plants partially parthenocarpic.

Extreme high and low temperatures may strongly affect reproductive members in floral buds. Such effects cannot be categorized under environmentally controlled changes in sex expression, but they are of considerable economic importance. In inflorescences of the dwarf Cavendish banana, which attain floral differentiation under low winter temperatures (below 12 °C), the embryonal bisexual flowers develop into abnormal female flowers, lacking one or more carpellar leaves. Such flowers result in fruits which are unacceptable on the market (FAHN et al., 1961).

RYLSKI, who studied the effects of environmental conditions on flowering, fruit-set and fruit-shape of pepper found overdeveloped and deformed ovaries in flowers which attained anthesis under low (8–10 °C) night temperatures (RYLSKI, 1973). Flowers with split ovaries and multiple styles were observed by NOTHMANN and KOLLER (1973 a) in eggplant plants growing under low temperatures, although such conditions did not increase the length of the style.

Cereals were also reported to be affected by extreme environmental conditions. GREGORY and PURVIS (1947) reported teratological development of anthers and pistil in barley following a very dry and hot period in the early summer; and MELETTI (1961) reported that conversion of anthers into carpellar structures occurred in a wheat variety kept under constant illumination and low temperature.

Floral abnormalities caused by very high or very low temperatures are of little economic importance as long as the crops are cultivated during their normal season, but with the increasing tendency to produce food out-of-season, these abnormalities should be considered seriously in both breeding work and crop production. The reader is referred to MEYER (1966) for a thorough review of this subject.

3.2.2.5 Chemical Agents

In this section we shall deal with the effects of the main groups of plant growth regulators on sex expression. We shall not present a comprehensive review of this subject because this has already been done quite thoroughly and critically (see HESLOP-HARRISON, 1972). Additional information on the effect of plant hormones on sex expression of several crops will be presented in Section 3.2.3. We shall thus confine ourselves mainly to some basic considerations and present a selected group of experimental results on this subject.

Experimental modification of sex in angiosperms by chemical agents was probably first reported by MININA (1938), then working in Kiev. She treated cucumber plants by the 'Klin' method: the greenhouse was sealed when the plants reached the 3 to 4-leaf stage and flue gas consisting mainly of CO was introduced, up to about 1 % of the greenhouse atmosphere. An 11 to 12-hour treatment caused a vast change in sex-tendency, increasing the ratio of female to male flowers. MININA'S results were verified in her own further studies as well as by HESLOP-HARRISON and HESLOP-HARRISON (1957). The latter reported on the effect of CO on the sex expression of *Mercurialis* and later attributed the sex-modifying effect of the flue gas to its ethylene component rather than to CO (HESLOP-HARRISON, 1959). This strong ethylene effect was indicated more directly some years later by MCMURRAY and MILLER (1968). These authors applied 2-chloroethylphosphonic acid (a compound which decomposes in the plant and releases ethylene) to monoecious cucumber plants. One 240-ppm spray of this compound resulted in 19 consecutive nodes bearing only female flowers. This vast change in sex expression seems to be quite outstanding, but basically at least a significant change of sex expression in cucumbers and related species has been indicated by other studies (see Section 3.2.3.1).

RESENDE and co-workers attempted to rationalize the effect of growth regulators on sex expression in plants (RESENDE, 1967). They argued that the floral members can be arranged according to their 'vegetative' and 'reproductive' character in the following order: calyx, gynoecium, corolla and androecium. Thus, growth regulators, such as auxins, which promote vegetative development, cause hypertrophy of the calyx and the gynoecium. In accordance with this assumption, RESENDE (1953) found that when IAA was injected in *Hyoscyamus niger* stems, flowering was retarded and in those floral buds which did develop, there was an outgrowth of the calyx and the gynoecium and a corresponding suppression of the corolla and androecium. Support for a parallel reaction of calyx and gynoecium was furnished by HESLOP-HARRISON (1959).

There are two basic questions about the response of plants having diclinous (unisexual) flowers to compounds affecting sex expression: (1) Is the change in sex expression realized by suppression of the reproductive members of the floral bud or by a direct promoting activity? (2) Do the growth regulators affect the embryonal floral buds themselves, thus causing them to develop further into either male or female flowers, or do these regulators act at an earlier stage, perhaps indirectly, thus inhibiting or promoting potential male or female floral buds from incipient development? Evidence was furnished in favor of contradictory answers for each of these questions; some of it will be discussed below in respect to sex expression of cucumber, but it may well be that there exists no general answer to these questions and different regulatory mechanisms exist in specific species and in response to specific regulators.

3.2.2.5.1 Auxins and Related Compounds

Cucumber was the first plant in which it was shown that auxin can alter sex expression (LAIBACH and KRIBBEN, 1949). Thus, disregarding unconfirmed previous reports on the effect of animal hormones on plant sex, cucumber can be regarded as the first plant in which sex expression can be modified by a defined chemical compound. The above authors applied α-naphthaleneacetic acid to monoecious cucumber plants and observed a substantial increase in the number of female flowers. Further studies by LAIBACH and KRIBBEN (Table

Table 3.10. The distribution of female flowers in the first seven leaf axils of monoecious cucumber plants treated with auxin

Treatment	Number of female flowers in successive leaf axils							Number of female flowers in the first seven leaf axils
	1	2	3	4	5	6	7	
Control	—	—	1	2	—	—	2	3
0.1 % 2,4 dichlorophenoxy acetic acid	—	—	—	—	2	—	—	3
0.1 % naphthaleneacetic acid	4	—	8	1	10	11	8	45
0.1 % indoleacetic acid	—	—	1	2	1	2	2	11

Data summarized from LAIBACH and KRIBBEN (1951); auxins were applied in lanolin paste to the cut surface of the petioles of the first two leaves; average of ten plants, rounded off to whole integers.

3.10) and several other authors confirmed the femaleness-promoting effect of auxins in cucumber as well as in other monoecious plants. Several dioecious plants such as hemp and papaya responded similarly to auxin application. WESTON (1960) reported the development of male flowers on genetic female hop plants after 2-chlorophenyl(thio)propionic acid application and suggested that, contrary to SD plants, LD plants such as hop respond to auxin application by an increase in maleness; but this suggestion was apparently not substantiated in further studies with other LD plants.

In several species with monoclinous flowers (e. g., *Silene*, *Hyoscyamus*, *Bryophyllum*) the relative development of the androecium was retarded and that of the gynoecium was promoted after auxin application. Nevertheless, in many hermaphrodite plant species no modification of sex expression could be induced by auxin application.

As mentioned above, the ethylene-releasing compound 2-chloroethylphosphonic acid[2] caused a marked increase in the number of female flowers in cucumber (see further information in Section 3.2.3.1). KENDER and RAMAILY (1970) applied this compound to grapes at the stage of bud burst and observed the development of fertile hermaphrodite flowers in grape stocks, which normally produce only staminate flowers. DE JONG and BRUINSMA (1974b) reported a promotion of pistil development after application of 2-chloroethylphosphonic acid to *Celome*

---

[2] This compound, having the structure $ClCH_2CHP{\overset{\displaystyle O}{\underset{\displaystyle OH}{<}}}OH$, appears in the literature under several commercial and trivial names: Ethrel, Amchem. 68-62, Amchem. 68-240, Ethephon.

spinosa, although it should be noted that these authors treated the plants several days after anthesis; thus, the observed effect can not be considered as a result of an external interference in the mechanism of sex determination. In wheat it was shown that 2-chloroethylphosphonic acid suppressed anther differentiation quite specifically and the application of this compound to wheat plants was suggested as a method to produce hybrid seeds on a small scale (BENNETT and HUGHES, 1972). The time of 2-chloroethylphosphonic acid application is very critical: only when applied before and very close to meiosis of the PMC does this treatment result in the required male sterility. Treatment at any other time either did not affect floret differentiation or resulted in female sterility (HUGHES et al., 1974). The use of 2-chloroethylphosphonic acid as a tool for hybrid seed production was also suggested in barley (LAW and STOSKOPF, 1973) and, as with wheat, application time was shown to be critical for reliable results. A similar suppressive effect on male fertility was observed after α-naphthaleneacetic acid application to the hermaphrodite LD plant *Silene pendula* (HESLOP-HARRISON and HESLOP-HARRISON, 1958a).

A suppression of stamen development by maleic hydrazide was reported to occur in several crops (e.g. maize, tomato, melon and watermelon), but the applicability of this growth suppressor for the attainment of male-sterile seed parents in commercial hybrid seed production is questionable.

The information available on the effect of triiodoacetic acid (TIBA) on sex expression does not allow any final conclusion to be drawn. This compound was regarded as an "anti-auxin" in the old literature; thus, the lack of its antagonistic effect relative to auxin caused NAPP-ZINN (1967) to question the role of auxins in sex expression in plants. It is now fairly well established that triiodoacetic acid retards the polar movement of auxin in plant tissue, and thus it may well cause accumulation of auxin at specific sites. Hence, under certain conditions, it may simulate auxin application.

From the consideration pointed out above and others (see HESLOP-HARRISON, 1972; DE JONG and BRUINSMA, 1974a), there is no obligatory relationship between the effect of externally applied plant regulators and the internal hormonal regulation of sex expression. We are concerned here mainly with the ability of chemical agents to modify sex rather than with the role of plant hormones in sex regulation; we shall, therefore, not elaborate on this question but point out that since there seems to be no safe 'short cut', the effects of auxins on the crops in question must be tested empirically.

3.2.2.5.2 Gibberellins

The role of gibberellins (GA) as endogenous regulators of growth, differentiation and enzymatic activities is now well established, and their various effects are being utilized advantageously in agriculture. The effect of gibberellin application on sex expression of plants was studied by numerous investigators, but in retrospect it now seems that a clear change in sex expression following gibberellin treatment was proven in a rather small number of species. Moreover, even among these species the effect was not unidirectional.

A substantial change in sex expression following gibberellin application to cucumber was found by WITTWER and BUKOVAC (1958), and GALUN (1959b). The latter found that heterozygous-female plants, which normally produced only female flowers in the upper nodes of their main shoots (and only female flowers in their side branches), could be 'converted' by repeated gibberellic acid (GA_3) treatment into plants having the flowering pattern of normal monoecious plants (Fig. 3.23). On the other hand, in another species of the same genus (melons), sustained efforts to cause a change in sex expression by gibberellin application were fruitless. We shall deal further with the effect of gibberellins in Cucurbitaceae in Section 3.2.3.1.

Control-no treatment

Treated for one week

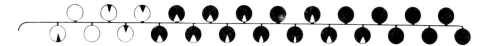

Treated for two weeks

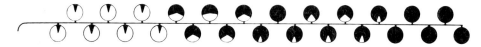

Treated for three weeks

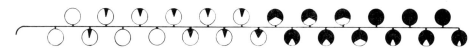

Treated for four weeks

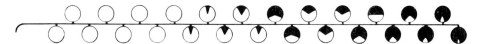

Fig. 3.23. Change in sex expression in heterozygous-female (st^+/st) cucumber plants treated with gibberellic acid. *Circles:* mean sex at specific nodes *clear areas:* proportion of nodes bearing staminate flowers *shaded areas:* nodes bearing female flowers (data from GALUN, 1959b)

In castor beans *(Ricinus communis)*, application of GA_3 was reported by
SHIFRISS (1961a) to increase the ratio of female to male flowers in the racemes, and repeated treatment caused some racemes to produce only female
flowers. This change in sex tendency was accompanied by a substantial reduction in the total number of flowers per raceme. SHIFRISS (1961a) attributed the
contradictory effects of gibberellins in cucumber and castor beans to their
different response to day length, since in respect to female flower development
these plants can be regarded as SD and LD plants, respectively; thus, gibberellins
were claimed to simulate the LD effect. Such a general explanation does not
conform with the results of WITSCH (1961), who used a *Xanthium strumarium*
line in which female flowers were promoted by SD as well as by gibberellin
application.

PHATAK et al. (1966) reported that anther differentiation and pollen production
in tomato occurred in a stamen-less (sl) mutant exposed to 10^{-4} M GA_3 in
its root medium. This gibberellin effect was restricted in time: flower buds
already visible at the beginning of treatment did not produce anthers, and
flowers which appeared three weeks after treatment cessation showed the
mutant phenotype. KUBICKI and POTOCZEK (1972) utilized this gibberellin effect
to self-pollinate 'gynoecious' (sl) lines and suggested accordingly a method for
F_1-hybrid seed production in tomatoes. The commercial applicability of such
a method still awaits verification (see also Section 3.4.3).

A change from female to hermaphrodite flowers was reported by RESENDE
and VIANA (1959) to occur in *Bryophyllum* after the application of GA_3 (0.005 %),
but a higher GA_3 concentration as well as the utilization of a related species
resulted in an opposite shift in sex expression.

Contradictory results were reported for the effect of gibberellins in hemp.
These will be detailed in Section 3.2.3.2.

Some reports indicate that gibberellins cause malformation of stamens, rendering them nonfunctional (e. g. in maize: NELSON and ROSSMAN, 1958; in rice:
NUNES, 1964), or of ovaries (e. g. in eggplant: NOTHMANN and KOLLER, 1973b),
but such effects should not be regarded as changes in sex expression and are
of little importance in breeding and crop production.

3.2.2.5.3 *Kinins*

Kinins differ from auxins and gibberellins in respect to their distribution in
the plant after external application. Gibberellins move rather quickly from the
application site to other parts of the plant. The distribution of applied auxins
is slower, partially polar, and is known to be controlled by internal mechanisms
of the plant tissues. Unless applied to the roots, kinins seem to move very
slowly in plants, and thus their effect after foliar treatment is probably restricted
to the vicinity of the application site. This restricted distribution led most investigators interest in the effect of kinins on sex expression to apply kinins directly
to the developing floral buds. An exception was the work of BOSE and NITSCH
(1970), who dipped *Laffa acutangula* seed in benzyladenine and reported a marked
change in sex expression toward femaleness in the developing plants. One could
assume that the benzyladenine or its active derivative persisted in the young

plant until the differentiation of the floral bud could be affected. The sex expression of no other plants of the Cucurbitaceae is known to be affected by kinins.

Kinin-induced changes of sex expression were reported in some other species (e. g. in *Bryophyllum* by CATARINO, 1964; in *Mercurialis* by DURAND, 1967), but detailed information on the effect of kinins on sex expression is available only for grapes *(Vitis vinifera)*. NEGI and OLMO (1966) dipped inflorescences of a male grape clone in 1,000 ppm of the synthetic kinin SD 8339 [6-(benzyl-amino)-9-(2-tetra-hydropyranyl)-9 H-purine] at about three weeks before anthesis and observed a total conversion of the male flowers into perfect flowers. Several other growth regulators used in the same way (e. g. auxins, TIBA and GA_3) failed to affect the sex expression of this clone. It should be noted that this male clone, a *sylvestris* variant of *V. vinifera*, is heterozygous to the dominant gene Su^F and even without kinin treatment there is a low frequency (10^{-4}) of hermaphrodite (perfect) flowers in this clone; the frequency seems also to be affected by external factors. In Su^F plants there is an initiation of the ovary and the differentiation of it continues up to the second meiotic division, but no functional embryo sac is formed and the pistil is underdeveloped. Kinin application restores normal development of the pistil and the embryo sac (NEGI and OLMO, 1972). IIZUKA and co-workers investigated the effect of kinins and related compounds in several male stocks of *Vitis* species (see HASHIZUME and IIZUKA, 1971). They found that not only synthetic kinins but also naturally occurring cytokinins such as zeatin cause the formation of functional perfect flowers in such male stocks (Table 3.11).

In some cases it is difficult to draw a boundary line between the control of male and female floral-member organogenesis and mere stimulation of the differentiated stamens and pistil. The results of DE JONG and BRUINSMA (1974a,

Table 3.11. Sex conversion in male *Vitis* plants following treatment by kinins

Species and treatment	Percentage of flowers			
	Total no. of flowers	With fully developed female parts	With partially developed female parts	Without female parts (male flowers)
Vitis thumbergii				
Control		0	0	100
Zeatin (4)	1,124	54.1	13.7	32.2
Dihydrozeatin (2)	263	10.3	61.2	28.5
6-BAR (10)	3,130	76.8	13.2	10.0
Vitis coignetiae				
Control		0	0	0
Zeatin (6)	144	24.8	36.6	38.6
Dihydrozeatin (5)	136	7.8	41.8	50.4
6-BAR (21)	622	21.0	42.2	36.8

Data from HASHIZUME and IIZUKA (1971). Inflorescences were dipped for 20 sec in 250–1,000 ppm solutions of the kinins, 1 to 40 days before anthesis; numbers in parentheses indicate number of treated inflorescences; 6-BAR = 6-benzylamino-9-β-D-ribofuranosylpur-ine.

b) with *Cleome* seem to indicate only promotion of the already differentiated pistil: zeatin stimulated pistil growth of in vitro-cultured flowers and slightly reduced female abortion in inflorescences treated daily.

3.2.3 Sex Expression in Some Economic Crops and its Application to Breeding and Crop Improvement

3.2.3.1 The Cucumber and Other Cucurbit Crops

Cucumber *(Cucumis sativus)*, melon *(Cucumis melo)*, watermelon *(Citrullus vulgaris)*, and three species of the genus *Cucurbita* comprise the main economic crops of the family Cucurbitaceae. Detailed descriptions and further references concerning cultivated species of this family may be obtained from WHITAKER and DAVIS (1962) and PURSEGLOVE (1968).

The cucumber is outstanding among these species in several respects. With the exclusion of some tropical regions, it is an important relish throughout the world, being distributed over almost all cultivated areas, where it is grown either in open fields or under cover (plastic sheets or glass). In no other crop has sex expression been so extensively investigated and utilized for breeding and crop improvement as in cucumber. Moreover, the sex regulation in *C. sativus* served in many investigations as a model system for the study of morphogenesis in higher plants. For these reasons as well as due to our personal familiarity with this crop, we shall handle the sex expression of cucumber in some detail. We shall also outline how this information can be utilized for hybrid seed production and better crop management.

3.2.3.1.1 Patterns of Sex Expression

The family Cucurbitaceae consists of about 100 genera with over 800 species, distributed in tropical, subtropical and temperate regions of the globe. Most of the species are native to the warm regions of East and South Africa and Madagascar.

The family excels in sex variability. This variability can be found both within genera and within species. True hermaphrodite species are rare or nonexistent (the taxonomy of this family is rather problematic; see: JEFFREY, 1967), but in some species hermaphrodite plants do occur *(Bryonia dioica)*. Thus, almost all species have either only diclinous flowers (i.e. male and female flowers), or both monoclinous and diclinous flowers. The first group includes dioecious and true monoecious species, while the second group consists primarily of species with andromonoecious plants; gynomonoecious plants are rather rare. Table 3.12 summarized the sex expression of 210 African species comprising 40 genera. In 13 of these genera only dioecious taxa were reported, while in another 14 genera no dioecism was observed.

The cucumber and the related 'wild' *Cucumis hardwickii* have several features in common. Both have seven chromosome pairs and cross freely, producing

Table 3.12. Distribution of sex expression in African Cucurbitaceae species

Sex expression within each species	Geographic distribution		
	East Africa	South Africa	Madagascar and Comoro
Monoecious only	56	40	30
Mono- and dioecious	3	1	—
Dioecious only	64	23	34
Total	123	64	64

Data compiled from information kindly communicated by C. JEFFREY, Royal Botanic Gardens, Kew, Richmond, U. K. Monoecious *sens. lat.*, including andromonoecism; some species occur in two or three geographic areas, and thus are not tabled horizontally.

fertile progeny (see LEPPIK, 1966; DEAKIN et al., 1971, for references). The cultigen *C. sativus* probably originated in northern India, where *C. hardwickii*, presumably an escapee from a primitive cucumber, is also found. These species stand apart from other *Cucumis* species having 12 or 24 pairs of chromosomes and are indigenous to tropical Africa. The geographical distribution of 26 *Cucumis* species is listed in Table 3.13.

Cucumbers are known from historical records starting about 4,000 years ago (Twelfth Egyptian dynasty). They then appeared in several separated cultures and were kept there in relative isolation for hundreds of years. It is thus not surprising that within *C. sativus* (in the broader sense, including ill-defined species such as *Cucumis sinensis* and *Cucumis flexuosis*) there is a great variation in respect to sex expression and fruit structure. This variability is expecially clearly manifested when two plants with different flowering pattern and different geogra-

Table 3.13. Sex expression and distribution of *Cucumis* species

Annuals		*Perennials*	
C. sativus	M; Asiat.	*C. prophetarum*	M; Afr. N. E. and SW Asiat.
C. hardwickii	M; Asiat.	*C. zeyheri*	M; Afr. S.
C. muriculatus	M; Asiat.	*C. ficifolius*	M; Afr. E.
C. hystrix	M; Asiat.	*C. figarei*	M; Afr. E.
C. sacleuxii	M; Afr. E., Ma.	*C. aculeatus*	M; Afr. E.
C. metuliferus	M; Afr. Trop., S.	*C. globosus*	M; Afr. E.
C. melo	M; Afr. Asiat. and Australasia	*C. meeusei*	M. Afr. S.
		C. rigidus	M; Afr. S.
C. humifructus	M; Afr. Trop., S.	*C. sagittatus*	M; Afr. S.
C. quintanilhae	M; Afr. S.	*C. callosus*	M; Asiat.
C. anguria	M; Afr. S.	*C. heptadactylus*	D; Afr. S.
C. dipsaceus	M; Afr. E., S.	*C. hirsutus*	D; Afr., Ma.
C. africanus	M; Afr. S., Ma.	*C. kalahariensis*	D; Afr. S.
C. myriocarpus	M; Afr. S.		

Data compiled and kindly communicated by C. JEFFREY, Royal Botanic Gardens, Kew, Richmond, U. K. M: monoecious (*sens. lat.*, including andromonoecism), D: dioecious, Afr.: Africa, Asiat.: Asiatic, Trop.: Tropical, S: South, E.: East, Ma.: Madagascar, N.: North, SW: South west.

phical origin are crossed. Thus, PANGALO (1936) reported that a single cross between a monoecious and an androecious plant resulted in an F_2 generation of 338 plants which could be divided into 15 different forms of sex expression. This segregation prompted PANGALO to suggest no less than ten main genes to be involved in sex expression of cucumber! As we shall see, sex expression even of the same genotype is strongly influenced by environmental conditions. Therefore, a familiarity with the basic flowering pattern and the physiology of sex differentiation is a prerequisite for solving the inheritance of sex expression in cucumber.

The embryonal flower bud of cucumber has both stamen and ovary initials (ATSMON and GALUN, 1960), but the flowers of most cucumber varieties mature as diclinous flowers. Most varieties are thus true monoecious. However, even within this definition there is a vast variability of flowering pattern, quite typical for each variety cultured under certain environmental conditions (SHIFRISS and GALUN, 1956). This variability is expressed theoretically by the distribution of male and female flowers in all the leaf axils of the plant. In monoecious cucumber plants, in practically all leaf axils (with the exclusion of a few basal nodes of the main shoot), there are either several male flowers or individual (rarely two or more) female flowers. Therefore, in such plants the distribution pattern of 'female-nodes' will faithfully describe their sex expression. In most cucumber varieties there is an increase in tendency to femaleness as we move acropetally along the shoots (CURRENCE, 1932). Such a change was observed in other cucurbits by WHITAKER (1931) and detailed by NITSCH et al. (1952) for *Cucurbita pepo* (Fig. 3.24). Thus, for the evaluation of the change in sex expression caused by external factors, the location of the first female flower on the main shoot is rather useful (SHIFRISS and GALUN, 1956). This criterion is not adequate for describing different sex genotypes because these may differ in ways other than the node location of the first female flowers (KUBICKI, 1969 a).

A few cucumber varieties are andromonoecious (e. g. Lemon, Richmond Green Apple). Plants of these varieties bear both diclinous (male) and monoclinous flowers. Such plants differ from monoecious cucumbers not only by 'replacement' of female by hermaphrodite flowers. We mentioned that female flowers are commonly borne individually in the leaf axils. Hermaphrodite flowers, on the other hand, are usually followed in the same axil by additional floral buds which may attain anthesis, and in the latter flowers there is an increased male tendency.

Cucumber flowers are variable in another respect: the relation between the receptacle to other floral parts (Fig. 3.25). Male flowers are *hypogynous* and female flowers are normally *epigynous*. In hermaphrodite flowers, all the grades between *hypogynous* through *perigynous* to epigynous are known (KUBICKI, 1969 a).

The components participating in forming the patterns of sex expression of cucumber may now be listed as follows:

1. *Sex Expression of the Floral Bud.* The embryonal floral bud is bisexual. Further development of the pistil, the stamens, or both pistil and stamens, will result in a female, a male, or a hermaphrodite flower, respectively.

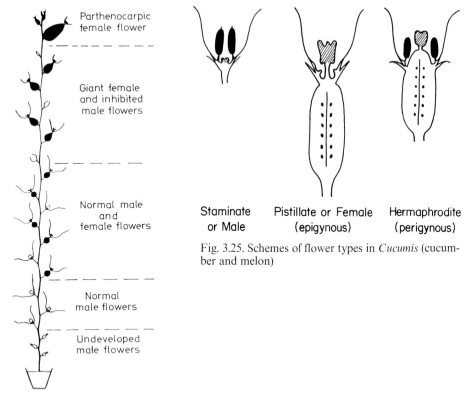

Staminate or Male | Pistillate or Female (epigynous) | Hermaphrodite (perigynous)

Fig. 3.25. Schemes of flower types in *Cucumis* (cucumber and melon)

Fig. 3.24. Schematic presentation of the sequence of flower types on the main shoot in a squash (cv. Acorn) (from NITSCH et al., 1952)

2. *Flowering Pattern in the Leaf Axils.* Due to the prevailing increase in maleness within each leaf axil, whenever the first floral bud in a leaf axil is male, additionally developing male flowers are male as well. Female flowers are borne usually individually, but if additional floral buds do occur and come to anthesis, these may be either female or male. Such male flowers were observed in genetic female plants after gibberellin treatment (FUCHS, HALEVY and ATSMON, unpublished). When the first flower in a leaf axil is hermaphrodite, subsequent flowers have an increased male tendency and true male flowers are quite common among the later developing flowers of such axils.

3. *Flowering Pattern along the Shoot.* Schematically, the main shoot of a monoecious cucumber has three phases in respect to flower sexuality (GALUN, 1973). In the most basal phase, there are only male nodes. Then, with the first female node, a mixed phase starts: female nodes are interspaced with male nodes. The nodal distance between the female nodes decreases acropetally until the attainment of the third (female) phase. In this phase, only female flowers occur. For this highly schematic pattern, although appropriate to almost all known monoecious cucumber varieties, there are obvious exceptions which we

shall not detail here. For andromonoecious varieties, a similar scheme can be described: these have a stronger male tendency, thus only the first two phases come to realization and, as mentioned above, instead of female nodes, andromonoecious varieties have nodes with one or more hermaphrodite flowers which may be followed by male flowers.

4. *Flowering Pattern in Different Shoots of the Plant.* As a rule, the main shoot of a monoecious cucumber has a stronger male tendency than the side branches. Thus, in some cases, the main shoot may be completely devoid of female flowers, and when the female tendency of a variety is strong, side branches may bear only female flowers. The distribution of female and male flowers among shoots of different orders, though strongly influenced by environmental conditions (as will be seen below), is a varietal character.

Most melon and watermelon varieties are andromonoecious. Their flowering pattern thus follows roughly that of andromonoecious cucumber varieties. Monoecious melon and watermelon varieties resemble monoecious cucumber varieties, but the former generally have a stronger male tendency and never attain the female phase. It should also be noted that all monoecious cucumber varieties have elongated ovaries, while many monoecious watermelon varieties have round ovaries resulting in spherical fruits.

Cultivated species of Cucurbita bear usually only diclinous flowers and are thus typically monoecious.

3.2.3.1.2 Effects of Day Length and Temperature on Flowering and Sex Expression

Floral induction in cucumber and other cultivated cucurbits is generally not considered to be regulated by specific thermoperiods. Flowering in these plants occurs under almost all climatic conditions which allow vegetative growth. Nevertheless, high temperatures (about 30°C) and long days (more than 16 h of light) tend to suppress anthesis and prevent the development of floral primordia in the basal leaf axils. Obviously thermal sensitivity varies between and even within species: sensitivity increases roughly from melons through watermelons and cucumbers to cultivated *Cucurbita* species. It should be noted that there are semi-wild lines of cucumber which seem to have a true SD requirement. The exact reaction of these plants to light and temperature was not investigated, but marginal conditions induce only male-flowering in these lines (FRIEDLENDER, ATSMON and GALUN, unpublished).

As pointed out above, the effect of climatic conditions on sex expression of cucumber and watermelon was observed many years ago (KNIGHT, 1819). Several authors reported environmentally induced changes in sex expression in cucurbits, but the first investigation undertaken under controlled conditions was that of NITSCH et al. (1952). They studied the effects of various temperatures and two day-length regimes (8 and 16 h) on the sex expression of a squash variety, a cucumber variety, and gerkins *(Cucumis anguria)*. The results showed clearly that within the range of 20°C to 30°C day temperature and 10°C to

30 °C night temperature, the lower temperatures promoted femaleness in both the cucumber and the squash varieties. It was also found that extension of day length promoted maleness in both monoecious (st$^+$/st$^+$) as well as in heterozygous gynoecious (st$^+$/st) cucumber plants (GALUN, 1959a). From information of many reports, most of which were not based on experiments conducted under controlled conditions, it may safely be concluded that generally speaking higher temperatures and longer photoperiods cause a shift in flowering pattern toward maleness in cucumber varieties. It may be noted that within a certain range, which is probably specific for each variety, the sex expression of monoecious cucumbers is affected similarly by either temperature or day-length changes (Tables 3.14 and 3.15).

Table 3.14. Effects of day length and temperature on sex expression of a monoecious cucumber variety (Bet Alpha)

Temperature (°C)	Day length (h)	Node location of first female flower on the main shoot
19	8	9.6
19	16	13.7
23	8	14.8
23	16	16.5

Unpublished data of LANG and GALUN.

Table 3.15. Effects of day length and temperature on sex expression of a Fushinari-type cucumber variety (Sagami-Hanziro)

Night temperature[a]	Day length (h)	Node location of first female flower on the main shoot	No. of female-flower-bearing nodes[b]
15	8	4.1	16.4
15	16	7.8	8.9
25	8	8.2	7.4
25	16	17.1	1.7

Data from ITO and SAITO (1960).
[a] Day temperature was not specified.
[b] Out of 25 nodes.

Extensive work in Japan with oriental cucumber varieties obtained originally from very different climatic conditions (mainly from China), is basically in line with the studies of European, Middle Eastern and American varieties. The Japanese studies were done at least in part under controlled conditions, and sex expression was evaluated in a rather clear way (see MATSUO et al., 1969, for reference sources). They indicated that sex expression is very variable among monoecious varieties and that the reaction of these varieties to temperature and light regimes is probably a genetic character—typical for each variety (Table 3.16). They also indicated that shortening daily illumination beyond a certain

Table 3.16. Varietal differences in sex expression and in response to temperature and daylength

Variety		Sex expression				
	Temper-ature (°C)	Photo-period (h)	Node location of first female flower	Number of male flower nodes[a]	Number of female flower nodes[a]
Kaga-Fushinari	17	8	1.0	0	30.0
	24	16	2.2	0.2	28.8
Shogoin-Fushinari	17	8	1.4	0	29.6
	24	16	3.3	1.1	27.4
Sagami-Hanziro	17	8	3.4	2.3	27.2
	24	16	15.8	23.2	4.6
Suyo	17	8	9.0	25.3	4.5
	24	16	25.8	27.8	1.2
Yamuto-Sanjyaku	17	8	15.6	25.6	4.1
	24	16	28.7	26.4	1.3
Tokyo-Aodai	17	8	12.7	26.7	2.3
	14	16	30.5	28.2	0.8
Taisen-Kema	17	8	20.0	28.4	1.3
	24	16	31.5	27.8	0.2

Data from SAITO and ITO (1964).
[a] Out of the first 30 nodes on the main shoot.

limit (e. g. 8 h) may suppress female flowers, and that the optimal temperature for promoting femaleness is a varietal characteristic entity.

A delay in appearance of hermaphrodite flowers and a less developed ovary in these flowers seem to occur in andromonoecious cucumber cultured under conditions of high temperatures and long days—but no precise information is available. A similar trend was observed in andromonoecious melon varieties, as well as in andromonoecious watermelon varieties. No reliable information on the effects of day length and temperature on the sex expression of monoecious melons and watermelons seems to exist.

3.2.3.1.3 Effects of Growth Regulators

Gibberellins seem to be the only group of growth regulators which induce development of male flowers in leaf axils otherwise occupied by female flowers in cucumbers. This observation was first reported by WITTWER and BUKOVAC (1958) for a monoecious cucumber variety. In studying self-pollination of female cucumber lines for hybrid seed production, GALUN (1959b) found (as mentioned above, see Fig. 3.23) that cucumber plants responded to repeated gibberellic acid (GA₃) treatment to such an extent that the continuous female phase could be prevented. This was confirmed for homozygous female cucumber lines by PETERSON and ANDHER (1960). Further studies showed that gibberellins differed in their effectivity to promote male flower production in cucumber: GA₄ and

GA_7 being several-fold more effective than GA_3 (WITTWER and BUKOVAC, 1962; CLARK and KENNEDY, 1969; PIKE and PETERSON, 1969). There are indications (e. g. KUBICKI, 1965a) that gibberellins do not directly promote stamen differentiation in the embryonal floral bud but merely suppress female flower formation and that, in the lack of the latter, male flowers ultimately develop. In confirmation of these indications, the direct application of GA_3 to in vitro-cultured cucumber floral buds did not promote stamen differentiation (GALUN et al., 1963). An interesting observation is the greater effectivity of gibberellins under regimes of low temperature and short days—conditions which promote femaleness (Table 3.17). Other observations indicated that the genetic background

Table 3.17. Effects of light and temperature regimes on the reactivity of monoecious cucumbers to GA_3 application[a]

Temperature[b] (°C)		Day length (h)	Node number of first female flower on the main shoot		Shift caused by GA_3
'Day'	'Night'		no GA_3	with GA_3	
19	19	8	9.6	30.1	20.5
19	19	16	13.7	21.3	7.6
23	19	8	11.7	26.8	15.1
23	19	16	14.5	22.0	7.5
19	23	8	16.2	22.6	6.4
19	23	16	15.8	21.4	6.6
23	23	8	14.8	24.3	9.5
23	23	16	16.5	18.6	2.1

Unpublished data of LANG and GALUN.
[a] GA_3 (300 ppm) was applied several times in 0.2-ml drops to young leaves.
[b] 'Day' temperature was for the daily 8 h of sunlight; 'Night' temperature prevailed for 16 h.

of cucumber lines controls the sensitivity to gibberellins (SAITO and ITO, 1963; SHIFRISS and GEORGE, 1964).

The information available on the effects of gibberellins on sex expression of cucurbits other than cucumber is rather limited. In melons, gibberellins strongly increase stem and tendril elongation, but their effect on sex expression in these plants is doubtful (KUBICKI, 1969b; ANAIS, 1971; RUDICH et al., 1972). A delay in the appearance of female flowers' development after GA_3 application to pumpkins was reported by SPLITTSTOESSER (1970), and a slight decrease in femaleness after gibberellin application on Luffa acutangula was reported by KRISHNAMOORTHY (1972).

Quaternary ammonium compounds and other growth retardants (Table 3.18) were reported to shift the sex expression toward femaleness (MITCHELL and WITTWER, 1962; HALEVY and RUDICH, 1967; LANG and GALUN, unpublished). The most effective among the plant retardants is probably SADH (RODRIGUEZ and LAMBETH, 1972). These retardants are known to suppress endogenous gibberellin synthesis. It is, therefore, not surprising that the female-promoting effect of plant retardants can be fully prevented by gibberellins (Tables 3.19 and 3.20)

Table 3.18. Chemical composition of plant retardants affecting sex expression of cucurbits

$CH_2Cl-CH_2-N^+(CH_3)_3 \cdot Cl^-$
(2-chloroethyl)trimethylammonium chloride
Trivial name: chlorocholine chloride—CCC

$(CH_3)_3N^+$ —[benzene ring with $CH(CH_3)_2$ and CH_3 substituents]—$O-C(=O)-N$—[cyclohexane ring]—Cl^-

2-isopropyl-4-dimethylamine-5-methylphenyl-1-piperidine carboxylate methyl chloride
Trivial name: Amo 1618

Cl—[benzene ring with Cl substituent]—$CH_2-P^+(C_4H_9)_3 \cdot Cl^-$

tributyl-2,4-dichlorobenzylphosphonium chloride
Trivial name: Phosphon D

$CH_2-C(=O)-NH-N(CH_3)_2$

N,N-dimethylaminosuccinamic acid (succinic acid-2,2-dimethylhydrazide)
Trivial names: B_{995}, B_9, SADH

Table 3.19. The effect of two plant retardants on sex expression in a monoecious cucumber line and their interaction with GA_3

Plant retardant treatment[a]	GA_3 treatment (ppm)			
	0	10	30	100
None	22 ± 0.5	26 ± 1.5	29 ± 0.2	37
Amo 1618 200 ppm	16 ± 0.4	22 ± 0.6	26 ± 1.2	31
Amo 1618 1,000 ppm	14 ± 0.3	21 ± 0.4	23 ± 1.1	35
CCC	16 ± 0.3	21 ± 0.4	25 ± 1.1	40

From GALUN (1973). Plant retardant treatment was by soil application; GA_3 was sprayed onto leaves; results are expressed as node location of the first female flower on the main shoot.
[a] Amo 1618 = 2-isopropyl-4-dimethylamine-5-methylphenyl-1-piperidine-carboxylate methylchloride; CCC = (2-chloroethyl)trimethylammonium chloride.

and it is noteworthy that CCC does increase femaleness in cucurbits (GOSH and BOSE, 1970), in which gibberellins do not affect sex expression.

Auxins were found by LAIBACH and KRIBBEN (1949) to induce a shift in sex expression toward femaleness. This pioneering finding, probably the first

Table 3.20. The effects of Amo 1618, GA$_3$ and day length on sex expression of monoecious cucumbers

Day length (h)	Without Amo 1618		With Amo 1618	
	without GA$_3$	with GA$_3$	without GA$_3$	with GA$_3$
16	29 ± 1.3	>42	9 ± 0.7	>40
8	9 ± 0.2	>43	9 ± 0.2	>42

From GALUN (1973). Amo 1618 and GA$_3$ were applied at 200 and 300 ppm, respectively, as detailed in Table 3.19. Results are expressed as node location of the first female flower on the main shoot.

reproducible report on a change in sex expression in plants by single defined compounds, was mentioned above (Section 3.2.2.5.1). The effect of auxins on the sex expression of cucurbits was confirmed by many authors and similar effects were reported for plants of other families. Auxin was even shown to be effective as a seed applicant (GALUN, 1956). Moreover, it was shown that in *Cucumis* auxins exert a direct effect on the development of the ovary of in vitro-cultured floral buds (GALUN et al., 1962, 1963; PORATH and GALUN, 1967; GALUN and PORATH, 1970). Studies with cucumber served actually as the main basis for general theories on the role of auxins in sex regulation of plants (e. g. KÖHLER, 1964a; HESLOP-HARRISON, 1972; GALUN, 1973).

The most dramatic shift in sex expression of cucurbits toward femaleness is caused by acetylene and ethylene. The effect of unsaturated hydrocarbon was observed by MININA many years ago and was utilized by her to increase greenhouse cucumber yields (MININA, 1938; MININA and TYLKINA, 1947). It took several decades until the effect of the ethylene-releasing compound 2-chloroethylphosphonic acid on sex expression of cucurbits was reported, almost simultaneously, in North Carolina (MCMURRAY and MILLER, 1968), Israel (RUDICH et al., 1969), New York (ROBINSON et al., 1969) and California (IWAHORI et al., 1969). This compound was found to affect both monoecious and andromonoecious cucumbers (Table 3.21). Several additional studies confirmed these reports in respect to cucumber as well as other cucurbits. Noteworthy among these was the finding that female flowers could be induced in genetic male cucumber plants after 2-chloroethylphosphonic acid treatment (AUGUSTINE et al., 1973). The female-promoting effect of this compound is now well-documented, and the ethylene release from 2-chloroethylphosphonic acid seems to have a decisive role in sex determination (BEYERS et al., 1972). Less agreement exists about the interaction of this compound with other growth regulators (e. g. IWAHORI et al., 1970; RUDICH and HALEVY, 1974).

There are reports on the effect of various additional chemicals, such as maleic hydrazide, abscisic acid and morphactin, on sex expression in cucurbits. Abscisic acid was found to be an endogenous growth regulator, and morphactin an inducer of parthenocarpy (ROBINSON et al., 1970). For various reasons, the practical utilization of all these compounds for regulating sex expression is questionable, but the use of morphactin—especially in combination with

Table 3.21. Effects of GA$_3$ and Ethrel on stem elongation and sex expression of two cucumber cultivars

Treatment	Length of first internode (cm)	Node location of first flower		Number of flowers on first 10 nodes	
		Female or hermaphrodite	Male	Female or hermaphrodite	Male
var. *Wisconsin SMR 18 (monoecious)*					
Control	7.7	8.0	2.0	2.0	33.4
GA$_3$, 2,000 ppm	16.0	above 17	1.4	0.0	33.0
Ethephon, 250 ppm	3.7	2.7	14.3	14.0	0.0
var. *Lemon (andromonoecious)*					
Control	4.6	above 17	3.0	0.0	26.6
GA$_3$, 2,000 ppm	8.3	above 17	5.7	0.0	18.0
Ethephon, 250 ppm	3.0	8.0	11.7	4.0	2.3

Data from ROBINSON et al. (1969). All data concern the main shoot only.

2-chloroethylphosphonic acid—for the attainment of parthenocarpic cucumbers seems now to be commercially applicable (CANTLIFFE and PHATAK, 1975).

We may summarize by indicating that it is possible now to shift sex expression by chemical treatment in either direction: genetic female cucumber plants can be induced by GA to produce abundant male flowers, and genetic male cucumber plants can be induced by 2-chloroethylphosphonic acid treatment to produce female flowers. As stated above, we are not primarily concerned in this book with the role of growth regulators in the control of sex expression. Nevertheless, it seems justifiable to remark that in spite of the interesting data on hand, additional information on the interaction of genetic sex types, on light and temperature conditions, and on the distribution and regulation of the levels of the endogenous growth regulators and their impact on the morphogenesis of floral buds is still required before a reliable general theory on sex regulation in cucurbits can be suggested.

3.2.3.1.4 Inheritance of Sex Expression and Breeding Procedures

In this section we shall deal first in some detail with cucumbers. For this crop we have more information on the genetic control of sex expression than for melon; moreover, the genetics of sex control in melon seems more complicated. We shall thus handle melons after discussing cucumbers, and then deal only briefly with the inheritance of sex in watermelons and other cucurbits.

The genetic control of sex expression in cucumber will be easily comprehended if we remind ourselves about two main considerations mentioned previously, in Section 3.2.3.1.1, when we listed the components participating in the flowering patterns in cucurbits.

1. The floral bud may develop either into a monoclinous or a diclinous flower; thus, we should look for a gene controlling this 'trigger' mechanism.

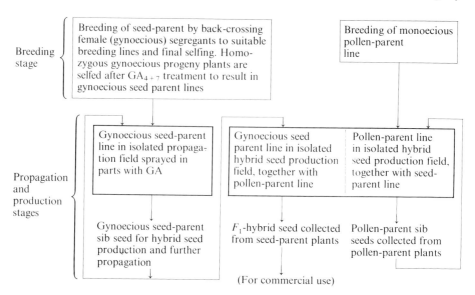

Fig. 3.26. F_1-hybrid seed production in cucumber based on a gynoecious (female) line as seed-parent

    In this method the pollination is carried out exclusively by bees and thus adequate care is necessary to assure ample bee activity. Complete isolation and purity of the female stock is also of utmost importance, and it is, therefore, good practice to renew the female stock and to check its purity continuously.
    This method has one major drawback. The resulting F_1 hybrid plants are st^+/st and thus may have different sex expression according to their genetic background and the environmental conditions. If they are almost complete females, it is necessary to add pollinators in the cucumber production field and the crop will not be completely uniform. On the other hand, if the F_1 plants are almost regular monoecious plants, one of the advantages of such hybrids is lost. In spite of the above mentioned disadvantage this method is successfully utilized for hybrid seed production in many countries and most new cucumber varieties are actually produced by this method. Complete femaleness of the F_1 plants will be of great advantage provided the plants bear parthenocarpic fruits. Thus, present breeding efforts are directed at combing genetic parthenocarpy into mechanically harvested F_1 hybrid varieties (see BAKER et al., 1973, for discussion). There are two main methods to assure complete femaleness in the F_1 hybrid, one of which is based on crossing two female lines. In this method the two lines are planted in the hybrid seed production field as in the previous method, but one of these lines is treated with GA and serves as pollinator. This method is quite expensive, as the two parental lines have to be propagated in isolation fields and a great amount of GA is required for the final F_1 hybrid production.
    The other method is based on a hermaphrodite line as pollen parent in the hybrid seed production field. Thus, both parental lines are st/st but one

has hermaphrodite flowers (i.e., it is m/m, st/st) and can therefore serve as pollinator and may also be propagated without GA treatment. This latter method was suggested by several authors (KUBICKI, 1970; PIKE and MULKEY, 1971; BAKER et al., 1973), but did not reach the stage of actually being used in large-scale commercial seed production.

The inheritance of sex expression in melons was first investigated by ROSA (1928) and POOLE and GRIMBALL (1939), and the main findings of the latter can be summarized as follows. Two main genes are involved in sex control: A/a and G/g. These are very similar to the genes M/m and st^+/st, respectively, of cucumber. According to POOLE and GRIMBALL, the main sex types are:

1. *monoecious* $A/-$, $G/-$; $(M/-, st^+/-)$;
2. *andromonoecious*   a/a, $G/-$; (m/m, $st^+/-$);
3. *gynomonoecious* $A/-$, g/g; $(M/-, st/st)$;
4. *hermaphrodite*   a/a, g/g; (m/m, st/st);

We put the designations for the cucumber genes in parenthesis to indicate the similarity in genetic sex control between cucumber and melon. Still, there are notable differences. In cucumber all $M/-$, st/st genotypes are female plants, whereas in melons POOLE and GRIMBALL defined $A/-$, g/g melons as gynomonoecious. In other studies (NIEGO and GALUN, unpublished) it was found that such gynotypes actually comprise a wide range of phenotypes from complete females to trimonoecious, which can be divided into 12 well separated sex types. Thus, while in cucumber M is absolutely dominant, in melons with an A/a genotype (i.e., parallel to M/m of cucumbers) hermaphrodite flowers do appear. Moreover, melon plants having the A/a, g/g genotype are strongly affected by environmental conditions: plants with the same genetic composition may have a strictly female phenotype under short days and relatively low temperatures, but be gynomonoecious when grown under long days and higher temperatures. Still, the variability is at least partially controlled by additional genetic modifiers (KUBICKI, 1969b). The genetic basis of sex expression is not yet fully resolved. This is probably mainly because of two reasons: similar to cucumber, there is no linkage map in melons, and genetic modifiers cannot be proven unequivocally; and, unlike in cucumber, no male flowers can be induced in female melons by GA, and thus such plants cannot be self-pollinated. For the latter reason, we do not know whether A/A, g/g plants resemble A/a, g/g plants. In addition to the above mentioned genes which affect directly sex expression in melons, another gene abrachiate (ab) should be considered (FOSTER and BOND, 1967). The ab/ab plants have only one shoot and no side branches; thus, abrachiate plants with a monoecious or an andromonoecious genotype never bear female or hermaphrodite flowers, respectively, as those should appear on the nonexisting side branches. Such plants are thus functionally strictly male.

Hybrid F_1 seed production of melons by hand pollination is a very laborious procedure, as most commercial varieties are andromonoecious and emasculation of the perfect flower should precede pollination. Labor can be reduced by introducing monoecism into the appropriate parents—though this procedure is complicated because the favorable spherical fruit shape is strongly correlated with bisexual flowers.

In order to eliminate the need for hand pollination, RUDICH et al. (1970) suggested to treat the seed parent in the hybrid seed production field with both 2-chloroethylphosphonic acid (ethephon) and N,N-dimethylaminosuccinamic acid (B-995). Such treatments prevented the development of male flowers on some monoecious lines but also reduced the number of female flowers and caused considerable damage to the plants. Moreover, environmental conditions may interfere and it is questionable whether chemical treatment alone will be applicable on a commercial scale. A combination of gynomonoecism with growth regulator treatment was suggested for the same purpose by ANAIS (1971), but it seems that the best practical solution would be to find genetic lines which are strictly female in certain seasons but will bear a sufficient number of hermaphrodite flowers in other seasons to enable the propagation of the line (Fig. 3.27). Experimental hybrids have been produced by this method (GALUN

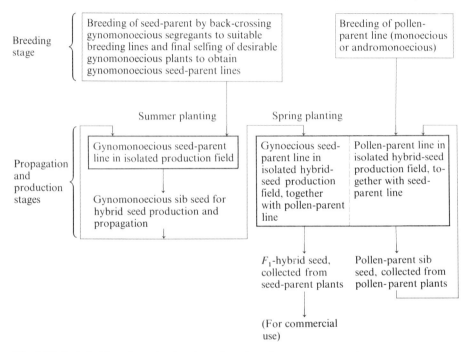

Fig. 3.27. F_1-hybrid seed production in melons based on gynomonoecism. The gynomonoecious line is of the type which bears only female flowers when planted in spring (relative SD and low temp.), but bears both female and hermaphrodite flowers in the summer

and NIEGO, unpublished). Finally, another approach should be mentioned. FOSTER (1968) suggested a combination of a seed parent segregating to male-sterile plants with appropriate morphological markers, with an appropriate pollen parent for the production of commercial F_1 hybrid seed.

Edible species of *Cucurbita*, i.e., *C. pepo, Cucurbita maxima, Cucurbita moschata, Cucurbita mixta*, are all monoecious and no femaleness was reported in them. ROBINSON et al. (1970) found that repeated 2-chloroethylphosphonic acid sprays (250 ppm) at the first-true-leaf stage eliminated male flowers from at least the first two nodes and increased femaleness, and is thus an appropriate method for hybrid seed production.

3.2.3.1.5 Scheme of Sex Expression in Cucumber

The information on the inheritance of sex expression and the effects of light, temperature and chemical agent on sex in cucumber are summarized in a schematic diagram (Fig. 3.28). At the bottom of this scheme the main shoot of a hypothetical plant is represented. It contains the three flowering phases: staminate (male), mixed, and pistillate (female). Shoot A represents a normal monoecious (st$^+$/st$^+$) plant having only the two first phases. Shoot B represents either the same genotype shifted by nongenetic factors to stronger femaleness, or a monoecious plant with modifying genes. Shoot C represents a typical st$^+$/st plant. Shoot D represents either the same genotype shifted to stronger femaleness by nongenetic factors, or a different genotype: either st$^+$/st with female modifiers or st/st with male modifiers. Shoot E represents a normal female plant (st/st). Obviously, effective treatment can shift sex expression over more than one step. Thus, st/st (E) plants after repeated GA applications may be shifted into a B or even an A phenotype, while st$^+$/st$^+$ (A) plants after appropriate 2-chloroethylphos-phonic acid application may be shifted into a C, D or even an E phenotype.

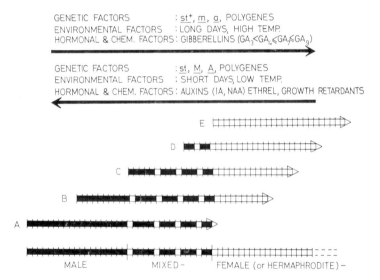

Fig. 3.28. Scheme of factors affecting sex expression in cucumber. *Bold arrow pointing to the right:* extension of the male phase *bold arrow pointing to the left:* shift of the female (or hermaphrodite) phase toward the base of the plant. The sections represent nodes: *shaded sections:* male *clear sections:* female (or hermaphrodite)

3.2.3.2 Hemp (Cannabis sativa)

Hemp is presently of marginal economic importance but it was a major fibre crop in Europe until several decades ago and its sex expression was then studied intensively. Thus, among the dioecious crops, hemp is still the one for which we have the greatest amount of information on the physiology of sex expression. This information was reviewed thoroughly by HESLOP-HARRISON (1957) and NAPP-ZINN (1967). It is noteworthy that in contrast to hemp, our knowledge on the sex regulation in the related *Cannabis indica* is negligible, although in the latter the inflorescence itself composes the harvested crop.

C. sativa is typically dioecious, but due to intensive breeding, varieties with different sex expression were produced; among the latter are monoecious varieties. Dioecism in hemp is accompanied by morphological dimorphism which is apparent only with the attainment of flowering. At this phase of development there is a considerable relaxation of elongation in the main shoot of female plants, and the bracts in sheathed female flowers are developed on short lateral shoots of a compact inflorescence. In male plants no such growth retardation occurs; consequently, a 'loose' terminal inflorescence is produced and the male flowers are typically borne in axils of the main shoot. This dimorphism in respect to flower location and inflorescence internode length is accompanied by a marked dimorphism in laminar shape (HESLOP-HARRISON and HESLOP-HARRISON, 1958b).

KÖHLER (1964a) agrees with WESTERGAARD (1958) that the inheritance of sex in hemp is probably the most complicated one among dioecious plants, but attempted to derive at a relatively simple general genetic interpretation, in the following way:

Normal male and female plants are X/Y and X/X, respectively. In addition there is an allele Xm which has a reduced femaleness-inducing ability. Thus, X/Xm plants do have a "female inflorescence" but may not be strictly females: depending on additional genetic and nongenetic modifiers they may tend to maleness, hence termed by KÖHLER subgynoecious. Xm/Xm plants still retain a "female inflorescence" but are functional males. From studies by several authors including results of crosses with polyploids, KÖHLER concluded that the genes promoting maleness are autosomal and balanced by the femaleness-promoting genes in the X chromosomes, while the Y chromosomes are 'empty' in respect to sex regulating genes, i.e., comparable to the mechanism of sex determination in *Rumex acetosa*.

From the above considerations we can draw the following tentative equations:

male plants with male-type inflorescence	X/Y, Xm/Y
male plants with female-type inflorescence	Xm/Xm
variable phenotypes from true females to monecious, but all with female-type inflorescence	Xm/X
Female plants with female-type inflorescence	X/X

Mutations and selection can alter the potency of both the autosomal male-inducing and the X-chromosome located factors. Consequently, great variability in sex types may be established.

It was shown many years ago that auxin treatment can induce femaleness in male hemp plants (HESLOP-HARRISON, 1956). More recently, a similar shift in sex expression was reported by MOHAM RAM and JAISWAL (1970, 1972b), after treatments with compounds which release ethylene. The strongest effect of this kind was caused by a 2-chloroethylphosphonic acid spray (480 and 960 ppm) to male plants grown under winter conditions, which are known to strengthen femaleness in hemp.

Morphactin was reported to affect sexual differentiation within the flower and even of the floral members (MOHAM RAM and JAISWAL, 1971). Spraying with 250 and 500 ppm delayed flowering, and in the delayed flowers stigmatic anthers and carpellate structures in anthers occurred.

HESLOP-HARRISON and HESLOP-HARRISON (1961) observed that GA_3 treatment strongly affected the stature of hemp plants, but they did not record a change in sex expression. Such a change was found by KÖHLER (1964b) in a certain female genotype and was more recently reported by MOHAM RAM and JAISWAL (1972a), who were probably not aware of KÖHLER's previous findings. MOHAM RAM and JAISWAL detected male flowers in female plants treated with GA_3, GA_{4+7} and GA_9 (Fig. 3.29). They also reported that this

Fig. 3.29 A–C. Changes in sex of hemp induced by gibberellic acid. (A) The upper part of a control female plant ($\times 0.3$). (B) Part of a gibberellic acid-treated female plant; note reversion to female flower on side branch ($\times 0.5$). (C) The upper part of a control male plant ($\times 0.2$) (from MOHAM RAM and JAISWAL, 1972a)

effect of GA could be prevented by simultaneous treatment with abscisic acid, and that abscisic acid alone had no effect on sex expression in hemp.

It may thus be concluded that both environmental factors and chemical agents affect sex expression in hemp rather similar to their effect in cucumber.

3.2.3.3 Maize *(Zea mays)*

Maize rates as one of the most important crops and its sex expression can be modified considerably by both genetic and nongenetic factors. Therefore, the sexuality of this plant deserves our attention. Sexual regulation is generally not considered to be of major importance in breeding and crop improvement of maize. Nevertheless, because of considerations outlined below, the modifications of sex expression in maize should not be ignored in agricultural and breeding practices.

Most of the research on sex expression of maize was undertaken quite long ago and will not be fully reviewed here. Readers interested in a thorough coverage of the literature on this subject should first consult the reviews of HESLOP-HARRISON (1957) and NAPP-ZINN (1967), was well as the article of HESLOP-HARRISON (1961).

3.2.3.3.1 Reproductive Morphology

Maize is a monoecious plant and its flowers are normally diclinous; male and female flowers are normally borne on different inflorescences.

The male inflorescence, termed tassel, is a terminal panicle, and its central axis is continuous with the axis of the main shoot of the plant. The tassel has usually a variable number of spirally arranged lateral branches. The spikelets are paired, one being sessile and the other pedicelled. Each spikelet has two flowers, an upper and a lower one, of which the first is advanced in its early development, and the latter may not attain maturity. The embryonal flower is monoclinous: after anther initials the gynoecium is initiated in form of an incomplete lip at the apical meristem of the flower, but further development stops when the gynoecium reaches a size of about 500 μm.

As the female inflorescence is less affected by external modifiers, it will not be detailed here. Good descriptions can be found in HECTOR (1936) and in BONNET (1940, 1948). This inflorescence ('ear') is the termination of a short lateral branch. The 'ear' is a modified spike, on the thickened axis of which are borne rows of paired spikelets, each having only one female flower. The interesting point here is that ontogenically in these flowers also, the stamen initials are first differentiated and attain a certain final size (ca. 250 μm). The later-appearing gynoecium then develops uninterruptedly. Thus, in spite of a great final difference between female and male flowers, they are morphologically indistinguishable up to a certain stage in their ontogeny (BONNET, 1940, 1948).

3.2.3.3.2 Environmentally Induced Sex Modification

Short photoperiods and low temperatures were reported by many authors to induce the development of female flowers on the tassel. These two environmental factors were analyzed in more detail by HESLOP-HARRISON (1961). Table 3.22 demonstrates that the plants had the normal flowering pattern under LD and a temperature of ca 22°C. Either lowering the night temperature (the effect of day-temperature was not reported by HESLOP-HARRISON) or shortening the daily photoperiod induced the appearance of female flowers in the tassel. These two environmental factors seem to be additive in respect to their effect on

Table 3.22. Effects of day length and night temperature on the number of leaves on the main stem and on the appearance of female flowers in the tassel of maize

Night temp.[a] (°C)	Number of leaves on the main stem		Number of female flowers in the tassel	
	8 h	21–22 h	8 h	21–22 h
	photoperiod		photoperiod	
22	9.00	12.17	5.67	0.00
10	8.33	13.33	12.50	1.17

Data from HESLOP-HARRISON (1961). All plants were kept in the greenhouse for 8 h daily (natural daylight; ca. 22 °C) and transferred to growth chambers for the rest of the day; supplementary light was provided by incandescent illumination (50–100 ft c).
[a] "Night temp." was applied during the 16 h in the growth chamber.

sex expression. Table 3.22 also indicates a quantitative photoperiodic effect in respect to flowering: the terminal inflorescence (tassel) appeared after a smaller number of foliar nodes under short rather than long photoperiods. These results are in full agreement with several previous reports based on less controlled experimental conditions. Exposure to short photoperiods and low temperatures during a critical period of anther differentiation also impaired the normal differentiation of those floral buds which were determined to have a suppressed gynoecium leading to nonfertile male flowers.

It is common that, when environmental conditions do affect the sex expression of the tassel, the promotion of the gynoecium is manifested mainly at the base of the inflorescence. Moreover, if sex modification is experienced in a pair of spikelets of the tassel, it is common that the sessile spikelet is more 'female' than the pedicellate one (Figs. 3.30A, B). On the other hand, in the much less environmentally affected ear—when long photoperiods and high temperature cause a change in sex expression—male flowers are restricted to the apical part. This would be regarded as a mere stronger manifestation of the normally existing gradient, because underdeveloped male flowers are quite common in young maize ears.

3.2.3.3.3 Sex Modification by Chemical Agents

Auxins seem to have a direct effect on sex regulation in maize (HESLOP-HARRISON, 1961), although a significant shift in floral bud differentiation from anther to pistil development could be obtained by α-naphthaleneacetic acid (NAA) treatment only when the plants were kept at 'marginal' environmental conditions (Table 3.23). Under these 'marginal' conditions female flowers were observed occasionally in tassels of untreated plants, although on the average less than one such female flower per tassel was observed in these plants. The effect of NAA on the tassel is pleiotropic; it also reduces drastically the branching. These results were confirmed by tests of endogenous auxin content made by

Table 3.23. Effect of α-naphthaleneacetic acid (NAA) on the formation of lateral branches and female flowers in the tassel of maize

Treatment	Number of branches		Number of female flowers	
	8 h	20 h	8 h	20 h
	photoperiod		photoperiod	
Control (water)	8.2	0.3	0.8	0.2
NAA (100 mg/l, 0.15 ml)	1.2	7.2	5.7	3.2

Data from HESLOP-HARRISON (1961), expressed as means per tassel; plants (Golden Bantam) were first raised in growth chambers (22°C, 75 % R.H.), with 10 h illumination (900–1,100 ft c) daily up to the stage of 5–8 cm length of the third leaf; they were then treated by injection in the vicinity of the shoot apex, and after 2 days were assigned to either 8- or 20-h photoperiods.

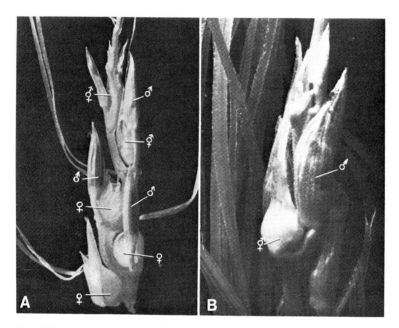

Fig. 3.30 A, B. Environmentally induced changes in sex expression in the tassel of maize. (A) Apical inflorescence of a plant maintained in short days. (B) An alicole of a terminal inflorescence showing a sessile hermaphrodite spikelet and a pedicilate male one (from HESLOP-HARRISON, 1961)

SLADKY (1969), who found a low auxin content in tassel initials and a high auxin content in ear primordia.

Gibberellin treatment was not reported to interfere with the sexual differentiation of normal maize plants unless applied at very high doses and thus causing strong retardation of floral development. Nevertheless, the role of endogenous gibberellin in the differentiation of maize inflorescences is indicated by two lines of evidence. The first of these comes from studies of NICKERSON (1960a) on maize mutants which are defective in tassel development. He found that sustained GA$_3$ treatment restored partially or fully the phenotypes having the dominant genes corn-grass (Cg) and teopod (Tp). Corn-grass is a pleiotropic gene located in chromosome 3, and among its effects are abundant tillering and nonfunctional male flowers, both of which characters were restored to normal by repeated GA$_3$ (125 ppm) application to the apical leaves; another Cg defect, nonfunctionality of the ear spikelets, was not restored. Teopod plants have strongly deformed inflorescences: the bracts, which are normally vestigial, grow enormously up to the size of blades; the functionality of both male and female spikelets is impaired; and, in addition, the plants' tillers are abundant. All these characters were changed toward a normal phenotype by GA$_3$ treatment. NICKERSON (see also 1960b) argues that these GA$_3$ effects are not a result of a direct regulatory role of gibberellin, but that at least some of them result from an interaction with the auxin balance of the plants.

Another line of evidence regarding the role of endogenous gibberellins comes from tests on gibberellin-like substance content in maize plants by SLADKY (1969), who found a high content of these substances in tassel primordia and a low content in ear primordia. Unfortunately, the bioassay determination used by SLADKY does not allow a rigorous identification of these substances or of unidentified inhibitors, the distribution of which also seems to be causally related to sex differentiation.

3.2.3.3.4 Inheritance of Sex Expression

During the years of intensive research on the genetics of maize a considerable number of mutants defective in inflorescence structure were detected, and the appropriate genes were identified and located on respective chromosomes (see EMERSON et al., 1935). Some of these, such as tassel-seed (ts) and silkless (sk), were mentioned above in Section 3.2.1.2, in the coverage of artificial dioecism in maize. These genes were useful in formulating a rational attitude to the inheritance of sex expression in plants, but have no direct relevance to maize breeding.

The implications of the above information for the practice of breeding and crop management of maize are manifold and quite obvious. We shall mention only one explicitly: the possible development of functional male flowers at the apices of ears of seed parents under certain environmental conditions may impair the attainment of F_1 hybrid seed from such plants. Moreover, one should be aware of the possibility that such and other modifications of sex expression are not only a result of environmental conditions but may also be affected by quantitative genetic factors.

3.3 Incompatibility

We shall use the term *incompatibility* to describe the phenomenon of an inability of plants having functional gametes to set seeds when either self-pollinated or crossed with some of their genetic relatives. BREWBAKER (1957) and most other authors accepted the term originally coined by STOUT (1917), but restricted their definitions to *self*-incompatibility, i.e., inability to set seed after self-pollination. As among all the systems of incompatibility known to us, none is restricted to self-incompatibility in the strict sense, we shall try to avoid the latter term.

Incompatibility systems of all the different kinds have one trait in common: they maintain a high level of genetic heterosigozity in a given plant population. HOGENBOOM (1975) stressed the difference between incompatibility and *incongruity*. As we shall handle here only the first of these mechanisms, we shall be very brief in describing incongruity. Thus, while in incompatibility we are faced with a situation of an intimate competence between the pollen and the pistil combined with a genetic regulation which prevents seed setting in specific combinations, no such basic competence exists between the pollen and pistil of plants separated by the incongruity barrier. Hence, while incompatibility is functioning within populations (and species), incongruity is operating between populations (and species).

Incompatibility was observed over 200 years ago. According to DARWIN (1877) it was KÖLREUTER who in the 18th century reported on *Verbascum* plants which did not set seeds when pollinated with their own, apparently fertile, pollen, but readily set seed when cross-pollinated. During the last century numerous cases of incompatibility in angiosperms were reported. These were reviewed by EAST (1940) and more recently by ROWLANDS (1964), LINSKENS and KROH (1967) and ARASU (1968), who also furnished historical sketches of this subject.

As we shall see, the genetic approach was rather successful and resulted in the revelation of genetic control mechanisms in all the main types of incompatibility. Studies of the histology, physiology and biochemistry of the pollen-pistil relationships revealed many interesting facts and explained many features of the incompatibility systems (see review of HESLOP-HARRISON, 1975). These facts led to numerous theories which were instrumental in the encouragement of further studies but are still short of providing an unequivocal explanation of the relationship between the genetic control and the structural and biochemical regulation of incompatibility. We shall not elaborate on the theoretical aspects of incompatibility. The reader is therefore referred to some of the reviews in which theories are either detailed or cited (e.g. SAMPSON, 1962; LEWIS, 1965; ARASU, 1968; PANDEY, 1970a; DE NETTANCOURT, 1972; LINSKENS, 1974) and to a discussion meeting fully reported in the Proceedings of the Royal Society (Ser. B), Volume **188**, no. 2 (1975).

The early authors who studied incompatibility were mainly concerned with its evolutionary implications and less with its impact on crop production and breeding. Later investigators pointed out the various problems encountered with a cultivated plant in which incompatibility occurs, and devised means to overcome these problems (e.g. incompatibility in clones of fruit trees and forage plants) and only recently were incompatibility systems exploited for the benefit of plant breeding, i.e., for the production of F_1 hybrid varieties and breeding of ornamentals with an extended flowering period due to the lack of seed setting.

3.3.1 Genetics of Incompatibility

Before the male gametophyte (the pollen) in angiosperms meets the female gameto-phyte, it has to make its way through the sporophytic tissue (the pistil) holding the female gametophyte (see Section 3.1.4). The pistil is thus the arena where, with a few exceptions to be mentioned below, the incompatibility phenomenon takes place. Moreover, since the pistil is a diploid sporophytic tissue, its compatibility reactivity is genetically controlled by its *sporophytic* genome. On the other hand, the male gametophyte (the pollen) carries with it at least some components from the sporophyte on which it developed (e.g. material invested into the exine cavities from the tapetum). It also carries with it the cytoplasm from its sporophytic parent. Thus the pollen compatibility reactivity can be genetically controlled either by the genome of the gametophyte itself or by the genome of its parental sporophyte. Accordingly the compatibility systems were divided into two main groups: *gametophytic* incompatibility and *sporophytic* incompatibility.

3.3.1.1 Gametophytic Incompatibility

In this group of incompatibility systems the pollen/pistil interaction is genetically controlled by the haploid genome of each pollen grain and the diploid genome of the pistil tissue. There are possible cases where at least part of the rejection mechanism is probably taking place in the embryo sac, which could point to a haplo–haplo incompatibility system (BOUHARMONT, 1960, cited by DE NETTANCOURT, 1972). Although cocoa *(Theobroma cacao)* may be such an exception, the genetics of its incompatibility is rather complicated and cannot be regarded as fully solved (KNIGHT and ROGERS, 1955; COPE, 1962). Additional cases, where the incompatibility probably takes place in the embryo sac, were reported in perennial geophytes of the genera *Freezia, Gasteria, Hemerocallis* and *Narcissus*.

3.3.1.1.1 One Multiallelic S Locus

The genetic control of this system, also termed the *Personata* type of incompatibility, can be demonstrated by the incompatibility in *Nicotiana sanderae* (EAST and MANGELSDORF, 1925). In this system one S locus with a large number of alleles: $S_1, S_2, S_3 \dots S_m, S_n$, controls the compatibility relationships. The pistil, being a diploid sporophytic tissue, always has two of these alleles, normally in a heterozygous state, e.g. $S_{12}{}^3$, S_{13}, S_{nm}. In each pollen grain (and tube) there is correspondingly only one allele, e.g. $S_1, S_2 \dots S_n$. Compatible pollination takes place only when a given allele in the pollen is met with *different* alleles in the pistil. In *N. sanderae* incompatible pollen is inhibited in the style, as shown schematically in Fig. 3.31. The pollination in A is incompatible because

---

[3] The heterozygote is assigned by some authors as $S_1 S_2$ and by other as S_{12}.

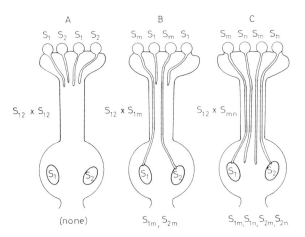

Fig. 3.31 A–C. Pollen-tube growth and fertilization in the one S locus gametophytic incompatibility system: three sources of pollen pollinating an S_{12} pistil. (A) pollen from an S_{12} plant: all pollen tubes are inhibited. (B) pollen from an S_{1m} plant: S_1 pollen is inhibited and S_m is compatible, resulting in S_{1m}, and S_{2m} zygotes. (C) pollen from an S_{nm} plant: both S_m and S_n pollen are compatible, resulting in S_{1m}, S_{2m}, S_{1n} and S_{2n} zygotes

both the S_1 and the S_2 pollen are matched in the pistil with identical alleles. Both B and C are compatible pollinations but in B half of the pollen (S_m) is not matched in the style and thus attains fertilization. Thus, B and C differ only in their progeny, segregating into two and four compatibility groups, respectively. Hence, legitimate pollinations always result in zygotes which are heterozygous in respect to the S alleles. The genetics of this system limits the minimal number of S alleles in an isolated population to three. Such a population will consist of three genotypes: S_{12}, S_{13} and S_{23}. When these are crossed (and selfed) in all nine possible reciprocal combinations, three (selfings) will be incompatible and in six only half of the pollen grains will be compatible with the respective pistil. An increase in the number of alleles in the population will obviously increase the ratio of compatible to incompatible pollinations. Thus, when the number of alleles in the population is 4, only 6/36 of the possible combination are incompatible; in 24/36 combinations half of the pollen grains are compatible; and in 6/36 combinations all the pollen grains are compatible. Table 3.24 demonstrates these compatibility relationships. Simple mathematic equations can obviously be constructed to acccount for an indefinite number of alleles by taking into account the prevalance of each allele in the population. Thus, if T is the total number of cross- (and self) pollination combinations in a given population and Ic is the number of incompatible combinations, the relation between these numbers will be $1 = \dfrac{Ic}{\sqrt{T}}$.

Another characteristic of a population with only three alleles in this system is a unidirectional compatibility in a given progeny when crossed to their maternal and paternal parents: when an S_{12} plant is pollinated by S_{13}, the progeny will be composed of S_{13} and S_{23} plants; both types of plants will be compatible

Table 3.24. Compatible and incompatible combinations among all possible reciprocal cross- and self-pollinations in populations having the one multiallelic S locus, gametophytic incompatibility system. The populations contain either three S alleles (A—plants are S_{12}, S_{13} or S_{23}) or four alleles (A + B—plants are S_{12}, S_{13}, S_{23}, S_{14}, S_{24}, or S_{34}). Numbers in parentheses represent fractions of compatible pollen grains in each cross-pollination

♂ ♀		A			B		
		S_{12}	S_{13}	S_{23}	S_{14}	S_{24}	S_{34}
A	S_{12}	—	(1/2)	(1/2)	(1/2)	(1/2)	(1)
	S_{13}	(1/2)	—	(1/2)	(1/2)	(1)	(1/2)
	S_{23}	(1/2)	(1/2)	—	(1)	(1/2)	(1/2)
B	S_{14}	(1/2)	(1/2)	(1)	—	(1/2)	(1/2)
	S_{24}	(1/2)	(1)	(1/2)	(1/2)	—	(1/2)
	S_{34}	(1)	(1/2)	(1/2)	(1/2)	(1/2)	—

pollinators to their maternal parent, but only S_{23} will be a compatible pollinator to its paternal parent.

With very few exceptions (e.g. Solanaceae) all the incompatible species belonging to the same family have a common incompatibility system. The one S locus multiallelic gametophytic incompatibility system occurs in several angiosperm families. Among these, detailed genetic studies were performed mainly in the following:

Leguminosae *(Trifolium)*
Solanaceae *(Nicotiana, Lycopersicon, Solanum, Petunia)*
Scrophulariaceae *(Antirrhinum, Nemesia)*
Rosaceae *(Prunus, Malus)*
Onagraceae *(Oenothera)*
Liliaceae *(Lilium)*

This system can thus be found in several genera having great agricultural or ornamental importance. In some cases the cultivated species themselves contain incompatibility (e.g. *Prunus avium*, *Trifolium repens*); in others, incompatibility is found in wild relatives of the cultivated species while the latter lacks this allogamic mechanism (e. g. *Lycopersicon peruvianum/L. esculentum; Nicotiana alata/N. tabacum*).

There is no recent detailed list of cultivated species and their wild relatives having the one S locus multiallelic incompatibility system. Those interested in detailed information are therefore referred to several reviews, some of them quite old: EAST, 1940; LEWIS, 1949a; WHITEHOUSE, 1950; BREWBAKER, 1957; CROWE, 1964; ARASU, 1968; LUNDQVIST, 1968; PANDEY, 1970b; DE NETTANCOURT, 1972).

The mutagenicity of the incompatibility gene S, and the change from incompatibility to compatibility due to polyploidization, were studied extensively. A detailed review of these studies and an account of the several hypotheses drawn from them are outside the scope of this book. Moreover, in many of these studies the incompatibility system was used as a tool to investigate fundamental gene action problems rather than to clarify the genetic control of incompatibility *per se*. In a recent discussion paper, DE NETTANCOURT et al. (1975) reviewed their own work on S gene mutation and function and related it to facts and hypotheses found in the extensive literature available on this subject, i.e., emerging from studies of A. J. BATEMAN, J. L. BREWBAKER, A. J. G. VAN GASTEL, D. LEWIS, F. W. MARTIN, K. K. PANDEY, and their associates. The above review can thus serve as an introduction to this subject. We shall hence confine ourselves here to the main facts and point out only some of the hypotheses which are accepted by most of the authorities in this field.

1. Mutations induced by X-ray radiation of flower buds result in two main types of mutations: *permanent* pollen part mutations rendering one of the S alleles to show a permanent loss of its incompatibility activity, and '*reversible mutations*' resulting in an unstable inactivity which permits the pollen tube to pass the incompatibility barrier in the pistil, while the normal incompatibility character reappears in the progeny. The exact nature of the latter type of mutation is not clear and it may be questioned whether or not it should be considered mutation in the common sense.

2. There are three types of permanent mutations: one affecting the expression of incompatibility in the pollen only; a second affecting the pistil only; and a third affecting both pollen and pistil.

3. Most authors noted that through X-ray-induced mutations no new S alleles could be isolated.

4. In several cases the loss of incompatibility in the pollen was linked to the appearance of a small chromosomal "centric fragment" which could be traced in one case *(Nicotiana alata)* to result from a chromosomal satellite, probably via a duplication-fragmentation process.

5. Diploid pollen, which carries two S alleles, and is produced in tetraploid plants resulting from induced chromosome duplication of incompatible diploid plants, can have one of three types of action. This pollen may act as the diploid pistil, e.g. the two alleles can function independently so that S_{12} pollen will be incompatible in all pistils carrying either or both S_1 and S_2. This is probably the situation common to species with this type of incompatibility belonging to families such as Commeliaceae, Bromeliaceae and Liliaceae (i.e., monocotyledons). In the second type of action the pollen may lose its incompatibility in respect to all pistil genotypes, probably by a kind of 'competition' between the two kinds of S alleles. Finally, one allele may be dominant over the other, so that if S_1 is dominant over S_2 the diploid S_{12} pollen will be compatible with a S_{23} pistil but incompatible with S_{12} or S_{13} pistils.

6. New S alleles may in some species *(Lycopersicon peruvianum)* arise following forced inbreeding for several generations.

Based on some of the above facts and on biochemical evidence to be handled below, LEWIS (1949b, 1960) proposed the tripartite structure of the S locus in this system. According to this hypothesis the S locus has a specificity segment, common to both pollen and pistil. This segment controls the type of S allele, i.e., whether it will be S_1, S_2, etc. A second segment controls the activation of the S activity in the pollen and the third segment controls the S activity in the pistil.

3.3.1.1.2 Two Multiallelic S Loci

This system of incompatibility was discovered by LUNDQVIST (1954) in rye and by HAYMAN (1956) in *Phalaris*. It was subsequently found in other species of the Gramineae. The review of LUNDQVIST (1975) should be consulted for previous work on this system. The incompatibility mechanism is based on two multiallelic S loci (S and Z) which are inherited independently but (with one exception — *Physalis* — to be mentioned below) interact in their incompatibility function. Allelic identity in respect to both these loci in the genotypes of the pollen and the pistil leads to incompatibility. Genetic identity in respect to only one of these loci allows compatible pollination. The compatibility reaction of the pollen is determined, in respect to both the S and Z alleles, by the genotype of the gametophyte only. An additional important characteristic of this system is the complete lack of competition or dominance between the two alleles of the same locus (typical in diploid pollen of dicotyledons of the *Personata* incompatibility mentioned above, and in the sporophytic incompatibility to be mentioned below). Thus, doubling the chromosomic number will not cause self-compatibility. There is also no epistasis between the S and Z alleles.

The following list of crosses will demonstrate the incompatibility of this system:

$$S_{11}Z_{33}(♀) \times S_{22}Z_{44}(♂) \rightarrow \text{compatible} \tag{1}$$

$$S_{11}Z_{33}(♀) \times S_{11}Z_{44}(♂) \rightarrow \text{compatible} \tag{2}$$

$$S_{11}Z_{33}(♀) \times S_{12}Z_{34}(♂) \rightarrow \text{compatible} \tag{3}$$

$$S_{12}Z_{34}(♀) \times S_{11}Z_{33}(♂) \rightarrow \text{incompatible} \tag{4}$$

From these examples it already becomes clear that reciprocal crosses such as (3) and (4) result in different compatibility responses. Further characteristics of this system will become clear from another example. Let us assume that a double heterozygous plant $S_{12}Z_{34}$ was obtained and, in spite of its self-incompatibility, a progeny was obtained by self-pollinating this plant (possible under certain conditions and was actually achieved). This I_1 progeny will consist of nine genotypes, each having a typical theoretical frequency. If all the plants in such a progeny are crossed (and selfed) in all possible combinations, the predictable results will be as presented in Fig. 3.32. The scheme of Fig. 3.32 indicates that one genotype, $S_{12}Z_{34}$ is self-incompatible and also incompatible with all other genotypes when used as seed parent, but compatible with all other genotypes when used as pollen parent. Other reciprocal differences in incompatibility can be seen clearly in Fig. 3.32: genotypes of this I_1 population as seed parents are incompatible with four or with two of the nine genotypes,

Pollen Parent \ Seed Parent	$S_{11}^{(1)}Z_{33}$	$S_{11}^{(2)}Z_{34}$	$S_{11}^{(1)}Z_{44}$	$S_{12}^{(2)}Z_{33}$	$S_{12}^{(4)}Z_{34}$	$S_{12}^{(2)}Z_{44}$	$S_{22}^{(1)}Z_{33}$	$S_{22}^{(2)}Z_{34}$	$S_{22}^{(1)}Z_{44}$
$S_{11}Z_{33}^{(1)}$	1	2	1	2	4	2	1	2	1
$S_{11}Z_{34}^{(2)}$	2	4	2	4	8	4	2	4	2
$S_{11}Z_{44}^{(1)}$	1	2	1	2	4	2	1	2	1
$S_{12}Z_{33}^{(2)}$	2	4	2	4	8	4	2	4	2
$S_{12}Z_{34}^{(4)}$	4	8	4	8	16	8	4	8	4
$S_{12}Z_{44}^{(2)}$	2	4	2	4	8	4	2	4	2
$S_{22}Z_{33}^{(1)}$	1	2	1	2	4	2	1	2	1
$S_{22}Z_{34}^{(2)}$	2	4	2	4	8	4	2	4	2
$S_{22}Z_{44}^{(1)}$	1	2	1	2	4	2	1	2	1

Fig. 3.32. The two multiallelic S loci incompatibility: compatibility relationships among progeny plants obtained by forced self-pollination of a $S_{12}Z_{34}$ plant. *Shaded and clear squares:* incompatible and compatible combinations, respectively *numbers in parentheses:* frequency (multiplied by 256) of a certain cross, provided all crosses are possible

respectively, when used as pollen parents. Actual results of this kind were found by LUNDQVIST (1954). The calculated value of the percentage of compatible pollinations in this example is 60.9 and the value found by LUNDQVIST was 59%. It becomes obvious that this percentage is higher than that obtained in a similar multiple crossing test of all the progeny of a forced self-pollination of an S_{12} plant in the one S locus incompatibility system. In the latter, only 37.5% of the crosses will be compatible.

Generally speaking, the strength of incompatibility in this system is much less than in the one S locus system. Thus, for example, while in those species of the genus *Petunia* which show incompatibility, pseudocompatibility is very rare; this latter phenomenon is common in species of the grass (Gramineae) family.

It should be noted that the assignment of S and Z alleles to a plant of unknown genotype is relatively simple when tester plants of known genotypes are available, but uncovering of the incompatibility alleles in a previously unknown system is a very complicated task. Hence, even experienced investigators may find it impossible to assign a specific incompatibility system to a species under their study. An example of such a case is the study of HAYWARD and WRIGHT (1971) on *Lolium perenne* in which the two S loci system typical to the Gramineae, could not be assured.

Finally, PANDEY (1957) reported that in *Physalis ixocarpa*, although there are two multiallelic incompatibility loci, identity at *either* locus in pollen and pistil is sufficient to lead to incompatibility. Moreover, there may be independent activity or epistatic interaction between the two loci.

3.3.1.1.3 Three or More S Loci

From the analysis of the two S loci handled above we deduced that unidirectional incompatibility is characteristic in cases where the pollen parent is heterozygous for one S locus for which the seed parent is homozygous. Thus, while the cross $S_a^{1.2}S_b^{3.3}(♀) \times S_a^{1.2}S_b^{3.4}(♂)$ is compatible the cross $S_a^{1.2}S_b^{3.4}(♀) \times S_a^{1.2}S_b^{3.3}(♂)$ is incompatible (note the change in designation of incompatibility loci: S_a and S_b rather than S and Z).

However, how about a case in which a certain plant (A) is compatible as pollinator with both plants (B) and (C), and plant (B) is compatible as pollinator with plant (C), but other crosses among (A) (B) and (C) are incompatible? If we try to design a scheme for such a case based on two loci for this three-level unidirectional incompatibility, we arrive at the following scheme (ØSTERBYE, 1975). (The arrow in this scheme denotes unidirectional compatibility).

$$
\begin{array}{cc}
S_a^{1.2} & S_b^{3.4} \\
\downarrow & \\
S_a^{1.1} & S_b^{3.4} \\
\downarrow & \\
S_a^{1.1} & S_b^{3.3}
\end{array}
$$

This scheme has a double homozygous plant at its bottom, but such a double homozygous plant would never result from legitimate crosses. From considerations of this kind, LUNDQVIST and his associates (LARSEN, 1974; LUND-QVIST et al., 1973; LUNDQVIST, 1975; ØSTERBYE, 1975) estimated the number of S loci in *Ranunculus acris* (Ranunculaceae) and *Beta vulgaris* (Chenopodiaceae). Working with several "families" of *R. acris*, these authors concluded that at least three S incompatibility loci are active in this species. *B. vulgaris* turned out to be even more complex and the data could be explained only by the operation of four S loci. Thus, when such incompatibility families in *B. vulgaris* were tested in a diallel cross, the results were compatible with the predicted pattern shown in Fig. 3.33, indicating four levels of unidirectional compatibility. A similar case may be *Papaver rhoeas* (LAWRENCE, 1975).

In the one multiallelic S incompatibility (*Personata*) system considered above, an increase in the number of *alleles* in a given population linearly increases the ratio of compatible to incompatible crosses. Increasing the number of incompatible *loci* would exponentially increase this ratio. Cases of unimpaired incompatibility on the polyploid level were reported in Chenopodiaceae as well as in Ranunculaceae. This retention of incompatibility by polyploid plants is the rule in all the monocotyledons, hinting to a similarity between the two S loci and the multiple loci incompatibility systems. Such and other considerations were elaborated by LUNDQVIST (1975) and interesting ideas on the evolution of incompatibility system were derived. These are outside the scope of this book. Nevertheless, it should be noted that the two families in which the multiple S loci systems were detected (i.e., Ranunculaceae and Chenopodiaceae) are considered to belong to a common dicot evolutionary branch, the basis of which is probably close to the point of origin of the monocot group. This may explain

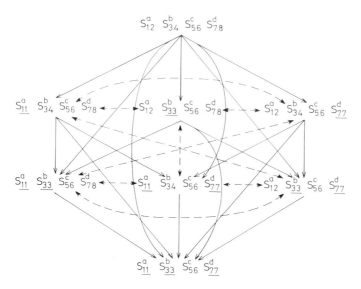

Fig. 3.33. Predicted incompatibility pattern in a family of *Beta vulgaris*. *Arrows:* direction of compatible pollinations (*arrowheads* point to the seed parent). *Double-headed arrows:* crosses connecting reciprocally compatible pollinations (after LARSEN, 1974)

the existence of two S loci in the Gramineae and more than two S loci in the former two families. On the other hand, the revelation of only one S locus in other monocot families must be justified by further evolutionary elaboration i.e., as a later reduction in the number of expressed S loci in isolated populations. One implication of such considerations for the plant breeder is that crossing plants from two widely separated populations may reveal 'hidden' S loci, possibly even in cases where either or both of these populations do not show any incompatibility because of independent mutations toward compatibility. The reader is thus referred to a detailed discussion of this subject by LUNDQVIST (1975).

3.3.1.2 Sporophytic Incompatibility

While the group of gametophytic incompatibility systems is characterized by its pollen reaction being determined by the genotype of the gametophyte, it is typical of the sporophytic incompatibility system that the pollen reaction is determined by the genome of the somatic tissue (of the sporophyte) in which this pollen developed.

For convenience we shall divide sporophytic incompatibility into *heteromorphic* and *homomorphic* incompatibilities. The designations 'heteromorphic' and 'homomorphic' are here limited to floral morphology. A further definition of this term in respect to incompatibility will be given below. It should, however, be noted that from various points of view the division of this incompatibility system, which is based on floral morphology, is not fully justified. We shall see that in contrast to the incompatibility systems mentioned above, e.g. one,

two or more S loci gametophytic incompatibilities, the heteromorphic inco
bility does not constitute a coherent group in respect to phylogenetic or evo
ary aspects. Moreover, we may see that in the same species, population
either or both heteromorphism and incompatibility coexist and that there may
be no genetic linkage between these two characters. It may even turn out that
in some heteromorphic species the incompatibility is gametophytic rather than
sporophytic. However, since in all heteromorphic incompatibility cases studied
up to now the genetic data revealed sporophytic determination, we included
heteromorphic incompatibility in the sporophytic incompatibility system.

3.3.1.2.1 Heteromorphic Incompatibility

The existence of variation in floral structure within the same species was detected
by several authors, but it was not until the publication of HILDEBRAND'S studies
in 1867 and a subsequent publication of DARWIN'S observations in 1877 (see
VUILLEUMIER, 1967, for review and detailed references) that a causal relation
between floral variations and the prevention of self-fertilization was fully appre-
ciated. The term heteromorphic incompatibility was coined by FISHER and
MATHER (1943).

Although heteromorphism may include a great number of features (e.g. anther's
height, pollen wall sculpture, pollen grain size, stylar length, stigmatic surface), it
is most clearly, but not necessarily always, manifested in the stylar length. The
term 'heterostyly' is therefore preferred over heteromorphism by many authors.
This heterostyly is usually expressed in either of two phenomena: (1) the species
(or population) is composed of two types of plants, one having short-styled,
and the other long-styled flowers (distyly); (2) there are three types of plants,
with short- medium-or long-styled flowers, respectively (tristyly). Pollinations
within the same floral form (morph) are termed illegitimate, and those between
different morphs are termed legitimate. Heteromorphic incompatibility hence
renders both self-pollination and intramorph (illegitimate) cross-pollination infertile.

VUILLEUMIER (1967) analyzed the literature and her own data and reported
on 23 families with 131 genera having heterostyly. These include both monocotyle-
don and dicotyledon families; only nine of these genera are tristylic, the others
being distylic. Contrary to what may be deduced from some texts, in most of the
heterostylic genera listed by VUILLEUMIER no incompatibility was reported,
although it may well exist in part of the latter genera. Heteromorphism which
lacks incompatibility may still strongly promote allogamy due to the mechanism
of insect pollinations which render intermorph pollination much more frequent
than intramorph pollination.

Heteromorphic incompatibility was reported in several ornamental plants
and is very rare in economic crops. Some of the former belong to the following
families: Linaceae (Linum), Plumbaginaceae (Plumbago, Ceratostigma,
Limonium), Lythraceae (Lythrum), Amaryllidaceae (Narcissus), Pontederiaceae
(Eichhornia), Primulaceae (Primula). Heteromorphism is correlated with ento-

mophily and the latter is associated with an attractivity of the flower to the pollinating insect (see Chapter 1.3). There thus seems to be some overlapping in the tastes of man and insects.

A now classical case of distyly was reported by DARWIN to exist in *Primula*. From his and several later studies (see BODMER, 1966, for references), the following picture emerges. The distylic populations of *Primula* species are composed of about equal numbers of plants with either long styles and short stamens ('pin') or short styles and long stamens ('thrum'). 'Pin' plants have also larger stigmatic cells and smaller pollen grains than 'thrum' plants, and in some cases additional characters were found to be strongly linked to this complex of floral heteromorphism. This complex is governed by a single locus which also regulates the incompatibility relation. Such populations are normally composed of Ss ('thrum') and ss ('pin') plants and the only fully compatible combinations are 'pin' (♀) × 'thrum' (ss × Ss) and 'thrum' (♀) × 'pin' (Ss × ss). Thus, both the S and the s pollen of 'thrum' plants are compatible on 'pin' pistils, but the s pollen of 'pin' plants is incompatible on 'pin' pistils, meaning that the compatibility reaction of the pollen is determined not by its own gametophytic genome but by the sporophyte on which it was produced. Although such a complex between sporophytic incompatibility and distyly is known in several other genera (e.g. *Linum*, *Limonium*), this complex is by no means obligatory within a genus neither within the same species. Thus, studies of DULBERGER (1964, 1970a) indicated that the heteromorphism in *Narcissus tazetta* (Amaryllidaceae) as well as in *Anchusa hybrida* (Boraginaceae) is not genetically linked to the incompatibility mechanisms of these species. This is clearly expressed by the normal fertility of "illegitimate" pollinations in spite of self-incompatibility. On the other hand, in a wide-range survey (from 11°N to 25°S and from 39°E to 168°E), LEWIS (1975) found within the species *Pemphis acidula* (Lythraceae) many distylic populations, which were probably intramorph-incompatible, and only one homomorph population, which was most probably self-compatible. In this case a loss of floral heteromorphism paralleled a loss of incompatibility. A further elaboration of evolutionary considerations concerning heteromorphism and incompatibility is outside the scope of this book. The above mentioned publication of LEWIS, as well as an earlier article of BAKER (1966), may serve as an introduction to these considerations.

Tristylic heteromorphism is known in three angiosperm families (Oxalidaceae, Lythraceae, Pontederiaceae). In tristylic populations every plant is characterized by having one type of flower. There are thus three 'morph' groups consisting of flowers with long, 'mid' or short styles. The anthers occupy typically (with notable exceptions) two out of three possible positions—low and 'mid', high and low, 'mid' and low—so that the third position is occupied by the stigma (Fig. 3.34). Investigation of the breeding behavior of *Lythrum salicaria* was already undertaken by DARWIN, and from his and subsequent studies the following may be summarized. Each of the three morphs contains one type of flower, similar to those shown in Fig. 3.34. Pollen from any given anther position is only compatible with a pistil having the same stigma position, which means that pollen from the two sets of anthers in one flower (e.g. long-styled) can fertilize flowers of two other complementary floral forms (e.g. short-styled and

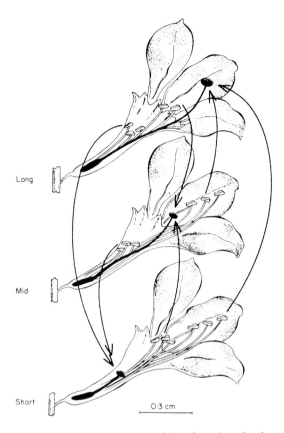

Fig. 3.34. Tristyly in *Lythrum junceum:* anthers and stigmata are positioned at three levels; compatible pollination takes place between anthers and stigmata of the same level (redrawn from DULBERGER, 1970b) Copyright (C) Blackwell Sci. Pub. Ltd. Reproduced by permission

'mid'-styled). The genetic constitution of the three morphs is based on two independent loci, M and S, with each having two alleles:

Long style – mmss
'Mid' style – Mmss or MMss
Short style – MmSs, mmSs, MMSs, MMSS or mmSS.

The dominant allele S conditions short style and is epistatic to M; the recessive s, in homozygous condition, interacts with M or m to give 'mid'- or long-styled forms, respectively. In addition to these morphological differences between the floral types additional differences, such as pollen form and stigma structure, also exist. The segregation of the two loci in *Lythrum salicaria* is complicated, as this species is an autotetraploid but basically the same compatibility relationships also exist in the true diploid tristylic *Lythrum junceum* (DULBERGER, 1970b) and are controlled by a similar genetic mechanism.

3.3.1.2.2 Homomorphic Incompatibility

Sporophytic control of pollen incompatibility was suggested already in 1912 by CORRENS for *Cardamine pratensis* (Cruciferae), but a satisfactory explanation of the experimental results was delayed for 38 years. Only then was the genetic control in this system fully explained for the two composite species *Crepis foetida* and *Parthenium argentatum* by HUGHES and BABCOCK (1950) and GERSTEL (1950), respectively. Since then, this system of incompatibility has been established in numerous species, many of them of great economic value. These belong mainly to the families Cruciferae, Compositae and Convolvulaceae. The genetic details of this incompatibility system were presented by BATEMAN (1952, 1954, 1955) and later information can be obtained from the general review on incompatibility of ARASU (1968). The main genetic features of this system may be summarized in the following:

1. Incompatibility is controlled by one S locus having several alleles; the number of these alleles is usually less than in the one S locus gametophytic incompatibility system, but in some species (e.g. *Brassica oleracea*) over 30 such alleles were revealed.

2. The reaction of the pollen is determined by the genotype of the sporophytic tissue in which it was formed, and thus it is controlled by two S alleles.

3. Resulting from the above type of pollen control, all the pollen of a plant has the same incompatibility reaction.

4. The two S alleles may react independently or they may interact by one being dominant over the other; these relationships may exist in one, both, or neither the pollen or the pistil; the recessive allele has then no activity.

5. Active allele identity in both pollen and pistil leads to incompatibility.

6. The dominance/independence relationships of the S alleles in the pollen and in the pistil may differ.

Such dominance/independence relationships may lead to rather complex incompatibility reactions. This will be demonstrated with an apparently simple cross: $S_{13}(♀) \times S_{12}(♂)$ in which the compatibility is expected to be the following:

Pollen reaction	Pistil reaction	Compatibility
Independent	Independent	Incompatible
S_1 dominant to S_2	Independent	Incompatible
S_2 dominant to S_1	Independent	Compatible
Independent	S_1 dominant to S_3	Incompatible
Independent	S_3 dominant to S_1	Compatible
S_1 dominant to S_2	S_1 dominant to S_3	Incompatible
S_1 dominant to S_2	S_3 dominant to S_1	Compatible
S_2 dominant to S_1	S_1 dominant to S_3	Compatible
S_2 dominant	S_3 dominant to S_1	Compatible

Figure 3.35 compares the sporophytic incompatibility with the gametophytic one. For simplicity the sporophytic incompatibility is represented by the case of independent activity of the S alleles in both pollen and pistil and the gametophytic incompatibility is represented by the one S locus control. Actually, among

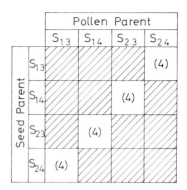

One S locus, multiallelic
gametophytic incompatibility

Homomorphic sporophytic
incompatibility with independent
S alleles activity in both pollen
and pistil

Fig. 3.35. Comparison between the one multiallelic S locus, gametophytic incompatibility and the homomorphic-sporophytic incompatibility (with independent S allele's activity in both pollen and pistil). The two schemes describe the theoretical results of diallele crosses among the plants of an F_1 progeny of a cross between two unrelated parental plants $(S_{12} \times S_{34})$. *Shaded squares:* incompatibility; *figures in parentheses:* number of zygote genotypes produced by each cross

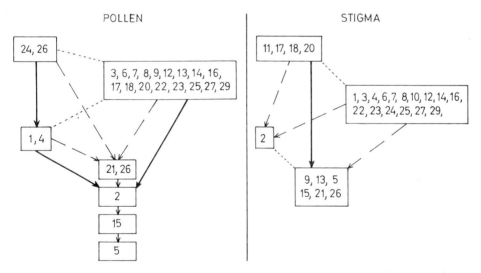

Fig. 3.36. Groups of S alleles arranged spatially in accordance with their dominance relationships in *Brassica oleracea. Whole lines with arrow:* dominance of upper over lower group *broken lines with arrow:* S alleles heterozygotes between groups show either activity of both alleles or alleles from upper group dominant over those from lower group *dotted lines:* both alleles active (from THOMPSON, 1968)

most of the species where sporophytic incompatibilities were studied in detail (e.g. several species of *Brassica*), independent activity of the S alleles is rather rare. Figure 3.36 gives an example of the dominance relationships of the S alleles in *Brassica oleracea* (THOMPSON, 1968). Additional information on the recognition of S alleles and their interrelation in the pollen and pistil can be obtained from THOMPSON and HOWARD (1959). From this scheme it may be deduced that the complicated dominance relationships which may exist in species with sporophytic incompatibility present a baffling problem. Fortunately, several methodologies and new approaches were recently developed to help solve such problems. As we shall detail in a subsequent section, various techniques are now available to overcome incompatibility barriers even in plants homozygous for an S allele, and thus forced self-fertilization is possible. Plants with a known pair of homozygous S alleles can then be used as efficient testers for the identification of new S alleles in genetically unidentified plants. Moreover, if a plant with known S alleles (say S_{12}) is self-pollinated and the progeny intercrossed, it is then rather simple to reveal the dominance relationship between the alleles concerned as shown in Table 3.25. The identification of the correct dominance/in-

Table 3.25. Expected compatibility relationships among the progeny of a (forced) self-pollinated S_{12} plant, when different dominance/independence relations between the S alleles are assumed. Plus and minus signs indicate compatible and incompatible crosses, respectively

Dominance/independence relationship of S alleles	Genotype of seed parent	Genotype of pollen parent		
		S_{11}	S_{12}	S_{22}
Independence in both pollen and pistil	S_{11}	−	−	+
	S_{12}	−	−	−
	S_{22}	+	−	−
Independence in pollen; S_1 dominant over S_2 in pistil	S_{11}	−	−	+
	S_{12}	−	−	+
	S_{22}	+	−	−
Independence in pistil; S_1 dominant over S_2 in pollen	S_{11}	−	−	+
	S_{12}	−	−	−
	S_{22}	+	+	−
S_1 dominant over S_2 in pollen and in pistil	S_{11}	−	−	+
	S_{12}	−	−	+
	S_{22}	+	+	−

dependence relationships of the S alleles can then be further verified by back-crossing to the parental plant and/or test-crossing to known testers.

Other, non-genetic techniques, applicable in the study of the incompatibility reaction, are also available. We shall return to these techniques while dealing with pollen/pistil interactions and with the utilization of incompatibility systems for plant breeding.

Contrary to what was found in most species having the one S locus gametophytic incompatibility, doubling the chromosome number in species with sporophytic incompatibility does not disturb the incompatibility reaction; neither does incompatibility disappear in natural polyploid species or variants of the latter type.

Finally, we shall mention the incompatibility of cocoa *(Theobroma cacao)*. The data for the incompatibility in this species are somewhat confusing. This species represents one of the cases in which the incompatibility occurs in the ovules: in incompatible crosses all or some of the male and female gametes do not fuse. Flowers in which the fusion of gametes was thus reduced, undergo early shedding. The incompatibility was found to have some characters of a sporophytically controlled system but gametophytic control cannot be completely excluded and probably other loci, in addition to the S locus, are involved (see COPE, 1962, for details).

3.3.2 Pollen–Pistil Interaction

The processes of pollen germination on the stigma, the growth of the pollen tube in the style, and the penetration of the tube into one of the synergid cells with the final discharge followed by the double fertilization, as described briefly above (3.1.4), obviously require a rather detailed and precise adjustment between pollen and pistil. Such adjustments were probably an integral component of the evolution of each species. Populations which ramified from the species but did not share the co-evolution of intimate pollen–pistil partnership, were thus probably separated and an incipient interspecific boundary evolved. Evidence is accumulating in favor of the notion that different control mechanisms govern pollen–pistil interactions of the above type and intraspecific incompatibility. It seems that the latter system is superimposed on a well-adjusted pollen–pistil interaction (see HESLOP-HARRISON, 1975, and HOGENBOOM, 1975). In the following we shall be concerned only with incompatibility in the restricted (intraspecific) sense.

   In our dealing with pollen–pistil interaction we shall mainly be concerned with structural and biochemical aspects of this interaction, bearing in mind their relevance to crop production and breeding. We shall mostly refrain from an evaluation of the factual information which may lead to a better understanding of the incompatibility. Those interested in the latter approach are referred to HESLOP-HARRISON et al. (1975) for further reading.

3.3.2.1 Pollen Cytology and Pollen–Stigma Interaction

Let us first look at the cytology of the pollen before it reaches the stigmatic surface. Following a survey of incompatibility systems, BREWBAKER (1957) pointed out the correlation between incompatibility system and number of nuclei in the pollen, noting that in general, species belonging to the gametophytic and sporophytic incompatibility system have two and three nuclei, respectively. BREWBAKER himself found exceptions to this correlation, but many more are known now. Thus, pollen of species with the two S loci gametophytic system has three nuclei, while at least in some of the plants with the heteromorphic–sporophytic incompatibility system the pollen has two nuclei. Furthermore, as detailed above, in *Ranunculus* a three S loci gametophytic system was revealed and the pollen has two nuclei, while in *Beta* belonging to a related family and having a similar incompatibility system, the pollen has three nuclei. On the other hand, in the two important families with sporophytic incompatibility—Compositae and Cruciferae—pollen with three nuclei is the rule.

A correlation between stigma structure and incompatibility system was suggested by HESLOP-HARRISON et al. (1975). These authors noted that generally (but not always) families with sporophytic incompatibility (e.g. Cruciferae and Compositae) have papillated dry stigmata (group II), whereas families with gametophytic incompatibility have either plumose stigmata with elongated receptive cells (group I, Gramineae) or 'wet' stigmata (groups III and IV). This correlation was concerned with families having either a gametophytic incompatibility system or a homomorphic-sporophytic one, but not with those having a heteromorphic incompatibility. In many plants belonging to this latter system, heteromorphism in respect to the stigmatic surface is notable: the stigmata may be either "papillate" or "cob" (almost smooth). Moreover, structural studies by DULBERGER (1975) on intermorph differences in the Plumbaginaceae showed that the pistils of the two morphs differ in several respects and at least in some distylic species (e.g. *Limonium meyeri*) stigmatic dimorphism is probably involved in the incompatibility mechanism. Thus, the A-type pollen, which is produced on plants with 'cob' stigmata and has a coarse sculpture, readily adheres to 'papillate' stigmata, and also to 'cob' stigmata, with which it is incompatible. On the other hand, B-type pollen, which is produced on plants with "papillate" stigmata and has a fine sculpture, readily adheres to 'cob' stigmata but not to the incompatible "papillate" stigmata (Fig. 3.37).

The pollen–stigma interactions in the homomorphic–sporophytic incompatibility system were clarified by several recent studies (see HESLOP-HARRISON, 1975; HESLOP-HARRISON et al., 1975; and HOWLETT et al., 1975, for review and details). Most of these were made with species of Cruciferae but supporting evidence came also from the Compositae. As noted in Chapter 3.1, it was found that soon after landing on either the stigma or an appropriate artificial surface, the pollen releases sequentially exine- and entine-containing substances. The former, proteins or glycoproteins, are released within minutes. When the exine exudate comes in contact with an incompatible stigma it causes there an immediate callose production. This exudate-induced callose production does not occur when its pollen source is compatible with the stigma. This pollen–stigma reaction is actually a cell-to-cell interaction: the callose production is restricted to the very same papillae which come in contact with the incompatible pollen grain (or the artificially applied exudate). As mentioned briefly above, the stigmatic surface of Compositae and Cruciferae is 'dry'. It was found to be covered with a hydrated overlayer of a proteineous *pellicle*. This is in contrast to many families with a gametophytic incompatibility where a copious fluid was detected on the stigmata. This relatively 'dry' surface in families with the sporophytic incompatibility may be responsible for the localization of the pollen–stigma interaction in the latter system. Production of callose is obviously not the only process resulting from the contact between pollen and papillae. Other enzymatic processes were detected, among them a cutinase activity which causes an erosion

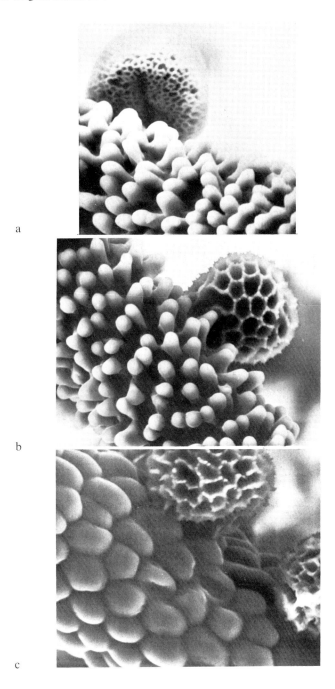

Fig. 3.37 a–c. Structural relations between pollen sculpture and stigma surface in *Limonium meyeri*. (a) Incompatible B pollen grain on a papillate stigma. (b) Compatible A pollen grain on papillate stigma. (c) Incompatible A pollen on a cob stigma. Scanning electron micrographs (× 1,000) of DULBERGER (1975) Reproduced by permission of The Royal Society

of the stigmatic cuticle. Observations in *Agrostemma githago* (Caryophyllaceae) indicated that such an erosion of the papillae, which is a prerequisite for pollen tube penetration, is a result of interaction between substances coming from the pollen and the papilae's own protein coat. Enzymatic removal of the proteins from the papilla prevented the pollen-induced erosion (HESLOP-HARRISON and HESLOP-HARRISON, 1975). The involvement of the cutinase activity in the compatibility reaction is nevertheless questionable, since in some cases (e.g. *Raphanus*) it occurs after both compatible and incompatible pollination. The early callose deposition may occur also on the protruding pollen tube, and with this deposition further germination is prevented. Although the exact mechanism of callose inhibition of pollen germination is not fully understood, it nevertheless is a specific phenomenon, restricted to incompatible pollinations in the sporophytic system, and can be readily used in this system to screen for the pollen-rejection response. Attempts to detect in the pollen, proteins which are specific for the compatibility genotype of this pollen's sporophytic parent, were disappointing. On the other hand, as we shall see below, serological differences between pollen genotypes differing in the S alleles, were detected in the gametophytic incompatibility system. Moreover, in contrast with the results with pollen, striking differences in the patterns of stigma antigens related to the S allele composition were detected in the plants of the sporophytic incompatibility. Whether or not these differences in stigmatic antigens represent pellicle-contained proteins which are involved in the rejection mechanism is still an open question, but this is a rather reasonable assumption. As we shall see below, once the pollen of Cruciferae and Compositae cross the stigma incompatibility barriers, it is not further inhibited and fertilization readily takes place even between an incompatible pollen and style. We may hence conclude our coverage of the pollen–pistil interaction in the homomorphic–sporophytic incompatibility system by stating that there now exists good evidence in favor of the claim that in this system the sporophytic response of the pollen is determined by the relatively late process of investment of components from the sporophytic tapetum into the exine of the pollen, rather than the response being determinated at a pre-meiotic stage and carried over to the gametophyte via its cytoplasm. Obviously one additional argument against the latter possibility is the "clearing" of cytoplasm, which was observed in the microcysts, and probably involves an 'erasing' of sporophytic information (see Chapter 3.1 above).

It now seems questionable whether a division between 'stigmatic' and 'stylar' incompatibility barriers is justified. Probably a more appropriate division will be between inhibition at the stigmatic surface and any later inhibition. We detailed above the former type of inhibition of incompatible pollen. A "stigmatic" inhibition of incompatible pollen occurs in *Oenothera organensis*, a species with gametophytic incompatibility. In this species both compatible and incompatible pollen germinate on the 'wet' stigmatic surface and the pollen tubes start to penetrate the stigmatic tissue and are then inhibited in their further growth. Nevertheless, it was argued that the incompatibility signalling actually started while the tube passed the viscous fluid coating the outer stigmatic cells (DICKINSON and LAWSON, 1975). Thus, the sites of the incompatibility triggering and its expression in the form of further tube growth inhibition, may be spatially separated.

3.3.2.2 Pollen Tube–Style Interaction

The differentiation between stigmatic and stylar inhibition of pollen tubes is problematic not only for the reasons pointed out above but also because in many species these two tissues cannot be separated precisely. Thus, in species with a very short style or with a pollen-receptive cell layer without an elaborate stigma, such a separation makes little sense. Moreover, in most cases where the incompatible pollen tube starts to penetrate the stigmatic tissue, the depth of penetration is far from uniform. Technically, it is rather simple to follow pollen tube growth in the pistil in either intact flowers (the pistils are then removed and analyzed) or by in vitro pollination of isolated flowers or pistils. We have therefore a wealth of information on this subject which is outside the scope of this book. The biochemical aspects of pollen development and pollen tube growth were reviewed comprehensively by MASCARENHAS (1975), while the text of STANLEY and LINSKENS (1974) gives additional general coverage on pollen biology, including applied aspects. The reviews of LINSKENS and KROH (1967), ARASU (1968), DE NETTANCOURT (1972) and HESLOP-HARRISON (1975) also deal with incompatibility aspects of pollen/pistil interactions.

It is noteworthy that many of the environmental effects on pollen tube growth in the style as well as the effects of pistil grafts were already observed by DARWIN (1877), whose work should still be the first reference for those interested in this subject.

The site of the incompatibility reaction in plants with gametophytic incompatibility may differ. Thus pollen incompatibility in *Oenothera* seems to be triggered in the stigma, leading to a subsequent retardation of pollen tube growth, while in *Lilium longiflorum* the incompatibility reaction obviously takes part in the style. The latter reaction is probably regulated by the transmitting tissue of the hollow style which is typical for the Liliaceae and related families. The exact nature of this inhibition is still obscure in spite of many biochemical approaches to this question for which the long *L. longiflorum* style is an especially favorable object.

Solanaceous plants have a "solid" style and the common incompatibility system in this family is the single, multiallelic S locus gametophytic one. In the genera of this family in which the pollen tube–style interaction was mostly studied (e. g. *Petunia*: LINSKENS and associates, *Lycopersicon*: DE NETTANCOURT and associates), the incompatible tube was found to be inhibited in the style. Several structural features were found to be correlated with incompatibility; many of them indicated a cessation of protein and/or polysaccharide synthesis in the incompatible tube. Ultrastructural studies in *Lycopersicon peruvianum* (DE NET-TANCOURT et al., 1973) indicated that the pollen tube wall is composed of two layers, the inner one of which undergoes degenerative changes in incompatible tubes, resulting in bursting of the tip at about one-third down the stylar length. These authors concluded that this bursting is an active (but probably premature) process, reminiscent of the programmed discharge of the pollen tube upon penetration in the synergid cell of the embryo sac, and that the incompatibility reaction is programmed by the vegetative nucleus of the pollen tube. A somewhat different process was observed in interspecific pollinations in which *L. peruvianum* was involved, although some of the features were common to the two types of incompatibility. Serological studies (initiated earlier by D. LEWIS) in species with gametophytic incompatibility were carried out in H. F. LINSKENS' laboratory, especially with *Petunia*. These studies reaffirmed that pollen with different S genes contain different antigens (as noted above, no such differences could be detected in species with sporophytic incompatibility). Moreover, styles of different S genotypes also contained specific antigens and pollen and styles of the same S genotype contain identical antigens. LINSKENS and co-workers also revealed an array of chemical compositions and biochemical activities to be related to incompatibility, but the exact control mechanism imposed by the genetic information on the incompatibility reaction in the style is still an unsolved question.

3.3.2.3 Pollen Tube–Ovule Interaction

This heading is listed only in order to mention that in some notable cases the incompatibility reaction is delayed until the pollen tube enters the ovule (e.g. *Theobroma cacao*, *Narcissus*). The mechanism of this type of interaction is still quite obscure. The possibility cannot be ruled out that even in such cases the pollen tube is triggered in respect to compatibility at an early stage but this triggering is manifested much later.

3.3.3 Incompatibility, Crop Production, and Breeding

In crops harvested for their vegetative biomass (e.g. sweet potatoes), especially if propagated vegetatively, as well as in ornamental plants (e.g. *Bougainvillea*), incompatibility may be advantageous or of little practical significance. On the other hand, in quite a number of crops, especially vegetatively propagated fruit trees (e.g. cherries), incompatibility complicates horticultural practice because the intraclonal incompatibility requires the addition of pollinators and renders the yield dependent on abundant pollen transfer among the trees. Incompatibility is also a disadvantage in the breeding procedure of many crops, when either a free intercross in a plant population or inbreeding is required. Conversely, incompatibility is of great practical importance for hybrid seed production in a number of cultivated plants. The plant breeder may therefore be interested either to eliminate incompatibility (permanently or temporarily), or to introduce incompatibility into a cultivar. The breeder may also be interested in ways and means of utilizing incompatibility for hybrid seed production. From the above considerations we can itemize the main possible objectives with which the agronomist, the horticulturist and the plant breeder, confronted with incompatibility, may be concerned.

1. Introduction of incompatibility into cultivars or breeding lines from either other lines or wild relatives.
2. Permanent elimination of incompatibility by genetic means.

3. Regulating, by either genetic or physiological means, the strength of incompatibility, thus enabling in a given line self-pollination, when required, but preventing it in other cases.

4. Development of breeding procedures which will enable the utilization of incompatibility for hybrid seed production.

Obviously, prior to changing the incompatibility reaction or utilizing it, the incompatibility system of the cultivar in question should be known. Although in numerous cultivated species the genetics of incompatibility was studied, the last survey was published several years ago (ROWLANDS, 1964). On the other hand, within a given family there is usually only one type of homomorphic-incompatibility system which is known for many of the economically important families (see previous sections of this chapter and ROWLANDS, 1964). The details of the inheritance of incompatibility and identification of the genotypes of the breeding lines will then have to be determined by test crossings, as outlined in Section 3.3.1 above. Several auxillary techniques were suggested to assist in this task. Among these are: serological characterization of the pollen and the pistil (LINSKENS, 1960); determination of the isozyme patterns of the somatic tissue and the pollen (BREDEMEIJER and BLAAS, 1975; NASRALLAH et al., 1970); utilization of marker genes linked to specific S genes (DE NETTANCOURT, 1972); in vitro pollination of isolated pistils (DEVREUX et al., 1975); and fluorescence microscopy of pollinated pistils. The references presented above in parentheses are by no means the only ones and further information on such methods can be obtained from DIXON (1968), DE NETTANCOURT (1972), HESLOP-HARRISON (1975), and STANLEY and LINSKENS (1974).

There are only a few general reviews on the impact of incompatibility on crop production and plant breeding (e.g. LEWIS, 1956; REIMANN-PHILIP, 1965; LUNDQVIST, 1968; DENNA, 1971; DE NETTANCOURT, 1972), most of which handle the subject mainly by considering basic concepts rather than by giving applicable suggestions. For more specific and up-to-date information, the reader is referred to the Incompatibility Newsletters (Assoc. Euratom, Wageningen).

3.3.3.1 Transfer of Incompatibility into Cultivars

As outlined above, the plant breeder may be interested in introducing incompatibility into cultivars, breeding lines or specific clones. In other cases he may want to strengthen an existing weak incompatibility. Cultivars which lack incompatibility may belong to a species in which other cultivars have strong incompatibility (e.g. *Brassica oleracea, Chrysanthemum morifolium*). In other cases, incompatibility exists in wild relatives or other cultivated and related species. Transfer of incompatibility may hence be achieved by the appropriate intra- or interspecific crosses. The lack of compatibility was in several cases reported to be controlled by specific fertility genes which are epistatic over the incompatibility ones. This transfer was actually achieved in *Petunia* by MATHER and in tomato by MARTIN (see DE NETTANCOURT, 1972, for references). Incompatibility can thus be transferred into a specific line by an appropriate breeding procedure. Transfer of incompatibility was also considered by DENNA (1971) for lettuce *(Lactuca)* and beans *(Phaseolus)*. We shall see in more detail below how the transfer of incompatibility is executed in *Brassica*, where hybrid seed production, based on incompatibility, is a well-advanced commercial practice. Such transfers are not devoid of pitfalls and difficulties. Hence, transfer of incompatibility from an allogamous species (e.g. *L. peruvianum*) into an autogamous one (e.g. *L. esculentum*) may be useless unless entomophily is also assured. The breeder should therefore be concerned also with floral characters which make the new incompatible line attractive for pollen vectors. The cross between a cultivar and its incompatible wild relative may involve several complications. Unilateral incompatibility is encountered in *Lycopersicon*, where *L. esculentum* pollen tubes are inhibited in *L. peruvianum* styles. This unilateral incompatibility is probably controlled by genes which are unrelated to intraspecific incompatibility. This subject was discussed in detail for *Lycopersicon* and other genera by HOGENBOOM (1972, and previous publications). The breeder faced with interspecific transfer of incompatibility will therefore have to account also for incongruity.

An interesting possibility for acquiring incompatibility is "haploidization" of cultivated tetraploids. Incompatible dihaploids were obtained from tetraploid potato varieties (HERMENSEN, 1973) and such a procedure may yield similar results in other species belonging to families with the one S locus gametophytic incompatibility.

3.3.3.2 Permanent Elimination of Incompatibility

There are three main ways to eliminate incompatibility from a breeding line:
1. doubling the chromosome number;
2. induction of compatible mutations;
3. transfer of compatibility into an incompatible line.

We mentioned above that doubling the chromosome number (e.g. by colchicine treatment) in many species with the one S locus gametophytic incompatibility system interferes with the incompatibility reaction of the pollen. Tetraploids of several genera such as *Trifolium, Nicotiana, Prunus* and *Petunia* showed this loss of incompatibility. It can therefore be assumed that polyploidization of incompatible cultivars belonging to families such as Leguminosae, Rosaceae and Solanaceae will lead in most cases to a loss of incompatibility. Permanent loss of incompatibility was achieved in several species with gametophytic incompatibility. Work on this subject was started by D. LEWIS and reviewed quite extensively by DE NETTANCOURT (1969). It is noteworthy that there exist variations in response to X-ray radiation. In some plants mutations leading to loss of incompatibility in both the pollen and the pistil were detected *(Trifolium repens)*, while in other genera *(Petunia, Oenothera)* loss of incompatibility was restricted to the pollen. Moreover, there seems to be a different radiosensitivity of the different S alleles. The usual practice is to irradiate the floral buds at the stage of pollen mother cells in the anthers. Because the common screening procedure for the detection of new S alleles is to apply the pollen from irradiated floral buds to styles with known genotypes—mutations with defective pollen incompatibility were usually isolated.

LEWIS' studies indicated that the frequency of spontaneous $S_1 \rightarrow S_f$ mutation in *Oenothera* was at the rate of 10^{-8} genes and that X-rays produced 1.6×10^{-8} mutations per one *r* unit. The induction of mutation by X-rays resulted in cherries in several mutations, all of which were stable. When a S_{12} tree produces mutated pollen and this is utilized for self pollination, S_{1f} and/or S_{f2} seedlings will result. It is obviously then better to make the cross:

$S_{12} \times S_{1f} \rightarrow S_{1f}$ and S_{f2} (all compatible)

than the cross:

$S_{13} \times S_{f2} \rightarrow S_{12}, S_{23}, S_{1f}$ and S_{3f} (two out of four compatible).

From theoretical considerations the screening of 'fertile' mutations in plants having the sporophytic incompatibility system should be complicated. When we take a case of S_{12} and the mutagenized pollen will be tested on S_{11} pistils, we should be able to differentiate between two possibilities: (1) if the sporophytic incompatibility is carried from the premeiotic stage in the cytoplasm, irradiating the flower bud at the stage of the early differentiation of the sporogenic tissue should result in individual mutated pollen grains; (2) if the incompatibility reaction is invested into the pollen by the tapetum, there would be very little chance of isolating such mutants.

The isolation of lines with a stable loss of incompatibility may be hampered by the occurrence of temporary compatibility and pseudo-compatibility. Thus, the selection of such lines is not a one-step process.

The degree of incompatibility within cultivars of the same species as well as among populations of wild species may vary considerably. This difference in the "strength" of compatibility may be the result of variation in the activity of the S alleles themselves, as was found among cultivars of *Brassica oleracea*, or a result of interference of (poly-) genes which in some cases may be epistatic over the S genes (e.g. *Abutilon*, *Petunia*). Actually, in most systems studied thoroughly, genes which interfere with strict incompatibility were found, but the heritability and the type of genetic control (e.g. epistatic, recessive, dominant or independent) do not seem to have a general trend and should be worked out for each specific case. Nevertheless, these genes can be transferred by appropriate crossing into incompatible lines. Obviously, if the breeder is interested in eliminating incompatibility of a cultivar in order to facilitate the isolation of pure lines, he will not be concerned with maintenance of incompatibility in his breeding material. On the other hand, if incompatibility is utilized for hybrid seed production (to be handled below) and seeds are the commercial yield of the F_1 hybrid, the breeder will be interested in having a 'fertility' restorer in one of his parental lines; this will make the F_1 less dependent on cross fertilization. The details involved in such fertility restoration and transfer of compatibility in the course of breeding procedures leading to successful hybrid varieties are unfortunately only rarely documented, due, probably, to commercial interests. We have therefore little information on this latter problem. RONALD and ASCHER (1975a) described the transfer of compatibility from garden clones of *Chrysanthemum morifolium* into incompatible greenhouse cultivars. From their results it appears that compatibility was dominant but maintained in a heterozygous state in the compatible garden clone. An appropriate breeding scheme could then serve two purposes: to facilitate the breeding of improved greenhouse *Chrysanthemum* cultivars and to establish lines to serve F_1 hybrid seed production. Similar procedures seem to be applicable in other ornamentals of the Compositae.

3.3.3.3 Surmounting the Incompatibility Barrier

Numerous methods were applied successfully to eliminate the incompatibility barrier and hence enable the obtention of seed from self- and cross-incompatible combinations. Most of these were reviewed by DE NETTANCOURT (1972) and LINSKENS and KROH (1967), and thus we shall only summarize them here.

3.3.3.3.1 Treatment of Pollen

Acute irradiation of pollen mother cells caused in several solanaceous species a temporary 'mutation'. Many of the resulting pollen grains were able to attain self-fertilization and among the progeny plants with either homozygous or heterozygous S alleles were obtained. Normal incompatibility recurred in this progeny. Other treatments of pollen, such as aging, did not result in surmounting the incompatibility barrier.

3.3.3.3.2 Bud Pollination

This is by far the most applicable method to overcome incompatibility. It can be utilized in species with either gametophytic or sporophytic incompatibility. It is usually performed by applying mature pollen to pistils a few days before anthesis. Methods used on a large scale and apparently very successful were described in detail by HARUTA (1966) for *Brassica* and *Raphanus*. In some species it was found that the application of ripe stigmatic fluid on the immature stigma further improved the results of bud pollination.

3.3.3.3.3 Delayed Pollination

The applicability of this method is controversial and there are conflicting reports on success even in the same species (e.g. *Brassica oleracea*). In some species incompatibility is significantly reduced at the end of the growth season or in very mature plants.

3.3.3.3.4 Heat Treatment

The exposure of pistils to temperatures of up to 60 °C eliminated the compatibility barrier in several genera (e.g. *Oenothera*, *Trifolium*, *Lycopersicon*). In several

Brassica crops, culture in relatively high temperature was found to break down the incompatibility, especially of 'weak' incompatibility S alleles.

3.3.3.3.5 Surgical Techniques

The application of pollen to the cut surface of the style after removal of the stigma (e.g. *Brassica*), or the introduction of pollen into the ovarian cavity *(Petunia)*, can result in successful self-pollination in plants with strong incompatibility. In *Petunia* a direct application of incompatible pollen to the placenta of in vitro-cultured pistils resulted in fertile seeds.

3.3.3.3.6 Double Fertilization

In several cases incompatible pollen achieved fertilization when applied a few hours after compatible pollen or when the two types of pollen were mixed.

3.3.3.3.7 Other Methods

Numerous other methods were suggested to surmount the incompatibility barrier, e.g.: treating flowers with carbon monoxide, injecting into the style substances which reduce the immune response in animals, applying an electrical potential difference (100 V) between stigma and pollen, treatment of the pistil with plant hormones or with protein inhibitors and steel brush pollination. The applicability of the above-mentioned methods for large-scale breeding purpose, i.e., maintenance by inbreeding of prospective parental lines for hybrid seed production, is obviously of prime importance for the plant breeder. Thus, methods which involve complicated surgical techniques or in vitro culture, as well as mixed pollination, may be applicable in physiological or genetic studies but are not recommended for breeding practice. Ideally, the most favorable technique for the breeder would be to propagate the parental lines by mass culture in isolation under thermal conditions which will enable self- and sib-fertilization of plants having the same S alleles genotype. Commercial hybrid seed production should then be done under different thermal conditions, which will assure strict incompatibility. Unfortunately, this goal can be achieved in very rare cases only, since there seems to be a strong positive correlation between the breakdown of incompatibility of an S allele at high temperature and the general incompatibility weakness of this allele. Thus, in *Brassica* the S alleles which are high in the dominance hierarchy and show strict incompatibility are very slightly (if at all) sensitive to thermal conditions, while those alleles which are lowest in the dominance hierarchy and are considered to induce only pseudo-incompatibility lose their incompatibility activity at high temperatures. Nevertheless, appropriate breeding and selection may overcome this undesirable correlation (RONALD and ASCHER, 1975b).

3.3.3.4 Hybrid Seed Production

Although the potential utilization of incompatibility was noted in the western literature during the first half of this century, its practical application was still questioned up to 20 years ago (LEWIS, 1956). In the West the first commercial hybrid based on incompatibility was probably the marrow-stem kale double cross hybrid *Maris Kestrel*, bred at the Plant Breeding Institute near Cambridge, U. K. (THOMPSON, 1959). Two Brussels sprouts hybrids followed in 1966; these were *Avoncross* and *Thor*, bred in England and The Netherlands, respectively. Breeding of commercial hybrids based on incompatibility started much earlier in Japan. It is difficult to trace the beginning of this breeding but soon after the second World War such hybrids were available from commercial companies (e.g. WATANABA, Sendai, and TAKII, Kyoto). These included several vegetable crops of the Cruciferae. As we shall see, hybrid production based on incompatibility, even when based on the most efficient and sophisticated methods, involves considerable labor. The commercial feasibility, in general terms, of an F_1 hybrid, can be considered by the following formulae:

$$(CHS—CCS) \ll (VHC—VCC) \tag{1}$$

$$\frac{CCS}{VCC} \to 0 \tag{2}$$

In these formulae all values are per unit area, and CHS = cost of hybrid seed, CCS = cost of conventional seed, VHC = value of hybrid crop, VCC = value of conventional crop. Thus, Eq. (1) states that the difference in cash value between a hybrid and conventional crop should be *much* higher than the difference between the costs of hybrid and conventional seed. Eq. (2) states that the ratio of seed cost to crop value should be as small as possible. From these considerations it is clear that the attractiveness of hybrid seed production increases with the crop value and decreases with the proportion of investment in seed relative to the gross income, as well as with the difference in cost between hybrid seed and conventional varieties. Without going into further details, these general considerations are not specific to hybrid seed production based on incompatibility. They can be applied also to hybrids based on sex expression and male sterility and should be considered in respect to any specific crop. Still, it should be noted that hybrid varieties have an additional attractiveness to the breeder. They assure his breeders' rights and consequently usually increase his willingness to invest long-range breeding efforts.

3.3.3.4.1 Characteristics of the Gametophytic and Sporophytic Incompatibility Systems which are Related to Hybrid Seed Production

Economically important crops bearing the one S locus gametophytic incompatibility system belong mostly to the families Rosaceae, Leguminosae, and Solanaceae. Of these, crops belonging to the Rosaceae include mostly vegetatively propagated perennials, in which F_1 hybrids in the usual sense are not applicable. The utilization of incompatibility for hybrid seed production in cultivated Solanaceae species is hampered by the fact that this allogamic mechanism is restricted almost exclusively to wild species. As noted above, incompatibility could, in several cases, be transferred to cultivars, but this involves difficult and long-range breeding work. The applicability of this incompatibility system was thus restricted to legumes, and among the latter actual F_1 hybrids were produced probably only in clover (Trifolium) (ANDERSON et al., 1972). The complex genetic control of the multiple S loci gametophytic incompatibility did not seem attractive to breeders concerned with F_1 hybrid seed production in cultivated Gramineae species, even though the potentiality of its utilization for hybrid seed production was suggested (ENGLAND, 1974).

Compositae and Cruciferae are the two economically important families having the homomorphic-sporophytic incompatibility. While incompatibility is known in several ornamental cultivars and wild species of the Composite, most cultivars of this family are self-compatible (see Table 1.5). Thus, as with Solanaceae, before incompatibility can be applied for hybrid seed production (e.g. in lettuce), it has to be transferred to the cultivars from incompatible wild relatives. Hence, in practice, there probably do not yet exist commercial hybrids of this kind. Contrary to the Compositae, in the Cruciferae many cultivars, especially of the genera Brassica and Raphanus, contain incompatibility and though the strength of this incompatibility varies and is in many cases not strong enough to be applicable for hybrid seed production, intra- and even interspecific hybridization is achieved with relatively little difficulty. Thus, species of these two genera are the main target of breeders trying to utilize incompatibility for commercial hybrid seed production, and considerable success has been achieved.

3.3.3.4.2 Hybrid Seed Production in Brassica and Raphanus

These two genera are typical representatives of the plants having the sporophytic incompatibility system. Thus, as mentioned above, the activity of the S alleles in the pollen as well as in the stigma can attain various relationships: independent, dominant or recessive. Moreover, the dominance hierarchy is quite complicated as exemplified for Brassica oleracea in Fig. 3.36. HARUTA (1966) studied the S allele relationships in several Brassica and Raphanus cultivars; his findings are summarized in Table 3.26. HARUTA'S findings should not be construed to mean that these are the only existing relationships in the cultivated species studied by him. Actually, although his tests involved thousands of selfings and crosses, the starting material in each crop was restricted to only a few cultivars and the total number of S alleles tested in each cultivar was small. Still, Table 3.26 shows clearly that within one crop there may be more than one type of incompatibility relationship. Thus, an actual collection of S alleles and the determination of their relationship is a project to be carried out separately for each crop before a reliable scheme for hybrid seed production can be worked out.

Table 3.26. Types of incompatibility relationship of S alleles in pollen and stigma found in several cruciferous cultivars[a]

Type of incompat-ibility relationship	Dominance relationship among S-alleles		Crops					
	In stigma	In pollen	Brussels sprouts	Head cabbage	Sprouting broccoli	Chinese cabbage	Turnip	Japanese radish
I	S_b dominant to S_a	S_b dominant to S_a	Senshu Komachi (A); Senshu Komachi (B).	Alde Copenhagen Market; Slow Bolling Flat Dutch; Nakato Kyoku Sosei.				Kuroha Mino Sosei (S_0B_3); Taibyo Mino Sosei (S_0MD_{11}).
II	S_b independent of S_a	S_b dominant to S_a		Surehead; Danish Ball Head; Copenhagen market.	Morse's Medium E.	Kyoto Hakusai; Matsushima Jun (S_0M_1); Matsushima Jun (S_0M_7).	Seigoin Okabu.	
III	S_b dominant to S_a	S_b independent of S_a		Maeda Banso.				
IV	S_b independent of S_a	S_b independent of S_a				Kyoto Hakusai; Nozaki Harumaki; Kintai Hakusai.	Sosei Kokabu.	Midoriha Mino Sosei (S_0Y_1); Midoriha Mino Sosei (S_0Y_2); Taibyo Mino Sosei (S_0MA_{12}); Tabiyo Mino Sosei (S_0ME_{11}).

[a] From HARUTA (1966).

Eliminating the details which should be taken into account with each of the *Brassica* and *Raphanus* crops, we shall now consider the main methods of hybrid seed production based on incompatibility.

1. *Single Cross.* This method is conceptually the simplest one. Two cross-compatible lines, each of them highly self-incompatible, are needed. The lines A (S_{11}) and B (S_{22}) are each propagated (e.g. by bud pollination or open pollination at elevated temperature), first by selfing and then by sib-pollination, in isolation. In the hybrid seed production field these lines are planted together to cross-pollinate each other. The hybrid seed is harvested from both parents. The scheme of the method will thus be:

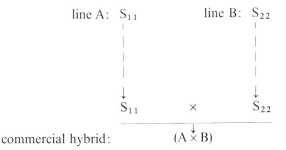

commercial hybrid: $(A \times B)$

Such hybrids are the most uniform ones but usually the production cost of the parental lines is prohibitive.

A slight variant of this method is the utilization of a B line with S alleles lacking incompatibility. The final cross will then be $S_{11} \times S_{ff}$. The advantage is then that only one incompatible parental line has to be maintained, and the disadvantage, that only the seed harvested from the S_{11} parental line is hybrid.

The single-cross method is favored only when crop uniformity is of prime importance and the value of hybrid seed is high enough to cover the expenses of parental lines production.

2. *Double Cross.* This method was utilized by THOMPSON (1959) for the production of *Maris Kestrel*—a marrow-stem kale hybrid—and later used with other cruciferous crops. It is based on the maintenance of four homozygous lines according to the scheme:

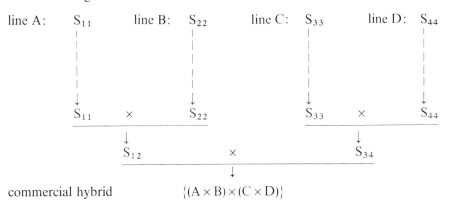

commercial hybrid $\{(A \times B) \times (C \times D)\}$

The great advantage of this method is the relatively inexpensive $A \times B$ and $C \times D$ parents needed for the final hybrid seed production. This method requires obviously more preparatory breeding work, both in respect to isolation of four self-incompatible lines and testings for combining ability. An additional disadvantage is the much less uniform character of the final hybrid, as compared with the single cross. Finally, in many cases the yield of double-cross hybrids is lower than that of single-cross hybrids. In order to combine the lower production cost of double crosses with the uniformity of single crosses, a method of utilizing isogenic lines can be used (NIEWHOF, 1968). According to this method, rather than using different A and B lines, different S alleles are bred into one line to give two isogenic A lines $(A' = S_{11}; A'' = S_{22})$; and similarly, instead of C and D, two isogenic B lines are produced. Such isogenic lines can be produced by starting with one plant which is heterozygous in respect to the S alleles. These alleles should both be as high as possible in the dominance hierarchy (e.g. $S_{24\,28}$). The plant is propagated by several generations of self-pollination, keeping only the S heterozygotes for further selfing until a uniform progeny is obtained, and among these only two S homozygotes are kept further as isogenic $(A' = S_{24\,24}; A'' = S_{28\,28})$ lines. The above is a highly schematic description of the breeding procedure. Incompatibility is obviously not the only concern of the breeder. Many other characters (yield, quality of the crop, resistance to diseases and pests, etc.) have to be taken in account. Thus, in practice, the production of parental lines for a prospective double-cross hybrid is a rather complicated procedure.

3. *Triple Cross*. In order further to reduce the cost of producing inbred parental lines, which involves expensive steps of bud pollination (or any other technique to assure self- or sib-pollination), THOMPSON (1964) suggested the "triple-cross" method:

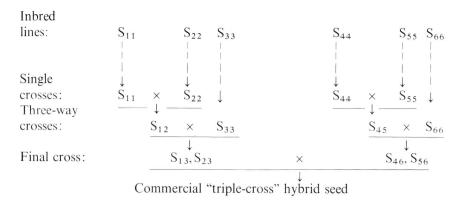

Commercial "triple-cross" hybrid seed

Such a method obviously reduces the cost of the final hybrid seed but has also obvious drawbacks. First, the final hybrid is far from uniform. It may be regarded almost as a synthetic variety. This is of no great concern if one is dealing with a forage crop, provided that there is still a substantial yield increase relative to conventional varieties. The need to 'build' six different parental lines which will include the appropriate S alleles so that the self-incompatibility will be strong enough in the single-cross and double-cross progenies, may be problematic. Finally, this method requires long-range preparation with very little breeding flexibility after the original six lines have been determined.

Several modifications of this, as well as of the previous method, were suggested (see GOWERS, 1975). Thus, S_f alleles can replace some of the incompatibility alleles in order to save part of the inbreeding, as mentioned above in dealing with the single-cross method. Another modification is to incorporate the isogenic-lines concept into the triple-cross technique. Finally, due to recent success in androgenesis in *Brassica* (see Table 3.1), the obtainment of homozygous S lines may be facilitated.

We have mentioned repeatedly the cost of the final hybrid in respect to expenses involved in propagation of homozygous lines. This can be made clearer by considering the seed-multiplication rates of self- or sib-pollinating a homozygous line vs cross-pollination. Obviously, different lines and different crops vary in this respect but whereas in self- and sib-pollination the seed-multiplication rate is usually between one and two orders of magnitude (e.g. from 10 mg to between 100 mg and 1 g), this rate for cross pollination is between three and four orders of magnitude (e.g. from 1 g to between 1 and 10 kg). This means that the difference between these two propagation methods can be estimated as two orders of magnitude. For each actual crop such calculations can be helpful in considering the method of hybrid seed production which should be adapted.

3.3.3.4.3 Problems in Breeding Aimed at Hybrid Seed Production

Hybrid seed production based on incompatibility is involved in several problems. The most important of these will be itemized in the following.

1. *New Incompatibility Specificities.* It was found in several species having the gametophytic incompatibility system (e.g. *Trifolium pratense*, *L. peruvianum*) that continuous inbreeding results in the appearance of new incompatibilities, thus causing self-compatibility. Plants with such new mutated specificities will therefore not serve the purpose of parents for hybrid seed.

2. *Depression by Continuous Inbreeding.* Such degeneration is well known among allogamous species and can lead in many *Brassica* crops to a complete loss of inbred lines. As there seems to be a variability in the degree of inbreeding depression, the breeder may overcome this problem by proper selection or by exchanging self- or sib-pollination by vegetative propagation.

3. *Pseudo-Compatibility.* The degree of incompatibility is very much dependent on the S alleles genotype as well as on additional, ill-defined genes. The breeder thus has to select his parental lines accordingly, incorporating into them the strongest available S alleles, but keeping in mind the propagation of homozygous lines.

4. *Reduction of Incompatibility by Environmental Conditions.* It was mentioned above that elevated temperatures may reduce incompatibility or even break it down. High humidity was also reported to reduce incompatibility. Thus, hybrid seed production under unfavorable conditions may result in a high proportion of 'sib' seed (30% or higher). Such seed is of no commercial value. On

self-pollinating crops. Substantial amounts of heterozygosity may be retained by genic male sterility (JAIN, 1961; LLOYD, 1975) and composite crosses can be established in self-pollinated plants. Thus, male sterility in artificial populations of inbreeding crops can be a useful tool in plant breeding. The application of nuclear male sterility to recurrent selection schemes, as in soybeans (BRIM and STUBER, 1973) or sorghum (DOGGETT, 1972), capitalizes on the fact that resulting lines need not carry restorer genes, as they would have to if cytoplasmic male sterility were to be used (GILMORE, 1964).

3.4.2 Inheritance of Male Sterility

Three categories of inheritance of male sterility are suggested by the genetic data in the literature: Mendelian (genic), maternal (cytoplasmic), and a combination of the two (gene-cytoplasmic). Table 3.27 details inheritance of the three categories in its simplest form, as follows:

With *genic male sterility* only one plasmatype (S) exists, interacting with a recessive plasmon-sensitive (ms) and a dominant fertility-restoring (Ms) allele.

With *cytoplasmic male sterility*, normal (N) and male-sterility-inducing (S) plasmatypes exist, but only recessive plasmon-sensitive alleles are present in the population.

With *gene-cytoplasmic male sterility*, normal (N) and male-sterility-inducing (S) plasmatypes exist, the latter interacting with a recessive plasmon-sensitive (ms) and a dominant fertility-restoring (Ms) allele.

Table 3.27. Mode of inheritance of genic, cytoplasmic, and gene-cytoplasmic male sterility (simplest model featuring two plasmatypes and one recessive plasmon sensitive allele—see text for details)

Plasmatype	Genic male sterility		Cytoplasmic male sterility		Gene-cytoplasmic male sterility	
	Fertile genotypes	Male sterile genotype	Fertile genotype	Male sterile genotype	Fertile genotypes	Male sterile genotype
Normal (N)	—	—	ms ms	—	Ms Ms / Ms ms / ms ms	—
"Sterile" (S)	Ms Ms / Ms ms	ms ms	—	ms ms	Ms Ms / Ms ms	ms ms

3.4.2.1 Genic Male Sterility (Mendelian Male Sterility)

Expression of genic male sterility may take a variety of forms ranging from abortion of anthers, malformation of anthers, abortion of pollen at various stages during microsporogenesis and indehiscense of anthers ("functional male sterility"), to highly extruded stigmata ("positional male sterility"); these expressions will be discussed later on. Generally, genic male sterile mutants appear to have a much more stable expression than cytoplasmic or gene-cytoplasmic male sterile types.

Sporophytic control of the trait is a general feature of recessive genic male sterility. Consequently, all gametes of the heterozygote are functional and only the female gametes of the homozygote recessive are functional. This, of course, permits survival of the male sterility alleles in natural populations. Selective abortion of microspores, sporophytically controlled by an alien locus, has been reported in the progeny of an interspecific tobacco cross (CAMERON and MOAV, 1957).

The overwhelming majority of analyzed cases of Mendelian male sterility show control by single recessive factors (see Section 3.4.1.2 above). Large numbers of non-allelic recessive male sterility alleles have been found within the same species. In all these cases, the presence of a single recessive gene in homozygote condition leads to the expression of male sterility. A few cases of double recessive male sterility have been reported (e. g. in cotton: WEAVER, 1968). Double recessive male sterility in monoecious plants is exemplified by the $babats_2ts_2$ and $skskts_2ts_2$ genotypes of maize (see Section 3.2.1.2).

A few cases of monogenic dominant male sterility have been reported, such as in *Solanum* (SALAMAN, 1910), *Coleus* (FORD, 1950) and *Streptocarpus* (ZEVEN, 1972). Although a few cases of dominant abortion of pollen during microsporogenesis have been reported (e.g. Ms_4 and Ms_7 in cotton: MURTHI and WEAVER, 1974), most dominant male sterile phenotypes appear to be "functional male sterile", leaving the option of forced pollen release open. Epistatic gene action between a dominant and recessive gene leads to male sterility in *Lactuca* (RYDER, 1963) and in *Origanum* (LEWIS and CROWE, 1956); in these cases male steriles are FFhh or Ffhh genotypes, the double recessives may or may not be lethal, and FFHH, FFHh, FfHH and FfHh are bisexual. The literature on genic male sterility has been reviewed by JAIN (1959) and GOTTSCHALK and KAUL (1974).

Genic male sterility has repeatedly been reported to occur spontaneously (see ALLARD, 1953; RICK, 1945; REIMAN-PHILLIP and FUCHS, 1971). Genic male sterility has also been induced artificially: two X-ray-induced recessive male sterile mutants of *Petunia* were studied by WELZEL (1954). An apetalous and antherless recessive mutant of *Tagetes erecta*, induced by X-rays (BOLZ, 1961), is presently used in hybrid seed production (HORN, 1974). Genic male sterility has also been induced by gamma rays in watermelon and tomato (WATTS, 1962; KWASNIKOW et al., 1970), and by chemical mutagen in peppers and pea (BREUILS and POCHARD, 1975; KWASNIKOW et al., 1970).

Pleiotropic effects of male sterility alleles (or close linkages with these alleles) have been reported in a number of cases. Reduced female fertility in the homozygous recessive (e.g. in an apetalous and antherless *Antirrhinum:* SINK, 1966), or reduced pollen fertility in the heterozygote (e.g. in hybrid tomato produced with ms_{41}: ANDRASFALVY, 1974), may cause problems in the production of hybrid cultivars. On the other hand, certain marker phenotypes associated with male sterility genes can lead to a more efficient hybrid seed production procedure (see Section 3.4.4.2.1 below).

3.4.2.2 Cytoplasmic and Gene-Cytoplasmic Male Sterility

Rare events, such as loss of restorer genes in the breeding population after initial wide crosses, or recent mutations toward a male sterile plasmatype, may serve as plausible explanations for the few cases reported of male sterility controlled solely by the maternal cytoplasm (see JAIN, 1959; BERNINGER et al., 1970). Furthermore, many cases of cytoplasmic male sterility proved, upon additional study, to be actually affected also by Mendelian genes, e.g. in corn (see RHOADES, 1933; DUVICK, 1965; GRACEN and GROGEN, 1974), sugarbeet (see OWEN, 1945), and *Petunia* (EDWARDSON and WARMKE, 1967; VAN MARREWIJK, 1969). The absence of a plasmon-genome interaction in cytoplasmic male steriles tends to make phenotypic expression of the trait more stable than in gene-cytoplasmic or genic male steriles. Lack of plasmon–genome interaction produces male steriles only in the progeny of male sterile plants.

In view of the dependence of various metabolic systems in the cytoplasm on nuclear attributes, the constancy of the genetic property of the plasmatype must be established. A change of the cytoplasm derived from the maternal parent may consequently be brought about by the action of nuclear genes. We know that nuclear genes may or may not inhibit or modify the action of the cytoplasm in a fertility-restored or non-restored genotype. In the former case the genetic property of the cytoplasm must remain latent in the restored genotype and show up again as a result of gene segregation to indicate genetic permanence of the maternal plasmatype. Furthermore, a contribution of the paternal cytoplasm to the zygote, as indicated in some cases could possibly cause a gradual change in the plasmatype upon successive backcrosses (see e.g. FLEMING, 1975, for paternal cytoplasmic influences on agronomic traits in maize; HAGEMANN, 1964, for status "paralbomaculatus" in *Pelargonium*, *Hypericum* and *Oenothera*). Such a possibility is considered graphically in Fig. 3.44. To date no change in the genetic properties of the maternal plasmatype by introgressive paternal cytoplasm has been reported for the male sterility trait (see e.g. MICHAELIS and MICHAELIS, 1948; BURK, 1960). Independence from Mendelian genes and constancy of the maternal transmission of the male sterile plasmatype have been repeatedly demonstrated by successive nuclear substitution experiments (e.g. JONES, 1956; DUVICK, 1965; FUKASAWA, 1967; SCHWEPPENHAUSER and MANN, 1968; see Fig. 3.45) or even by androgenesis (see e.g. GOODSELL, 1961). In contrast with all cases of genic male sterility, where expression of male sterility is controlled entirely sporophytically, gametophytic control on pollen function in heterozygous fertility-restored plants carrying male sterile cytoplasm has been reported in a number of cases. Data from maize populations expected to segregate indicate the occurrence of some type of selective fertilization in plants bearing the (S) cytoplasm and heterozygous for the fertility-restoring genes, resulting in a higher frequency of fertile plants than expected (BRIGGLE, 1957). BUCHERT (1961) showed that in the (S) plasmatype, only microspores which contain a dominant allele for fertility restoration, Rf_3, produce functional gametes, whereas male gametes containing the recessive allele, rf_3, abort. The female gametes of the (S) plasmatype all remain functional. In contrast to this, all microspores are functional in heterozygous plants, Rf_3rf_3, with the (T) plasma-

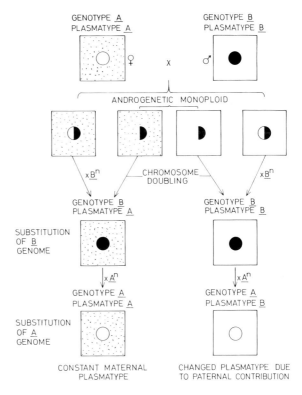

Fig. 3.44. Proof for the constancy of maternal plasmatype by substitution and restoration crosses. Dots in cytoplasm *(rectangle)* indicate male sterility inducing factors, dark nucleus *(circle)* indicates plasmon sensitive nuclear factors. *Left side:* constancy of maternal *right side:* constancy of paternal plasmatypes

type irrespective of the allele. The rf gametes may not function due to abortion or due to elimination when in competition with Rf gametes, as a result of differential pollen germination or growth (JOSEPHSON, 1964). In the case of a recessive restorer gene to the (S) cytoplasm, a suppressor of this gene, S^{Ga}, also operates gametophytically (SCHWARTZ, 1951).

Variation of pollen production in heterozygous fertility-restored plants and shifts of expected segregation ratios toward fertility have been reported in a number of plants. Studies by BARHAM and MUNGER (1950) showed reduced pollen production in heterozygous restored onion. Male sterile (S) ms ms plants did not produce any pollen; fertile homozygous plants—irrespective of the plasmatype—(S) Ms Ms or (N) Ms Ms produced about 82% viable pollen; and pollen production in the heterozygote (S) Ms ms was only 64%. These results have been interpreted as due to incomplete dominance. Deviating segregation ratios in onion crosses between male steriles and (relatively undefined) genotypes of pollen donors with a shift toward fertility have been interpreted as due to the action of modifier genes (NIEUWHOF, 1970). However, the malfunction of the plasmon-sensitive gamete in the (S) cytoplasm cannot as yet be ruled out in these cases.

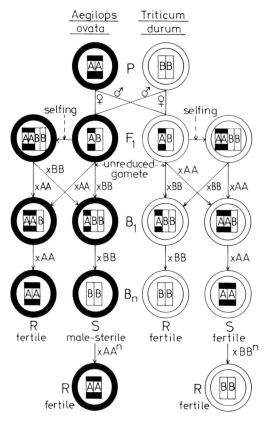

Fig. 3.45. Diagrammatic representation of the substitution and restoration of genome comple-
ments between *Aegilops ovata* and *Triticum durum. Inner circle:* nucleus and *outer zone:*
cytoplasm. *Rectangle with A or B in inner circle:* haploid set of ovata and durum genome.
S: substitution, R: restoration (after FUKASAWA, 1967)

The interacting mechanisms of the nuclear and cytoplasmic hereditary systems,
causing gene-cytoplasmic male sterility, provide for a finer balance of fertility
in the different genotypes than is possible with the purely nuclear system. Basic
differential fertility levels, as found in natural and artificial gynodioecious populations, are further varied by environmental influences causing cyclic or seasonal
fluctuations in the degree of male sterility (see e.g. DUVICK, 1965; PETERSON,
1958) usually not encountered with genic male sterility. Thus, the plasmon–
genome–environment interaction produces a delicate balance in the expression
of the trait.

The genetics of gene-cytoplasmic male sterility is abundantly documented
in the literature for about 150 species and will not be detailed here (see reviews
by EDWARDSON, 1956, 1970; CHOWDHURY and DUVICK, 1966; VARGHESE,
1968). "Restorer genetics", dealing with the interaction between plasmon-sensitive
alleles and male-sterility-inducing plasmatypes, has become an important field
of applied genetics due to its application in commercial hybrid seed production.

To be useful in hybrid seed production, genes restoring fertility of "male sterile" plasmatypes should be dominant. From an evolutionary point of view dominance of restorer genes is expected and, as a matter of fact, dominance of restorer genes is found with most "male sterile" plasmatypes [e.g. *C*, *S*, and *T* plasmatypes in maize (GRACEN and GROGAN, 1974); *S* and *T* plasmatypes in onion (VAN DER MEER, 1970; SCHWEISGUTH, 1973); timopheevi plasmatype in wheat (MILLER et al., 1974; BAHL and SAGE, 1972; MAAN, 1973)]. The presence of fertility restoring genes in populations carrying the "male sterile" plasmatype should be expected, whereas the recessive plasmon sensitive allele may occur in high frequency in populations carrying the normal plasmatype. Genetic studies, e.g. with onions (LITTLE et al., 1944; DAVIS, 1957; PIENAAR, 1958; KOBABE, 1958; BERNINGER, 1965; VAN DER MEER, 1970), crucifers (THOMPSON, 1972; BONNET, 1975), peppers (OHTA, 1972, 1973; FRANKEL and SHIFRISS, unpublished), and rice (SHINJYO, 1972a, 1972b), confirm correspondence of (*S*) plasmatypes with restorer genotypes in some populations while (*N*) plasmatypes can be found with plasmon-sensitive and/or restorer genotypes, the plasmon-sensitive allele often occurring with a very high frequency in (*N*) plasmatypes populations. We may then reach the conclusion that a non-sensitive cytoplasm may be combined with both sensitive and non-sensitive alleles in certain plant populations, and non-sensitive alleles may be combined with both sensitive and non-sensitive cytoplasm in other plant populations.

3.4.3 Structural, Developmental, and Biochemical Characterization of Male Sterility

We have so far categorized male sterility according to the inheritance pattern of the net effect, i.e., absence of viable pollen for fertilization. Characterization of male sterility by its final result makes possible generalization of a large number of cases of common hereditary, evolutionary and practical consequences. The male sterile phenotype may be described in any developmental stage and at different levels of details and compared with microgametogenesis in the normal phenotype. Structural, developmental and biochemical characterization of the androecium may help in understanding the sequence of events leading to male sterility and their control by hereditary elements. Absence of viable pollen for fertilization is often accompanied by additional phenotypical expressions of morphological or physiological nature, indicating pleiotropic effects of the male sterility genes or their close linkage with other genes controlling these expressions. We will try to integrate the present knowledge on genetically controlled events in microgametogenesis responsible for male sterility.

3.4.3.1 Developmental Modifications Leading to Breakdown in Microgametogenesis

Normal developmental organization of the androecium and of microgametogenesis has been described in Chapter 3.1.1 above. Comparative developmental studies of normal vs male sterile anthers reveal definite stages and modes of disturbances resulting in the absence of functional pollen. Figure 3.46 gives a schematic outline of probable events in microgametogenesis responsible for male sterility. The ontogenetic sequence of developmental stages at which the male sterile phenotype can first be identified, consists of the differentiation of androecium and microsporangium, microsporogenesis, microgametogenesis and anther dehiscence and will be discussed in that order. For a detailed review of the literature on stages and modes of disturbances leading to breakdown in microgametogenesis, the reader is referred to the paper of LASER and LERSTEN (1972) for cases of cytoplasmic male sterility and to the paper of GOTTSCHALK and KAUL (1974) for cases of genic male sterility.

3.4.3.1.1 Modifications in the Structural Differentiation of the Stamen

As we have seen in Chapter 3.2, in most cases the potentiality of sex suppression appears to be under hereditary control, whereas the degree of the expression depends largely on environmental conditions. Consequently, cases of male sex suppression in normally hermaphroditic flowers of monoecious plants vary in degree and are often vaguely defined as stamenless, antherless or simply as producing rudimentary, vestigial, warty, minute staminoida instead of normal stamens.

Failure of orderly stamen differentiation has been reported in a number of mutant genotypes of cultivated plants, e. g. in sorghum (KARPER and STEPHENS, 1936), maize (BEADLE, 1932), rice (CHANDRARATHNA, 1964), tobacco (RAEBER and BOLTON, 1955), snapdragon (SINK, 1966), pea (KLEIN and MILUTINOVIC, 1971); Origanum vulgare (LEWIS and CROWE, 1952), cucumber (BARNES, 1961), tomato (RICK, 1945b; LARSON and PAUR, 1948; BISHOP, 1954; HAFEN and STEVENSON, 1955, 1958), squash (SHIFRISS, 1945), and cotton (ALLISON and FISHER, 1964). All cases of faulty differentiation of the stamens, except cotton, appear to be controlled by single recessive genes, but their norm of reaction seems to be widely influenced by the environment. A good illustration of environmental influence is provided, on the one hand, by cases of artificial stamen suppression or sex reversion e. g. by morphactin in Capsicum annuum (JAYAKARAN, 1972) and in Cannabis sativa (MOHAN RAM and JAISWAL, 1971), respectively and, on the other hand, by artificial restoration of normal stamen differentiation e. g. by gibberellin in stamenless tomato (PHATAK et al., 1966). The relationship between level of gibberellin and the mutant genotype has recently been further clarified by SAWHNEY (1974). Male sex suppression often takes the form of differentiation into earlier sterile lateral members of the floral bud. Petals or petalloid structures are usually produced instead of stamens; this phenomenon is utilized in the production of "double flowered" cultivars of ornamentals such as stock (Matthiola incana) or begonia (Begonia semperflorens). An example of petaloidy is presented in Fig. 3.47 (E), where male suppression is the result of interaction between the Nicotiana undulata plasmatype and plasmon-sensitive alleles. Male sex suppression is sometimes associated also with the suppression of the petals. Antherless and apetalous phenotypes have been reported, e. g.

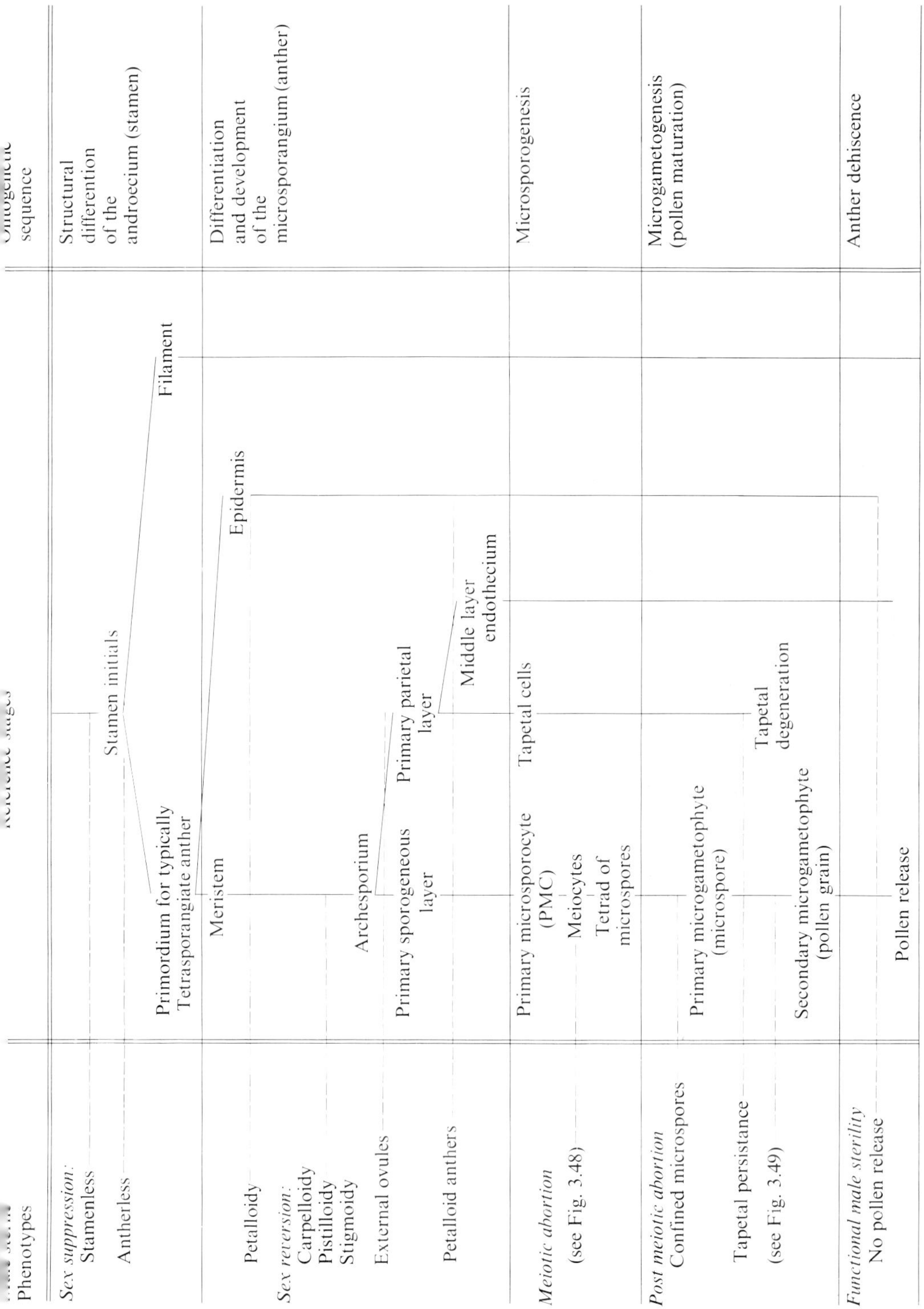

Phenotypes

Sex suppression:
Stamenless
Antherless

Petalloidy

Sex reversion:
Carpelloidy
Pistilloidy
Stigmoidy

External ovules

Petalloid anthers

Meiotic abortion
(see Fig. 3.48)

Post meiotic abortion
Confined microspores

Tapetal persistance
(see Fig. 3.49)

Functional male sterility:
No pollen release

Reference stages

Stamen initials

Primordium for typically Tetrasporangiate anther
Meristem
Filament
Epidermis

Archesporium
Primary sporogeneous layer
Primary parietal layer
Middle layer
endothecium

Tapetal cells
Primary microsporocyte (PMC)
Meiocytes
Tetrad of microspores

Primary microgametophyte (microspore)
Tapetal degeneration
Secondary microgametophyte (pollen grain)

Pollen release

Ontogenetic sequence

Structural differention of the androecium (stamen)

Differentiation and development of the microsporangium (anther)

Microsporogenesis

Microgametogenesis (pollen maturation)

Anther dehiscence

Fig. 3.46. Schematical outline of probable events in microgametogenesis responsible for male sterility

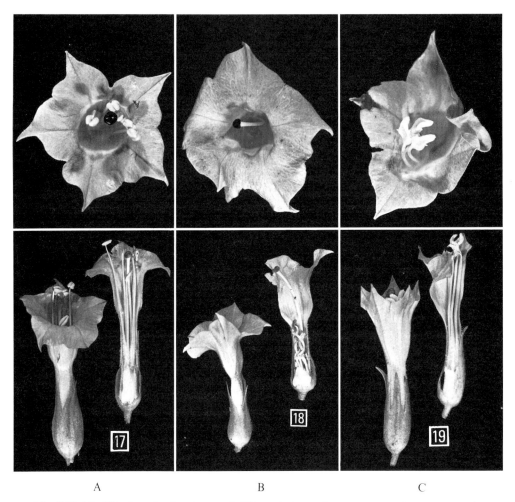

A   B   C

Fig. 3.47. Modifications in the structural differentiations of stamen or anthers as a result of interaction between *Nicotiana* plasmatypes and plasmon sensitive alleles in a *N. tabacum* × *N. glutinosa* hybrid: (A): *N. tabacum* plasmatype, cv "Hicks": phenotype normal, (B): *N. suaveolens* plasmatype, cv "Hicks": phenotype pistilloid, (C): *N. bigelovii* plasmatype, cv "Hicks": phenotype petalloid anthers, (D): *N. megalosiphon* plasmatype, cv "Hicks": phenotype carpelloid functional, (E): *N. undulata* plasmatype, cv "Hicks": phenotype petalloid, (F): *N. debnyi* plasmatype, cv Red Russian: phenotype stigmoid and split corrola

in zinnia (GOLDSMITH, 1968), snapdragon (SINK, 1966), thyme (APPL, 1933) and petunia (SINK, 1973).

3.4.3.1.2 Faulty Differentiation of the Anther

Aberrant anther differentiation most often indicates sex reversion which ranges from stigma-like extensions of the connective tissue in otherwise normal anthers to *stigmoid* male sterile anthers, *pistilloid* male sterile anthers to the development

D   E   F

of carpels instead of anthers with typical stigma and style, bearing ovules capable
of development into normal seed *("Carpellody")*. A large number of cases of
sex reversion can be related to plasmon–genome interaction (see PORTER et al.,
1965; MEYER, 1966; FISHER and SYMKO, 1973). Figure 3.47 (F) shows stigmoid,
pistilloid [Fig. 3.47 (B)], and carpelloid [Fig. 3.47 (D)] phenotypes based on
Nicotiana debnyi, *N. suaveolens* and *N. megalosiphon* plasmatypes respectively
interacting with plasmon-sensitive alleles in a *N. tabacum* × *N. glutinosa* hybrid.

Sex reversion may also be expressed by the formation of external ovules
on abnormally thickened staminal tubes [e. g. in interspecific hybrids of cotton
(MEYER and BUFFETT, 1962) and tobacco (BURK, 1960) or in genic "stamenless"
tomato (SAWHNEY and GREYSON, 1973) and castor bean (STEIN, 1965)].

Aberrant differentiation of the epidermal and primary parietal layers, and
sometimes even of the primary sporogeneous layer often lead to petaloid anthers.
If the primary sporogenous layer is also altered, such anthers are usually male
sterile as, e. g., in a *Nicotiana bigelovii* plasmatype interacting with plasmon-sensit-

ive alleles in a *N. tabacum* × *N. glutinosa* hybrid [Fig. 3.47 (C)], or in *Brassica nigra* plasmatype interacting with plasmon-sensitive alleles from *Brassica oleracea* (PEARSON, 1972).

3.4.3.1.3 Breakdown in Microsporogenesis

Breakdown during microsporogenesis may involve a number of aberrations (see Fig. 3.48). The onset of irregularities seems to be related to faulty timing

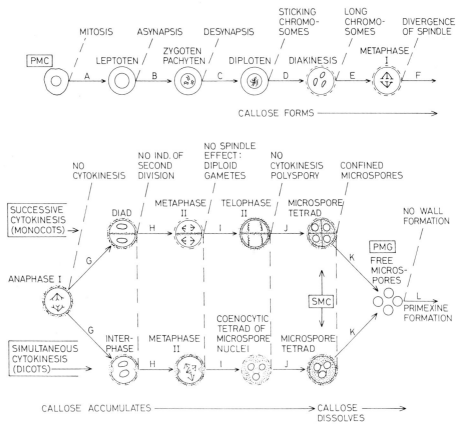

Fig. 3.48. Breakdown stages during microsporogenesis. A: Induction of meiosis, B: Pairing 1, C: Pairing 2, D: Terminalisation of chiasmata, E: Spiraling, F: Spindle effect, G: Cytokinesis (successive type), H: Induction of second division, I: Spindle effect, J: Cytokinesis, K: Microspore release, L: Primexine formation, PMC: Primary microsporocyte (2n), SMC: Secondary microsporocyte (n), PMG: Primary microgametophyte

and coordination of developmental sequences between sporocytes and tapetal cells since most cases show no definite stage but a range of abnormal behavior extending over part or all of the meiotic stages or even cover pollen maturation. The relative speed with which sporocytes undergo meiosis, the gradients in the degree of deviation, moving from the upper to the lower microsporangium and from the periphery of the PMCs to the middle of the locule, make critical evaluation of the exact stage of arrest in microsporogenesis difficult. Cells located in the upper anther lobes or near the tapetum appear usually less abnormal than cells in the lower anther lobes or farther away from the tapetum. Such observation, as well as the common occurrence of desynchronized groups of sporocytes, imply nutritional difficulties connected with tapetal development and unsynchronized activity.

The abortive process, though ordinarily not confined to definite stages, is recognized usually as an arrest centered on the early or final stages of meiosis [e.g. in tomato (see RICK and BUTLER, 1956) and in maize (see BEADLE, 1932)]. This may be so because of the relative duration of the meiotic stages or the relative genetic and metabolic autonomy of the meiocytes between late prophase and microspore tetrad stage as compared with the early and final stages of meiosis (see above).

In some male sterile genotypes, the primary (diploid) microsporocytes may not enter meiosis (Fig. 3.48 A); they may simply develop vacuoles and disintegrate [e.g. in winter squash (SINGH and RHODES, 1961) or tomato ms_3 (RICK, 1948)] or undergo synchronous mitosis instead of meiosis [e.g. in maize (PALMER, 1971) or raspberry (CRANE and THOMAS, 1949)].

Absence of pairing of homologous chromosomes during the first meiotic division (Fig. 3.48 B) has been reported in asynaptic male sterile mutants of maize (BEADLE, 1930) and tomato (ms_{10} RICK and BUTLER, 1956). Precocious separation of homologous chromosomes between pachytene and diplotene (i.e., desynapsis, Fig. 3.48 C), as found in male sterile alfalfa (CHILDERS, 1952) and pea (see GOTTSCHALK and KAUL, 1974), may be correlated with an observed chiasmata failure. Lack of terminalization of chiasmata (Fig. 3.48 D), resulting in sticky chromosomes, has been observed in male sterile violet (CLAUSEN, 1930), cowpea (SEN and BHOWAL, 1962) and tomato, ms_{18} (RICK and BUTLER, 1956). A few cases of faulty contraction of chromosomes (Fig. 3.48 E) have been reported for mutants of sweet pea (FABERGÉ, 1937), pea (see GOTTSCHALK and KAUL, 1974) and maize (BEADLE, 1932), arresting meiosis between diakinesis and metaphase I, exhibiting a "long chromosome phenotype".

In reviewing cases of male sterility showing irregularities during the first meiotic prophase, we may note that these are frequently associated with various degrees of impairment to the fertility of the female gamete. On the other hand, cases of male sterility showing irregularities in later stages only, never reveal impairment of female fertility. This may indicate that early in the sequence chromosomal pairing and gene recombination in both male and female sporocytes are controlled by the sterility genes, whereas a separate hereditary mechanism involves only microsporogenesis through metabolic upsets and nutritional deficiencies following the first metaphase. It may be of significance that meiotic irregularities after prophase I are always associated with deviant tapetal behavior (CHILDERS, 1952) and it may not be incidental that genic male steriles involving more than one locus and the majority of gene-cytoplasmic male steriles show abortion after the first meiotic division and characteristically do not affect female fertility.

Few cases of male sterility can be placed as being due to a breakdown between metaphase I and telophase II. Deficient callose formation has been found to be associated with abortion of meiocytes of cytoplasmic male sterile petunia (FRANKEL et al., 1969; IZHAR and FRANKEL, 1971) and sorghum (ERICHSEN and ROSS, 1963; WARMKE and OVERMAN, 1972). Divergent spindle orientation or action (Fig. 3.48 F) may lead to the formation restitution nuclei after the first meiotic division, [e. g. in potato (ELLISON, 1936)]. Irregular spindle orientation after metaphase II (Fig. 3.48 I) is evident in genic male sterility of muskmelon (BOHN and PRINCIPE, 1964) and peppers (NOVAK et al., 1971). Cytokinesis may fail after telophase I in monocots (Fig. 3.48 G) featuring the successive type of cell plate formation, [e. g. in maize mutants ms_8 and ms_9 (BEADLE, 1932)]. Quadripartition in the successive or simultaneous type of cyto-kinesis (Fig. 3.48 J) may not be completed and proper tetrads are not formed [e. g. in maize mutants ms_4 and ms_{10} (BEADLE, 1932) and pea mutants 395 (see GOTTSCHALK and KAUL, 1974)].

Release of microspores from the tetrad bindings (Fig. 3.48 K) appears to be one of the most critical stages for the breakdown of microsporogenesis in male sterile mutants. Lack of enzymatic digestion of the callose envelope surround-ing the young microspores in the tetrad seemingly starves the microspores and inhibits proper pollen production (FRANKEL et al., 1969; IZHAR and FRANKEL, 1971). The sequel to progressive confinement of microspores within the callose envelope is arrest or disorganization of wall formation and mitosis of the microga-metophyte and degenerated, sticky pollen grains [e. g. in *Aegilotricum* (FUKASAWA, 1953), *Hordeum* (MIAN et al., 1974; CHAUHAN and SINGH, 1966), *Dactylis* (FILION and CHRISTIE, 1966), *Zea* (MODJOLELO et al., 1966), *Solanum* (ABDALLA and HERMSEN, 1972), *Capsicum* (HORNER and ROGERS, 1974; HIROSE and FUJIME, 1975), *Petunia* (EDWARDSON and WARMKE, 1972; IZHAR and FRANKEL, 1971), *Cucumis* (CHAUHAN and SINGH, 1968), *Lupinus* (PAKENDORF, 1970), *Brassica* (COLE, 1959), and *Citrullus* (KIHARA, pers. comm.)]. Release of microspores without subsequent primexine formation upon the primary microgametophyte (Fig. 3.48 L) has been mentioned as the first significant deviation from normal pollen maturation in male sterile mutants [e. g. in barley (MIAN et al., 1974) and tomato (RICK, 1948)].

3.4.3.1.4 Abortion of the Microgametophyte

After release from the quartet callose capsule, the primary microgametophyte normally enters into a rapid stage of growth and develops vacuolation and exine (Fig. 3.49 M). Degeneration centered at this stage in microgametogenesis is very common in gene-cytoplasmic male steriles, e. g. in onion, beet, carrot and some crucifers (LASER and LERSTEN, 1972), and is frequently associated with delayed degeneration of the tapetum or other abnormal development of the tapetum or endothecium. Development of the microgametophyte may reach the stage of normal appearing vacuolate microspore but remain mono-nucleate, i.e. amitotic, and degenerate (Fig. 3.49 N). Such degeneration is known in particular in monocots (see LASER and LERSTEN, 1972; GOTT-

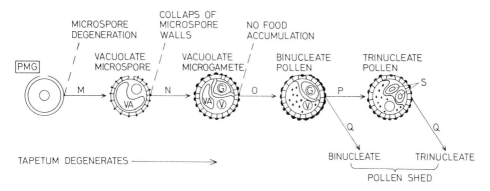

Fig. 3.49. Pollen abortion stages during microgametogenesis and functional male sterility. M: Exine and vacuole form, N: Microspore mitosis, O: Food reserve accumulation, P: Gamete mitosis, Q: Anther dehiscence, VA: vacuole, G: generative nucleus, V: vegetative nucleus, S: sperm cells

SCHALK and KAUL, 1974) and also is found correlated with abnormal tapetal activity and development (OVERMAN and WARMKE, 1972; PRIT-CHARD and HUTTON, 1972; HERVÁS, 1974; RAMANNA and HERMSEN, 1974). Amitosis is followed usually by collapse of the mononucleate microspore walls. Following microspore mitosis the vacuolate microgamete accumulates starch and other food reserves (Fig. 3.49 N). Cessation of further development associated with lack of food accumulation has been reported for gene-cytoplasmic male sterile grasses (e. g. for wheat—DE VRIES and IE, 1970; barley—SCHOOLER, 1967; sorghum—BROOKS et al., 1966; maize—KHOO and STINSON, 1957), and for genic male sterile tomato, ms_{13} (RICK, 1948), pea, mutant 195 B (GOTTSCHALK and KAUL, 1974), cucumber (WHELAN, 1972) and squash (FRANCIS and BERMIS, 1970). Non-degenerate tapetum appears to be correlated physiologically with the lack of build up of food reserves of the microgamete.

A review of the multitude of male sterile phenotypes reveals the most critical stages of the breakdown in microsporogenesis: clusters are evident at the early meiotic stages (stages A–E in Fig. 3.48), at the tetrad stage (stages J–K in Fig. 3.48), and at microspore maturation (stages M–N in Fig. 3.49). Sensitive phases in microgametogenesis appear to be related to concurrent events in the normal cataclysmic development of the tissue surrounding the sporocytes, through which nutritive material must either pass through or be metabolized: the tapetum (ECHLIN, 1971). Meiocytes and microspore tetrads are surrounded and isolated by a relatively impenetrable layer of callose which also undergoes normal cataclysmic build up and breakdown (WATERKEYN, 1964). Faulty timing of metabolic events during the short duration of many of the developmental stages in microgametogenesis (see e. g. FINCH and BENNETT, 1972; BENNETT and KALTSIKES, 1973; IZHAR and FRANKEL, 1973) must seriously affect the three mentioned phases in the development of the microgametophyte. The intimate metabolic relationship of an active tapetum with the early meiocytes obviously regulates nutrition and build up of the callose layers. The apparent enzymatic degradation of the callose envelope by glucanase activity of the declining and senescent tapetum is required to release the young microspore to the nutritive environment provided in part by degradation products of tapetal cells (see VASIL, 1967; MEPHAM and LANE, 1969; FRANKEL et al., 1969; ECHLIN, 1971; IZHAR and FRANKEL, 1971; STIEGLITZ and STERN, 1973). Thus, disturbance in the required physiological correlation between tapetal cells and developing microgametes must cause male sterility.

A number of generalizations can be made for male sterile phenotypes at the three critical phases of breakdown:

Early meiotic phase:	tapetum-deficient or disorganized callose—lack of formation or precocious digestion.
Tetrad phase:	tapetum—precocious or delayed degeneration callose—delayed digestion or persistence
Microspore maturation phase:	tapetum—persistent

The distribution of male sterile mutants studied (LASER and LERSTEN, 1972; GOTTSCHALK and KAUL, 1974) among phenotypes of the three phases appears to be different for plants of the successive versus the simultaneous type of cytokinesis and for gene-cytoplasmic versus genic male steriles (Table 3.28).

Table 3.28. Percentage distribution of male sterile phenotypes among three broad classes of abortion (based on LASER and LERSTEN, 1972; GOTTSCHALK and KAUL, 1974; and additional cases not covered by the two reviews)

Genetic type	Taxa	Breakdown phase		
		Early meiotic	Tetrad	Microspore maturation
Gene-cytoplasmic	all taxa	15	57	28
	dicots	27	58	15
	monocots	4	57	39
genic	all taxa	30	52	18
	dicots	32	56	12
	monocots	7	14	79

Due to gaps in the information available and to ranges of abnormal behavior often extending over several stages, assignment of phenotypes to a specific developmental phase is often quite imprecise. Nevertheless, the substantial comparison of the count of 62 cases of gene-cytoplasmic male sterilities (LASER and LERSTEN, 1972) and of 93 genic male sterilities (GOTTSCHALK and KAUL, 1974) seems to be indicative for quite significant differences: early meiotic breakdown is more frequently found among dicotyledonous plants (about 30 %) than among monocotyledonous plants (about 6 %). Very few tetrad breakdown types are found among monocotyledonous genic male steriles (14 %) as compared with dicotyledonous genic male steriles, monocotyledonous and dicotyledonous gene-cytoplasmic male steriles (all about 57 % of the cases). The very high concentration of cases of genic male sterile monocots at the microspore maturation phase (79 %) is of particular interest.

In quite a number of important cultivated plants, belonging to the Amaranthaceae, Chenopodiaceae, Compositae, Gramineae, and Umbelliferae, pollen is shed trinucleate, i. e., its generative cell undergoes the additional mitosis before the pollen is shed (Fig. 3.49 P; KREMP, 1965; BREWBAKER, 1967). Trinucleate, as compared with binucleate pollen is regarded as more sensitive to dehydration, radiation and long storage, possibly because of its less pronounced exine and deprivation of the autonomous pollen grain of its reserves by the second mitotic division. The difficulties involved in observation of pollen mitosis may be responsible for the lack of exact information on the breakdown of what appears to be mature pollen grains (e. g. in some male sterile mutants of maize and wheat).

3.4.3.1.5 Functional Male Sterility

A number of male sterility genes prevent pollen release by modifying normal differentiation or functioning of the stomium and thus prevent release of otherwise fertile pollen through failure of anther dehiscence (Fig. 3.49 Q). Functional male sterile mutants leave the option of forced pollen release open and thus are found not only as recessive traits (as in tomato—see CLAYBERG et al., 1966) but also as dominant traits (see Chapter 3.1.2.1). Non-dehiscent anthers may be due to the absence of stomium (e. g. in eggplant—JASMIN, 1954; barley—ROATH and HOCKETT, 1971; rice—HOFF and CHANDRAPANYA, 1973); stomal cells may be tenacious and pressed by hooked petals which do not allow unfurling of corrolla (e. g. in ps mutant of tomato—LARSON and PAUR, 1948); and anthers may be reflexed or hard and filaments not elongating (e. g. in maize—SPRAGUE, 1939; clover—ATWOOD, 1944; tomato—CLAYBERG et al., 1966; grape—OLMO, 1943; and rubber—RAMAER, 1935).

3.4.3.2 Biochemistry of Male Sterility

The biochemistry of angiosperm pollen development was recently reviewed in detail by MASCARENHAS (1975). Therefore, we will restrict ourselves here to a few remarks on comparative biochemistry of the developmental phases in the male sterile and fertile microsporophylls. Differences in the level of a number of substances, their accumulation or disappearance during development, characterize normal versus disturbed biochemical communication between microsporocytes and the surrounding tissue (tapetum) in normal versus male sterile anthers. Such information describes effects, not causes. The hereditary elements responsible for male sterility may act autonomously in certain tissues or through a distant trigger, and regulation of their timing further conceals the source of the disturbed metabolism in male sterile anthers.

Several comparative studies of the free amino acid pool during microsporogenesis indicated differences between fertile and male sterile anthers. Many studies showed a characteristic deficiency of proline or cystine and accumulation of asparagine in sterile anthers, compared with fertile anthers (e. g. in wheat—FUKA-SAWA, 1954; RAI and STOSKOPF, 1974; maize—KHOO and STINSON, 1957; FUKASAWA, 1954; sorghum—SILVOLAP, 1968; KERN and ATKINS, 1972; petunia—IZHAR and FRANKEL, 1973). Other studies of sterile anthers indicate different alterations from the normal free amino acid pool, and phase-specific changes in the amount of some of the free amino acids are indicated in many studies. Since the balance of free amino acids in anthers may be subject to the action of several factors (e. g. protein and amino acid breakdown and synthesis, translocation into and out of the anther), alteration of the amino acid profile in sterile anthers may reflect either anabolic or catabolic processes. Thus, abnormally high asparagine accumulation may be associated with breakdown of the tissue (MCKEE, 1962; IZHAR and FRANKEL, 1973) or with some block in the metabolic pathway between asparagine and ornithine in the TCA cycle (FUKUSAWA, 1962).

Changes in the amino acid pool within the anther locule may in part account also for normal or abnormal changes of the pH in the anther locule. Such changes, associated with asparagine or glutamine accumulation in male sterile anthers of petunia, have been correlated with a drop in pH altering the timing of the phase-specific callase activity, causing breakdown in microsporogenesis by delayed or precocious action of the enzyme in genic male sterile or cytoplasmic male sterile mutants of petunia, respectively (IZHAR and FRANKEL, 1971, 1973). The dependence of callase activity on pH, lack of glucanase activity in the microsporocytes at all stages of meiosis, and its presence in the tapetum and anther locules at the time of the normal degradation of the callose matrix of the microspore quartet (ESCHERICH, 1961; STIEGLITZ and STERN, 1973), lead to the belief that faulty activation of the enzyme in the anther locule may be a primary factor in microspore abortion.

The rationale of exact timing of the callase activity, external to the microsporocytes, becomes even more evident in the view of the normal dynamics of callose and primexine synthesis which is at least in part internal to the microsporocytes. Studies of plastids during microsporogenesis suggest that polysaccharide reserves of starchy plastids disappear in microspores during callose and cellulosic primexine synthesis (DUPUIS, 1974). Recurrent, transitory formation of callose walls around the normal generative cell of the microgametophyte (GORSKA-BRYLASS, 1967; HESLOP-HARRISON, 1968; MEPHAM and LANE, 1970) after the orderly dissolution of the callose matrix of the tetrad capsule, accentuates the significance of the exact dynamics of synthesizing and degrading enzymes during microgametogenesis. The role of the callose wall, as providing a molecular filter and thereby establishing a certain degree of transitory autonomy for meiocytes, quartets of microspores, and perhaps also to the generative cells of the microgametophyte, has been proposed (HESLOP-HARRISON, 1964) and questioned (MASCARENHAS, 1975). Regardless of the true role of the callose walls, unquestionably any abnormal and untimely formation or dissolution has been shown to cause male sterility (FRANKEL et al., 1969; IZHAR and FRANKEL, 1971; HAYWARD and MANTHRI-

RATNA, 1972; NOVAK, 1972; OVERMAN and WARMKE, 1972; PRITCHARD and HUTTON, 1972; HORNER and ROGERS, 1974).

We have seen that genetic elements control breakdown in microsporogenesis and microgametogenesis locally within the anther through upsets in the biochemical communication between microsporocytes or microgametophytes and surrounding tissue. Genetic elements causing modification of the normal structural differentiation of the microsporophyll appear to act through hormonal metabolism at the stamen primordia level, causing sex suppression or reversion. Although we are not aware of direct comparative studies on hormone production or receptors during differentiation of male sterile and fertile flowers, responses to exogenous growth regulators, such as auxins, gibberellins and kinins (see Section 3.2.1.5 above), leave no doubt about the hormonal control system of sex suppression or reversion.

3.4.3.3 Male Sterility Genes and Their Action

3.4.3.3.1 Site of Male Sterility Factors

Male sterility genes are transmitted biparentally, uniparentally or both. Reversible suppression of uniparentally–maternally inherited male sterility by the action of chromosomal fertility restoration testifies not only to the constancy of the genetic determinants, but also to the independent production of RNAs and proteins by the nuclear and extranuclear genome of the plant. BOGORAD (1975) asserts that "compartmentalization of functions in membrane-limited nuclei, mitochondria and plastids is the hallmark of eukaryotic life, and understanding the genetic and metabolic interactions of organelles and the nuclear-cytoplasmic system (is) a central problem of cell biology". In this sense, male sterility and fertility restoration resulting from intergenomic cooperation may be conceived as a result of *one* gene product composed of two or more complementary subunits or of *two or more* structural and regulatory gene products coded by each of the two hereditary systems.

Uniparental transmission of the extrachromosomal factor can be understood if we assume that the sterility factor is located in a DNA-containing cytoplasmic fraction which is contributed to the progeny exclusively by the maternal parent. Mitochondria and plastids are considered the most likely candidates to carry the genetic information for cytoplasmic male sterility for they are membrane-limited organelles accommodating independent DNA, transfer RNA, ribosomes, and suitable synthetizing enzymes (see BORST, 1972; KIRK and TIENEY-BASSETT, 1967 respectively for mitochondria and plastids). Absence of mitochondria or plastids from the male germ cells, or some mechanism preventing male organelles from entering the egg cells, must be responsible for the exclusive maternal transmission of mitochondrial or plastid male sterility genes.

Studies indicate that most plastids and mitochondria of the angiosperm male gametophyte become incorporated into the vegetative cell, and the generative cell and resulting sperm cells receive very few if any plastids, although they

supported by the fact of graft and seed transmission of some plant viruses causing various degrees of reduced male fertility (e. g. barley stripe mosaic virus—McKINNEY and GREELEY, 1965; tobacco ring spot virus—VALLEAU, 1932; DESJARDINS et al., 1954; broad bean wilt virus—CROWLEY, 1957; OHTA, 1970; tobacco mosaic virus—McKINNEY, 1952; OHTA, 1970). The relation between virus infection and gene-cytoplasmic male sterility has been corroborated in experiments with peppers indicating that the action of the nuclear fertility-restoring alleles is actually reduced by virus infection (OHTA, 1975).

Graft-transmitted cytoplasmic male sterility behaved generally as a cytoplasmically inherited trait in successive generations but in one case the transmission resulted in Mendelian inheritance of male sterility (FRANKEL, 1962). This case has been explained as being due possibly to chromosomal association or integration of the extrachromosomal factor transmitted through the graft union. As such, male sterility would be due to alternative genetic factors analogous to the behavior of episomes in bacteria (FRANKEL, 1971). In other words, bits of genetic information for male sterility could be distributed and only transitorily compartmentalized among genomes of a variety of cell entitites (e. g. nuclei, mitochondria, plastids, symbionts, pathogens) and "gene distribution mutants" (BOGORAD, 1975) may occur. A model of intergenomic and metabolic cooperation responsible for the mutant structure and function of cell organelles, the breakdown in microgametogenesis and associated phenotypic effects is indeed consistent with present knowledge of the genetics of male sterility and fertility restoration as well as of genetic continuity, distribution and exchange of male sterility factors among genomes of cell entities.

3.4.3.3.2 Pleiotropic Effects

Developmental modifications leading to the breakdown in microgametogenesis may be related developmentally to pollen abortion, but modifications of physiological traits leading to divergent agronomic performance among normal, male sterile and restored versions of otherwise isogenic material must be due to pleiotropic or linked effects of nuclear and extranuclear male sterility loci. Pollen sterility per se in non-restored genotypes may have a secondary effect and in maize shows yields about 2 % higher than those of restored genotypes (DUVICK, 1965). Reduced intraplant competition for assimilates by tassels during early development of the plant may explain such higher yields of non-restored hybrids (CRISWELL et al., 1974). However, such yield increases can be considered only a partial compensation for yield reductions of 2–4 % among restored hybrids carrying T cytoplasm (NOBLE and RUSSELL, 1963). Plant stature and leaf number are generally higher in hybrids with normal cytoplasm than with T cytoplasm (DUVICK, 1965). Hybrids having restored T cytoplasm show earlier silk emergence and delayed pollen shedding when compared with normal versions (GROGAN and SARVELLA, 1964; SARVELLA and GROGAN, 1965). The disastrous southern corn leaf blight epidemics in the USA (see above) can be accounted for by comparative grain yield studies of corn hybrids in normal and T male sterile cytoplasm infected with Helminthosporium maydis (e. g. LIM et al., 1974).

ATKINS and KERN (1972) evaluated normal, cytoplasmic male sterile and restored versions of sorghum lines and hybrids, but did not find differences of practical importance. Studies of the effect of two tobacco male sterile plasmatypes *(Nicotiana megalosiphon* and *N. suaveolens)* on agronomic performance and chemical characteristics indicated adverse effects which could not be overcome by fertility restorer genes (MANN et al., 1962; POVILAITIS, 1972; HOSFIELD and VERNSMAN, 1974; respectively). Male sterile (A) lines of wheat carrying *Triticum timopheevi* cytoplasm show high preharvest seed sprouting whereas their fertile (B) counterparts do not sprout at all (DOIG et al., 1975). Five *Aegilops* plasmatypes induce parthenogenesis at high frequencies in common wheat (TSUNEWAKI, 1974). Suppression of the development of the flower beyond the stage in which the corolla emerges (blindness), bud blasting, and extreme delay of flowering are usually associated with male sterile plasmatypes of petunia; this makes such plasmatypes useless for hybrid seed production, although the adverse effects may be partly overcome by gibberellin treatment (IZHAR, 1972).

3.4.4 Utilization of Male Sterility in Plant Breeding

Superiority of F_1 hybrids over the better of their two parents is a common phenomenon in both cross- and self-pollinated crops. Such superiority may be expressed in the heterotic phenotype by increased growth, height, leaf area, dry mytter accumulation, early flowering and higher total yields (see SINHA and KHANNA, 1975), as well as in uniformity and agricultural homeostasis of the cultivar population. In addition to the superiority of the hybrids per se, there are advantages in breeding F_1 hybrids over open-pollinated cultivars in speeding up programs (by parallel assembling desirable dominant traits in either of the two parents of the hybrid) and reducing problems of inbreeding depression and undesirable linkages of recessive genes in parents.

Economic benefits for the seed producers (based on proprietory monopoly, novelty value, etc.) no doubt contributed to the promotion of F_1 hybrid cultivars. Commercial F_1 hybrid cultivars become increasingly important for food, fiber and ornamental crops (see WITTWER, 1974; GABELMAN, 1974; HORN, 1974).

In cross-fertilized species the naturally imposed breeding system assures cross-fertilization, whereas in self-fertilized species selfing is favored by floral morphology. Hence, in cross-pollinated species problems arise particularly in the *inbreeding phase* providing suitable parents for the hybrid, whereas in self-pollinated species, they arise in the *crossing phase* of hybrid seed production. The particular problems in the inbreeding phase will not be dealt with here: these problems are related to natural mechanisms of incompatibility (making inbreeding difficult) and to inbreeding depression. To produce hybrid seed economically, the restrictions of controlled cross-fertilization caused by flower morphology, especially of perfect (hermaphrodite) flowers, must be overcome. The female parent should be prevented from self- or intraline fertilization. Moreover, pollen of the male line must effectively pollinate the female line, which requires an efficient natural pollen dispersal mechanism in the male, or artificial pollination. Elimination of self- or intraline fertilization of the female line requires andro-self sterility.

Such sterility can be produced by hand emasculation (castration), chemical emasculation, or manipulation of genetic male sterility or self-incompatibility.

Large-scale production of hybrid corn is done by detasselling the female parent, but large-scale emasculation of species with perfect flowers such as tomato, sorghum, etc., is usually economically unfeasible. Factors influencing the economics of hybrid production by hand emasculations are ease of emasculation, number of seeds produced per flower (per pollination), number of seeds sown per unit area and the upper limit of seed price in relation to crop production costs.

Chemical emasculation has been shown to be unreliable, so far. Therefore, genetic male sterility is of special interest for hybridizing crop plants having perfect flowers with few seeds per flower and where seed prices cannot cover the cost of extra expenses involved in hand emasculation. Thus, it happened that onion was the first crop in which genetic male sterility was clearly defined (JONES and EMSWELLER, 1937) and developed for production of hybrid cultivars (JONES and CLARKE, 1943; JONES and DAVIS, 1944). The crop to follow was field corn (JONES and EVERETT, 1949), and at present cytoplasmic male sterility serves in the production of hybrid seed of field corn, sweet corn, sorghum, pearl millet, sugar beet, alfalfa, onion, carrot, and radish, and may become useful in the production of hybrid wheat, rice, orchard grass sunflower, flax, cotton, soybean, field bean, *Crotolaria*, tobacco, garden beet, pepper, petunia, tuberous rooted begonia, columbine and other plants (GABELMAN, 1956, 1974; DUVICK, 1959, 1966; REIMANN-PHILIPP, 1964, 1974; HORN, 1974). Genic male sterility is used today for hybrid seed production of barley, tomato, pepper, marigold, zinnia, snapdragon, begonia and *Ageratum*, and is potentially useful in the production of hybrid cotton, lettuce, bicolor sweet corn and other crops.

3.4.4.1 Comparison of Hybrid Production Using Genic, Cytoplasmic and Gene-Cytoplasmic Male Sterility

The following lines are involved in hybrid seed production (FRANKEL, 1973):

A line (female parent). The female parent line which has to be male sterile in the seed production plots.

B line (maintainer). The function of this line is to maintain the A line, and with the exception of the male sterility factor, it should be isogenic.

C line (male parent). The male parent line, must also contribute (when required) fertility restoration factors to the offspring.

3.4.4.1.1 Genic Male Sterility

Genic male sterility is usually recessive and monogenic. Hence, fertility restoration in the hybrid and the crossing scheme are relatively easy.

The scheme shown in Fig. 3.50 indicates that removal (roguing) of fertile (heterozygous) segregates (Ms ms) is required in seed production plots and that pure-breeding male-sterile lines can not be maintained, unless fertility is restored

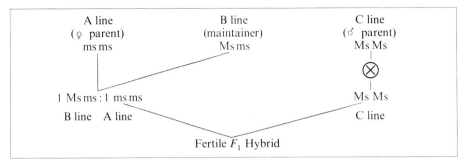

Fig. 3.50. A scheme for maintenance of parent lines and hybrid seed production using genetic male sterility

by a modified environment. Since the environmental conditions for fertility restoration in the A line are difficult to define, the hybrid is likely to contain many plants resulting from selfing. When the A line (but not the C line) is homozygous for an appropriate recessive marker gene, non-hybrid plants can be removed in the nursery before transplantation (in crops where this is feasible). Maintenance of the A line requires identification and removal of the heterozygotes before anthesis; this could be achieved by marker genes closely linked to the male sterility locus. However, insufficient linkage would result in recombination between marker and male sterility genes and thus contaminate the line.

When reproductive parts serve as the agricultural product, we have to be sure that recessive male sterility genes are not present in the C line. Fortunately, the frequency of male sterility genes is low and virtually all F_1 hybrid plants are fertile.

3.4.4.1.2 Cytoplasmic Male Sterility

Cytoplasmic male sterility is based solely on plasmagenes transmitted maternally. Thus, fertility in the hybrid cannot be restored. Consequently, the system is useful only in plants where seed production is not important. Production of the hybrid seeds and maintenance of the parent lines are shown in Fig. 3.51.

3.4.4.1.3 Gene-Cytoplasmic Male Sterility

Here the interaction between the sterile plasmatype and fertility restoration genes permits utilization of a breeding system most favorable for hybrid seed production. On the one hand, pure breeding male sterile A lines can be maintained and on the other hand, fertility in the final hybrid can be restored. Figure 3.52 outlines a scheme for the maintenance of parent lines and the production of hybrid seed utilizing gene-cytoplasmic male sterility.

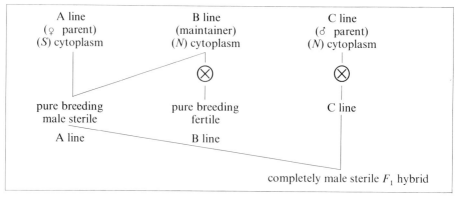

Fig. 3.51. A scheme for maintenance of parent lines and hybrid seed production using cytoplasmic male sterility

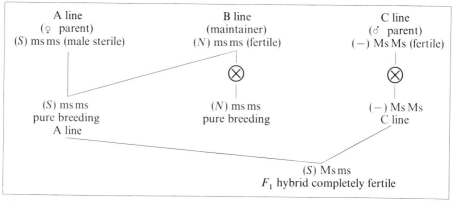

Fig. 3.52. A scheme for maintenance of parent lines and hybrid seed production using gene cytoplasmic male sterility

3.4.4.2 Application of Genic Male Sterility

Most efficient use of clever genetic manipulation of male sterility may be required to overcome a basic shortcoming in the use of genic male sterility, namely, the difficulty in obtaining economically a 100 % genic male sterile stand. Consequently, the application of genic male sterility is still restricted to hybrid seed production of those cultivated plants in which cytoplasmic male sterility has not been found (e.g. in autogamous crops such as tomatoes and barley) or those in which the male sterile plasmatype shows inferior agronomic performance.

3.4.4.2.1 Genetic and Field Management Programs to Provide a Homogeneous Stand of the Genic Male Sterile Seed Parent

The customary maintenance scheme for the female parent of the hybrid (A line) will produce 50 % heterozygous fertile plants (see Fig. 3.50), but a number of field management and genetic programs can be devised to provide an homogenous stand of the male sterile seed parent:

Selfing of Recessive Male Sterile Mutants. Many male sterile mutants will restore male fertility under certain natural or artificial environmental conditions (see e. g. JUSTUS and LEINWEBER, 1960; HANSCHE and GABELMAN, 1963; RICK and BOYNTON, 1965). Such conditions could be defined accurately and applied for the increase of pure A line seed. Functional male sterile mutants [see Chapter 3.4.3.1.5] produce, but do not release, pollen and can be selfed. However, due to the variable phenotype and unpredictable environmental conditions, reliable sterility of such mutants can often not be kept in the hybrid seed production field. To overcome this difficulty it may be useful to have in the A line recessive marker genes not present in the C line (δ parent line). This way, the degree of contamination of the hybrid seed lot can be monitored and, in some instances, the selfed progeny of the A line can be removed in the nursery before transplanting.

Clonal Propagation of Male Sterile Mutants. A solid stand of male sterile plants can be established by vegetative propagation. Such a method may be economical in plants easily propagated vegetatively, such as brassicas (GABELMAN, 1975), cotton (SANTHANAM et al., 1972) and ornamentals (GOLDSMITH, 1968).

Physical Removal of Fertile Heterozygotes. In mutants which can be forced to self, stability of male sterility is usually not dependable under field conditions. Consequently we must use stable mutants and resort to removal of the heterozygous fertiles from the seed production field. Double density planting should be practised to insure a full stand in the field (see HORN, 1974; LAPUSHNER and FRANKEL, 1967). Normally, fertile plants must be removed before anthesis, since pollen shedding obviously endangers the production of pure hybrid seed. In many cases reliable identification of fertile plants also requires microscopical examination. However, in plants that are essentially self pollinating, where no natural vectors for pollen transfer exist, and where hand pollination is practised (e. g. in tomatoes), we can wait for the first fruit to set upon selfing and then remove all plants bearing fruits.

The removal of heterozygous fertile plants may be facilitated by pleiotropic effects of the recessive male sterility gene, such as deviant structural differentiation of the corolla, the stamens or the anthers [see Chapter 3.4.3.1.1 and 3.4.3.1.2]. In the absence of pleiotropic effects, marker genes closely linked to the male sterility locus with a clear phenotype expression may be used. Seedling markers are of special value since roguing at the seedling stage in the nursery is cheaper than removal of adult plants in the field. Some successful attempts have been made in this direction (in tomato CLAYBERG, 1966; SORRESSI, 1968; PHILOUZE, 1974; cabbage—SAMPSON, 1970; cucurbits—WHELAN, 1974; WATTS, 1962). Physiological markers, such as resistance to phytocides, can be used to leave only the homozygous male steriles in the hybrid seed production field (e. g. in barley— HAYES, 1959; WIEBE, 1960). Height differentials, resulting from linkage of stature with male sterility genes, can be used for roguing by way of pre-anthesis cutting of the taller heterozygous phenotypes, thus leaving a pure stand of the shorter homozygous male sterile (A line) phenotypes (WIEBE, 1968).

SINGLETON and JONES (1930) proposed to use close linkages between endosperm color and male sterility genes in maize to insure a pure stand of the male sterile parent in the hybrid seed production field. Since no close linkages were available the scheme was not found useful. However, the recent proposal of GALIANT (1975) for hybrid seed production of bicolor sweet corn, appears to resolve the problems of recombinants when linkage is fairly close; the scheme may potentially be exploited as well for other crops having a conspicuous endosperm coloration. The program makes use of a homozygous seed parent and a heterozygous maintainer (in the coupling phase) for white endosperm and male sterility (A line y–ms/y–ms; B line Y–Ms/y–ms). The male parent of the hybrid must be homozygous for fertility and yellow endosperm (C line Y–Ms/Y–Ms). The white seed from the maintenance field is separated by an electronic eye seed sorter to furnish A line seed mixed with a small proportion of recombinants. Separation of yellow seed from the hybrid seed production field removes all selfed seed of the recombinants (which will be white).

Use of Extra Chromosomes. A number of possibilities to obtain an 100 % genic male sterile stand are made available by the breeding behavior of balanced tertiary trisomics or plants carrying an extra homoeologous chromosome derived from related species. Since chromosome segments on both sides of the breakpoint of the extra chromosome or the extra homoeologous chromosome are not transmitted through the pollen, fertility and marker genes located on them can aid in the identification of the male sterile genotypes. Schemes for utilizing balanced tertiary trisomics for hybrid seed production of diploid crops were outlined by RAMAGE (1965) and are presently used for the production of hybrid barley seed. These schemes use height, phytocide susceptibility, and seed size or shape differentials to selectively combine harvest or separate diploids, trisomics and F_1 hybrids. DRISCOLL (1972) proposed the use of genic male sterility of the crop plant and a corresponding fertility factor on an homoeologous alien chromosome. The principles of such an efficient scheme for hybrid wheat seed production are outlined in Fig. 3.53.

3.4.4.2.2 Pollination Control

Sufficient natural cross pollination in combination with male sterility would be desirable for economic hybrid seed production, but utilization of natural cross pollination in the production of hybrid seed on genic male sterile seed parents faces two types of problems: The highly self-fertilizing nature of many of those species where genic male sterility for hybrid seed production is of potential importance, and the very nature of male sterile flowers to be unattractive to pollen-gathering insects. Consequently, we may often find that pollen transfer from the C line to the A line by wind or insects does not result in sufficient seed set and that the activity of discriminating insect vectors may be limited to C line flowers. Even in flowers in which nectar is the main attractant, natural cross pollination of male sterile flowers often yields much less seed than hand pollination (see e.g. BREUILS and POCHARD, 1975).

Sometimes natural cross pollination may be improved by various means, such as: manipulating the ratio of pollen donors to seed parents, manipulating the spacing and location of pollen donors in relation to seed parents (planting

	X Line	Z Line	Y Line	C Line
chromosomes	21″ wheat + 1″ alien	21″ wheat	21″ wheat + 1′ alien	21″ wheat
genotype	ms/ms Ms−Hp/Ms−Hp	ms/ms	ms/ms Ms−Hp	Ms/Ms
phenotype	(Hairy neck fertile)	(Normal neck male sterile)	(Hairy neck fertile)	(Normal neck fertile)

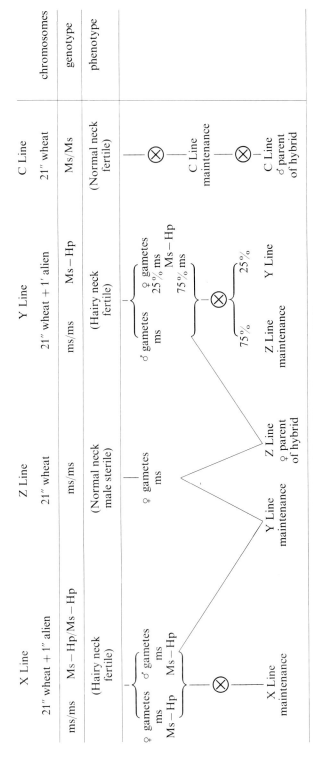

X Line:
♀ gametes {ms, Ms−Hp}   ♂ gametes {ms, Ms−Hp}   ⊗   X Line maintenance

Z Line:
♀ gametes ms   Z Line ♀ parent of hybrid   Y Line maintenance

Y Line:
♂ gametes ms   ♀ gametes 25% ms, 75% Ms−Hp   ⊗   75% Ms−Hp ms (Z Line maintenance), 25% Y Line

C Line:
⊗   C Line maintenance
⊗   C Line ♂ parent of hybrid

Fig. 3.53. The XYZ system of producing hybrid wheat (adapted from DRISCOLL, 1972)

design), choice of pollen donor (extent of pollen shedding, attraction to vectors, flower morphology), choice of male sterile seed parent (flower morphology, ovule and egg number per flower), choice of appropriate season and regions, and proper nicking of A and C lines. Some of these subjects have been discussed in Sections 1.3 and 2.3.2.3.

If natural cross pollination cannot be improved sufficiently, it can only be replaced by hand or otherwise artificially forced pollination, which may still be economical with high-value or widely spaced crops. The details of the production procedures depend on the flower biology of the species. As an example, Table 3.29 compares tomato hybrid seed production detailing the production

Table 3.29. Steps in the production of tomato hybrid seed (Lapushner and Frankel, 1967)

	Production procedures		
	Use of male fertile female parents	Use of male sterile female parents and artificial pollination	Use of male sterile female parents and natural cross-pollination
Maintenance of male sterile line	−	+	+
Identification and removal of heterozygotes	−	+	+
Emasculation	+	−	−
Pollen collection and storage	+	+	−
Hand pollination	+	+	−
Marking of flowers	+	−	−
Removal of fruits from uncontrolled pollination	+	−	−
Selective harvesting of fruit	+	−	−

steps required when hand-emasculated and -pollinated male fertile seed parents, hand-pollinated male sterile seed parents, or naturally cross-pollinated male sterile seed parents are used. Breakdown of time expenditures for efficient production methods with fertile seed parents (about 40 % for emasculation, 15 % for pollen collection and storage, 30 % for hand pollination, 5 % each for marking flowers, removal of fruits from uncontrolled pollination and selective harvesting of fruit), and a higher fruit set and seed count when using male sterile seed parents, attest that expenditures in maintenance of the male sterile line and the removal of heterozygotes in the hybrid seed production field are much more than offset by savings in time. Thus, when hand pollination is necessary in high-value autogamous plants (e. g. tomato, pepper, eggplant and flower crops), genic male sterile parents become a definite asset in hybrid seed production and sometimes may be decisive in its economics.

3.4.4.3 Use of Gene-Cytoplasmic Male Sterility

Gene cytoplasmic male sterility has already found large-scale application in the production of hybrid seed. In the USA, production based on gene-cytoplasmic male sterility reached about 90 % of the cultivated area for hybrid sorghum, about 60 % for hybrid sugar beet, and up to 85 % for hybrid field corn (DUVICK, 1966). Hybrid cultivars of vegetable crops, produced by means of gene-cytoplasmic male sterility, comprise 94 % of the sweet corn, 27 % of the onion, and 7 % of the carrot acreage in the United States (GABELMAN, 1974). In other highly developed countries the statistics are similar. The reason for such large-scale use of gene-cytoplasmic male sterility lies in the straightforwardness of the procedures providing for pure breeding male sterile A lines and fertility restoration in the hybrid. Male-sterility-inducing plasmatypes of "Autoplasmic" origin in allogamous plants (Fig. 3.42) and of "Alloplasmic" origin in autogamous plants (Fig. 3.43) have been found almost universally. Past experience showed that detection of a male-sterility-inducing plasmatype in a crop plant was always followed by the discovery of additional, distinct cytosteriles. Thus, it appears probable that cytoplasms causing male sterility and fertility restorer genes could very likely be found for any crop plant by combining various plasmatypes and genotypes. However, successful application of gene-cytoplasmic male sterility to F_1 hybrid seed production cannot be achieved by male sterile plasmatypes and restorer genes alone, but a number of obstacles have to be overcome.

Insufficient cross pollination to produce the hybrid seed is a problem mainly with autogamous plants. Flowers may stay closed (as in flax) or not expose their stigmata sufficiently to receive pollen (as in wheat). Obviously cleistogamous flowers cannot serve as pollen donors, and genetic control of anther extrusion may become important (see e. g. DE VRIES, 1973; SAGE and ISTURIZ, 1974). In entomophilous plants, effective transfer of pollen from the C line to A line may be hampered by preference of insects for the pollen fertile C line (see WILLIAMS and FREE, 1974). To overcome limitations in cross pollination, phenotypes with appropriate flower morphology, planting design, choice of season and region for growth, manipulation of pollen vectors, etc., could be of help (see above and DE VRIES, 1971; STOSKOPF and RAI, 1972; MILLER et al., 1974; MOFFETT and STITH, 1972). Concurrent receptivity of styles of the female parent and pollen shedding by the male parent ("nicking") are of utmost importance and must be manipulated genetically or by cultivation techniques, as outlined in Chapter 3.3.1.

Inbreeding depression is a problem with allogamous plants and results in low seed production of inbred parent lines. This may be due also to partial female sterility as a result of a pleiotropic effect of the male sterile plasmatype. A solution to the problem may be sought in the production of double-cross hybrids, as is general practice with field corn (RICHEY, 1950; DUVICK, 1965). Various combinations of male sterile cytoplasm and fertility restoration genes can be used in double-cross hybrid production schemes (see Fig. 3.54). Alternatively, when fertility of the hybrid is required, non-restored double-cross hybrid seed may be mixed (in cross-pollinated crops) with seeds of the same cross combination, using emasculation procedures.

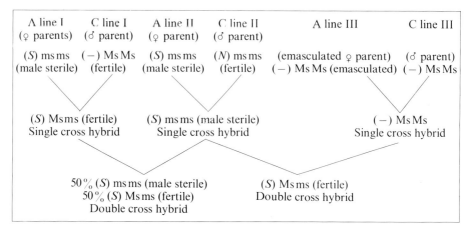

Fig. 3.54. Examples of alternative schemes for double-cross hybrid seed production using gene-cytoplasmic male sterility and fertility restoration

Unreliable stability of male sterility may make the desired genotype useless as seed parent or necessitate additional breeding work (e. g. in cotton—MEYER and MEYER, 1965; and alfalfa—BARNES and GARBOUCHEVA, 1973). The polyploid nature of species (e. g. wheat) and undesirable linkages may cause *complications in fertility restoration* and, as discussed in Chapter 3.4.3.3.2 above, *undesirable pleiotropic effects of the plasmatype* may cause inferior agronomic performance of the hybrid. In maize, yield advantage of hybrids produced with cytoplasmic male sterility is in some instances partly due to the absence of pollen (DUVICK, 1965), but restoration of pollen production removes this advantage. Mixing seed of an identical hybrid combination made with detasselling and with male sterility may improve yields of hybrids.

Where male fertility of the hybrid is of no importance ploidy combinations can be utilized. Triploid hybrids of root crops (e. g. sugar beets—KINOSHITA and NAGAO, 1968) or ornamental plants (e. g. begonia—HORN, 1971) can be conveniently produced by a tetraploid male sterile A line and a diploid C line. Female fertility in tetraploids is generally better than male fertility and haploid pollen is far more functional than diploid pollen.

We observe gynodioecy in natural populations, recognize multitudes of male-sterility-inducing plasmatypes, nuclear male sterility and fertility-restorer genes, and see large-scale application of male sterility for hybrid seed production. Therefore we may expect that the cross pollination mechanism based on male sterility will indeed increasingly serve plant breeding and mankind.

References

ABDALLA, M. M. F., HERMSEN, J. G. TH.: Plasmons and male sterility types in *Solanum verrucosum* and its interspecific hybrid derivatives. Euphytica **21**, 209–220 (1972)

AKKOC, E.: Zytogenetische Beiträge zur Geschlechtsvererbung beim Spinat *(Spinacia oleracea)* im Hinblick auf die Züchtung triploider Hybridsorten. Z. Pflanzenzücht. **53**, 226–246 (1965)

ALLARD, R. W.: A gene in lima beans pleiotropically affecting male sterility and seedling abnormality. Proc. Am. Soc. Hort. Sci. **61**, 467–471 (1953)

ALLARD, R. W.: Natural hybridisation in lima beans in California. Proc. Am. Soc. Hort. Sci. **64**, 410–416 (1954)

ALLARD, R. W.: Principles of Plant Breeding. New York: John Wiley 1960

ALLARD, R. W.: Genetic systems associated with colonizing ability in predominantly self-pollinated species. In: The Genetics of Colonizing Species, BAKER H. G., STEBBINS, G. L. (eds.). New York–London: Academic Press, 1965, pp. 49–76

ALLARD, R. W., JAIN, S. K., WORKMAN, P. L.: The genetics of inbreeding populations. Advan. Genet. **14**, 55–131 (1968)

ALLARD, R. W., KAHLER, A. L., WEIR, B. S.: The effect of selection on esterase allozymes in a barley population. Genetics **72**, 489–503 (1972)

ALLEN, C. E.: Sex-inheritance and sex determination. Am. Naturalist **66**, 97–107 (1932)

ALLEN, C. E.: The genotypic basis of sex-expression in angiosperms. Botan. Rev. **6**, 227–300 (1940)

ALLISON, D. C., FISHER, W. D.: A dominant gene for male sterility in upland cotton. Crop Sci. **4**, 548–549 (1964)

AMICI, G. B.: Observations microscopiques sur diverses éspèces des plantes. Ann. Sci. Nat. Bot. **2**, 41–70 and 211–248 (1824)

AMICI, G. B.: Note sur la mode d'action du pollen sur le stigmate. Ann. Sci. Nat. Bot. **21**, 329–332 (1830)

ANAIS, G.: Nouvelles orientations dans la sélection du melon *(Cucumis melo)*. Utilisation de la gynoecie-modification de l'expression du sexe par traitements chimiques (gibberellines-ethrel) et par greffage. Ann. Amélioration Plantes **21**, 55–65 (1971)

ANDERSON, M. K., TAYLOR, N. L., KIRITHAVIP, R.: Development and performance of double-cross hybrid red clover. Crop Sci. **12**, 240–242 (1972)

ANDRASFALVY, A.: Genes, mutations and cytoplasms of special interest in modern vegetable breeding. Proc. 19th Inten. Hort. Congr. **3**, 457–463 (1974)

APPL, J.: Artkreuzungen, Geschlechtsvererbung und Nondisjunction bei Lippenblütlern aus der Thymiangruppe. Blätter Pflanzenzücht. **11**, 51–56 (1933)

ARASU, N. T.: Self-incompatibility in angiosperms: a review. Genetica **39**, 1–24 (1968)

ARNY, D. C., WORF, G. L., AHRENS, R. W., LINDSEY, M. F.: Yellow leaf blight of maize in Wisconsin: its history and reaction of inbreds and crosses to the inciting fungus *(Phyllostica* sp.), Plant Disease Reptr. **54**, 281–285 (1970)

Association of Official Seed Certifying Agencies: Specific requirements for the certification of plant materials under the AOSCA system. AOSCA Handbook No. 23 (1971) revised 1973, pp. 9–17

ATANASOFF, D.: Viruses and cytoplasmic heredity. Z. Pflanzenzücht. **51**, 197–214 (1964)

ATANASOFF, D.: The viral nature of cytoplasmic male sterility in plants. Phytopathol. Z. **70**, 306–322 (1971)

ATKIN, J. D., DAVIS, G. N.: Altering onion flowering dates to facilitate hybrid seed production. Calif. Agric. Exp. Stn. Bull. **746** (1954)

ATKINS, R. E., KERN, J. J.: Cytoplasm effects in relation to agronomic performance of grain sorghums, *Sorghum bicolor* (L.) Moench. Crop Sci. **12**, 777–780 (1972)

ATSMON, D., GALUN, E.: A morphogenetic study of staminate, pistillate and hermaphrodite flowers in *Cucumis sativus*. Phytomorphology **10**, 110–115 (1960)

ATSMON, D., GALUN, E., JAKOB, A., STEIN, H.: New methods of hybrid seed production based on studies in sex expression. U. N. Conf. Appl. Sci. Technol. for the Benefit of Less Devel. Areas. E/Conf. 36/c 189, 1–4 (1962)

ATWOOD, S. S.: Oppositional alleles in natural populations of *Trifolium repens*. Genetics **29**, 428–435 (1944)

AUGUSTINE, J. J., BAKER, L. R., SELL, H. M.: Female flower induction on androecious cucumber, *Cucumis sativus*. J. Am. Soc. Hort. Sci. **98**, 197–199 (1973)

AYYANGAR, G. N. R., PONNAIYA, B. W. X.: Cleistogamy and its inheritance in sorghum. Curr. Sci. **8**, 418–419 (1939)

BADR, S. A., HARTMANN, H. I.: Effect of diurnally fluctuating vs. constant temperatures on flower induction and sex expression in the olive *(Olea europaea)*. Physiol. Plantarum **24**, 40–45 (1971)

BAHL, P. N., MAAN, S. S.: Chromosomal location of male fertility restoring genes in six lines of common wheat. Crop Sci. **13**, 317–320 (1973)

BAKER, H. G.: Infection of species of *Melandrium* by *Ustilago violacea* and the transmission of the resultant disease. Ann. Botany **11**, 333–348 (1947)

BAKER, H. G.: Reproductive methods as factors in speciation in flowering plants. Cold Spring Harbor Symp. Quant. Biol. **24**, 177–191 (1959)

BAKER, H. G.: The adaptations of flowering plants to nocturnal and crespuscular pollinators. Quart. Rev. Biol. **36**, 64–73 (1961)

BAKER, H. G.: Evolutionary mechanisms in pollination biology. Science **139**, 877–883 (1963)

BAKER, H. G.: The evolution, functioning and breakdown and heteromorphic incompatibility systems. 1. The Plumbaginaceae. Evolution **20**, 349–368 (1966)

BAKER, H. G., HURD, P. D.: Intrafloral ecology. Ann. Rev. Ent. **13**, 385–414 (1968)

BAKER, L. R., SCOTT, J. W., WILSON, J. E.: Seedless pickles—a new concept. Res. Rep. Mich. Sta. Univ. Agric. Exp. Stn. **227**, 1–12 (1973)

BARHAM, W. S., MUNGER, H. M.: The stability of male sterility in onions. Proc. Am. Soc. Hort. Sci. **56**, 401–409 (1950)

BARNES, D. K., CULBERTSON, J. O., LAMBERT, J. W.: Inheritance of seed and flower colors in flax. Agron. J. **52**, 456–459 (1960)

BARNES, D. K., GARBOUCHEVA, R. A.: Intra-plant variation for pollen production in male sterile and fertile alfalfa. Crop Sci. **13**, 456–459 (1973)

BARNES, W. C.: A male sterile cucumber. Proc. Am. Soc. Hort. Sci. **77**, 415–416 (1961)

BARRATT, D. H. P., FLAVELL, R. B.: Alterations in mitochondria associated with cytoplasmic and nuclear genes concerned with male sterility in maize. TAG **45**, 315–321 (1975)

BARRETT, H. C.: A new pollen collection technique for plant breeders. Hort. Res. **9**, 153–155 (1969)

BARRETT, H. C., ARISUMI, T.: Methods of pollen collection, emasculation and pollination in fruit breeding. Proc. Am. Soc. Hort. Sci. **59**, 259–262 (1952)

BARRONS, K. C., LUCAS, H. E.: Production of first generation in hybrid tomato seed for commercial planting. Proc. Am. Soc. Hort. Sci. **40**, 395–404 (1942)

BATCH, J. J., MORGAN, D. G.: Male sterility induced in barley by photoperiod. Nature (London) **250**, 165–167 (1974)

BATEMAN, A. J.: Contamination of seed crops. I. Insect pollination. J. Genet. **48**, 257–275 (1947a)

BATEMAN, A. J.: Contamination of seed crops. II. Wind pollination. Heredity **1**, 235–246 (1947b)

BATEMAN, A. J.: Contamination of seed crops. III. Relation with isolation distance. Heredity **1**, 303–336 (1947c)

BATEMAN, A. J.: Self-incompatibility in angiosperms. I. Theory. Heredity **6**, 285–310 (1952)

BATEMAN, A. J.: Self-incompatibility in angiosperms. II. *Iberis amara*. Heredity **8**, 305–332 (1954)

BATEMAN, A. J.: Self-incompatibility systems in angiosperms. III. Cruciferae. Heredity **9**, 53–68 (1955)

BATESON, W., SAUNDERS, E. R., PUNNETT, R. C.: Male sterility in *Latyrus odoratus*. Rep. Evol. Comm. Roy. Soc. London **4**, p. 16 (1908)

BEADLE, G. W.: Genetical and cytological studies of Mendelian asynapsis in *Zea mays*. Mem. Cornell Univ. Agr. Exp. Stn. **129**, (23 pp.) (1930)

BEADLE, G. W.: Genes in maize for pollen sterility. J. Genet. **17**, 413–431 (1932)

BEATTIE, A. J.: The floral biology of three species of *Viola*. New Phytologist **68**, 1187–1201 (1969)

BÉNARD, G., MALINGRAUX, C.: La production de semences selectionées de palmier à huile à l'IRHO. Principe et réalisation. Oléagineux **20**, 297–302 (1965)

BENNETT, M. D., HUGHES, W. G.: Additional mitosis in wheat pollen induced by ethrel. Nature (London) **240**, 566–568 (1972)

BENNETT, M. D., KALTSIKES, P. J.: The duration of meiosis in a diploid rye, a tetraploid wheat and the hexaploid triticale derived from them. Can. J. Genet. Cytol. **15**, 671–679 (1973)

BERGAL, P., CLEMENCET, M.: The botany of the barley plant. In: Barley and Malt, LOOK, A. H. (ed.). New York–London: Academic Press 1962, p. 13

BERGH, B. O., STOREY, W. B.: Character segregation in avocado racial-hybrid progenies. Calif. Avocado Soc. Ybk. **48**, 61–70 (1964)

BERNINGER, E.: Contribution à l'étude de la stérilité mâle de l'oignon (*Allium cepa* L.). Ann. Amélioration Plantes **15**, 183–199 (1965)

BERNINGER, E., MARCHAT, C., NORTH, C., KOBABE, G., BERTHELEM, P., VALDEYRON, G.: Cas de stérilité mâle. In: La stérilité mâle chez les plantes horticoles. Versailles: Eucarpia Meeting 1970, pp. 205–206

BERTHELEM, P.: Essai d'utilisation d'insectes pollinateurs (*Bombus* Labr.) dans l'amélioration de la féverole (*Vicia faba* L.). Ann. Amélioration Plantes **16**, 101–115 (1966)

BEYERS, R. E., BAKER, L. R., SELL, H. M., HERMER, R. H., DILLEY, D. R.: Ethylene: a definitive role in the sex expression of *Cucumis melo*. Proc. Nat. Acad. Sci. **69**, 717–720 (1972)

BHAT, N. R., KRISHNAMOORTHI, T.: A male sterile mutant in *Nicotiana tabacum*. Curr. Sci. **25**, 297–299 (1956)

BIANCHI, F.: Transmission of male sterility in *Petunia* by grafting. Genen en Phaenen **8**, 36–43 (1963)

BISHOP, C. J.: A stamenless male sterile tomato. Am. J. Botany **41**, 540–542 (1954)

BLACKBURN, K. B.: Sex chromosomes in plants. Nature (London) **112**, 687–688 (1923)

BLAKESLEE, A. F., BELLING, J., FARNHAM, M. E., BERGNER, A. D.: A haploid mutant in *Datura stramonium*. Science **55**, 646–647 (1922)

BODMER, W. F.: The genetics of homostyly in populations of *Primula vulgaris*. Phil. Trans. Roy. Soc. Ser. B **242**, 517–549 (1966)

BOGORAD, L.: Evolution of organelles and eukaryotic genomes. Science **188**, 891–898 (1975)

BOHN, G. W., PRINCIPE, J. A.: A second male sterility gene in muskmelon. J. Heredity **55**, 211–215 (1964)

BOLING, M., SANDER, D. A., MATLOK, R. S.: Mung bean hybridisation method. Agron. J. **53**, 54–55 (1961)

BOLZ, G.: Genetisch. züchterische Untersuchungen bei Tagetes II. Herstellung, Genetik und Verwendung röntgeninduzierter Mutanten in der Züchtung. Z. Pflanzenzücht. **45**, 121–142 (1961)

BOLZ, G.: Monohybride Vererbung der plasmatisch-genetisch bedingten Pollensterilität bei *Beta vulgaris*. Z. Pflanzenzücht. **60**, 219–234 (1968)

BOND, D. A., HAWKINS, R. P.: Behavior of bees visiting male sterile field beans (*Vicia faba*). J. Agr. Sci. **68**, 243–247 (1967)

BONNET, A.: Introduction et utilisation d'une stérilité mâle cytoplasmique dans des variétés précoses européenes de radis. *Raphanus sativus* L. Ann. Amélioration Plantes **25**, 381–397 (1975)

BONNETT, O. T.: Development of the staminate and pistillate inflorescences of sweet corn. J. Agr. Res. **60**, 25–37 (1940)

BONNETT, O. T.: Ear and tassel development in maize. Ann. Missouri Botan. Garden **35**, 269–280 (1948)

BONNETT, O. T.: The oat plant: its histology aned development. Univ. Illinois Agric. Exp. Stn. Bull. **672** (111 pp.) (1961)

BOROWSKY, M. I.: (The transfer of character of male sterility in maize by vegetative hybridisation). Proc. Darwin Jubil. Conf. Kisinev 1960: 209–212 (Russian)

BORST, P.: Mitochondrial nucleic acids. Ann. Rev. Biochem. **41**, 333–376 (1972)

BORTHWICK, H. A., EMSWELLER, S. L.: Carrot breeding experiments. Proc. Am. Soc. Hort. Sci. **30**, 531–533 (1933)

BORTHWICK, H. A., SALLY, N. J.: Photoperiodic responses of hemp. Botan. Gaz. **116**, 14–29 (1954)

BOSE, T. K., NITSCH, J. P.: Chemical alteration of sex expression in *Luffa acutemgula*. Physiol. Plantarum **23**, 1206–1211 (1970)

BOURGIN, J. P., NITSCH, J. P.: Obtention de *Nicotiana* haploides a partir d'étamines cultivées in vitro. Ann. Physiol. Vegetale **9**, 377–382 (1967)

BOWDEN, J. E., NEVE, R. A.: Sorghum midge and resistant varieties in the Gold Coast. Nature (London) **172**, 551 (1953)

BRANDENBURG, W.: Broad beans: causes of poor yields sought. New Zealand J. Agr. **102**, 277–280 (1961)

BREDEMEIJER, G. M. M., BLAAS, J.: A possible role of a stylar peroxidase gradient in the rejection of incompatible growing pollen tubes. Acta Botan. Neerl. **24**, 37–48 (1975)

BREUILS, G., POCHARD, E.: Essai de fabrication de l'hybride de piment "Lamuro-Inra" avec utilisation d'une stérilité mâle génétique (ms 509). Ann. Amélioration Plantes **25**, 399–409 (1975)

BREWBAKER, J. L.: Pollen cytology and incompatibility systems in plants. J. Heredity **48**, 271–277 (1957)

BREWBAKER, J. L.: The distribution and phylogenetic significance of binucleate and trinucleate pollen grains in the angiosperms. Am. J. Botany **54**, 1069–1083 (1967)

BRIDGES, C. B.: Cytological and genetic basis of sex. In: Sex and Internal Secretions. ALLEN, C. (ed.) London: Baillier 1939, pp. 15–63

BRIGGLE, L. W.: Interaction of cytoplasm and genes in a group of male sterile corn types. Agron. J. **49**, 543–547 (1957)

BRIM, C. A., SHUTZ, W. M.: Inter-genotypic competition in soybeans. II. Predicted and observed performance of multiline mixtures. Crop Sci. **8**, 735–739 (1968)

BRIM, C. A., YOUNG, M. F.: Inheritance of a male sterile character in soybeans. Crop Sci. **11**, 564–567 (1971)

BRIM, C. A., STUBER, C. W.: Application of genetic male sterility to recurrent selection schemes in soybeans. Crop Sci. **13**, 528–530 (1973)

BRITTINGHAM, W. H.: Type of seed formation as indicated by the nature and extent of variation in Kentucky blue grass, and its practical application. J. Agr. Res. **67**, 255–264 (1943)

BROOKS, M. W., BROOKS, J. S., CHIEN, L.: The tapetum in cytoplasmic-genetic male sterile *Sorghum*. Am. J. Botany **53**, 902–908 (1966)

BROWN, G. K.: Pollination Research Discussion. Date Growers Inst. Rept. **43**, 29 (1966)

BUCHERT, J. G.: The stage of the genome-plasmon interaction in restoration of fertility of cytoplasmically pollen-sterile maize. Proc. Nat. Acad. Sci. **47**, 1436–1440 (1961)

BUISHAND, T.: The crossing of beans. Euphytica **5**, 41–50 (1956)

BUKOVAC, M. J., HONMA, S.: Gibberellin-induced heterostyly in the tomato and its implication on hybridisation. Proc. Am. Soc. Hort. Sci. **91**, 514–520 (1967)

BURK, L. G.: Male sterile flower anomalies in interspecific tobacco hybrids. J. Heredity **51**, 27–31 (1960)

BURK, L. G.: Haploids in genetically marked progenies of tobacco. J. Heredity **53**, 222–225 (1962)

BURK, L. G.: Green and light-yellow haploid seedlings from anthers of sulfur tobacco. J. Heredity **61**, 279 (1970)

BURTON, G. W.: Artificial fog facilitates *Paspalum* emasculation. J. Am. Soc. Agron. **40**, 281–282 (1948)

BURTON, G. W.: Natural sterility maintainer and fertility restorer mutants in Tift 23A, cytoplasmic male-sterile pearl millet, *Pennisetum typhoides* (Burn) Stapf. and Hubb. Crop Sci. **12**, 280–282 (1972)

CAMERARIUS, R. J. 1694. De sexu Plantarum epistola: (transl. and ed. M. MÖBIUS) Ostwalds
Klassiker der Exakten Wissenschaften Nr. 105. Leipzig: Wilhelm Engelmann 1899

CAMERON, D. R., MOAV, R.: Inheritance in *Nicotiana tabacum* XXVII. Pollen-killer, an
alien locus inducing abortion of microspores not carrying it. Genetics **42**, 326–335
(1957)

CAMERON, J. W., SOOST, R. K.: Nucellar lines of citrus. Calif. Agr. **7**(1), 8, 15, 16 (1953)

CAMERON, J. W., SOOST, R. K., FRONT, H. B.: The horticultural significance of nucellar
embryony in citrus. In: Citrus Virus Disease. WALLACE, J. (ed.) Berkeley: Univ. Calif.,
Div. Agric. Sci. 1959, pp. 191–196

CANTLIFFE, D. J., PHATAK, S.: Use of ethephon and chloroflurenol in a once-over pickling
cucumber production system. J. Am. Soc. Hort. Sci. **100**, 264–267 (1975)

CARLSON, P. S.: The use of protoplasts for genetic research. Proc. Nat. Acad. Sci. **70**,
598–602 (1973)

CARLQUIST, SH.: The biota of long distance dispersal. IV. Genetic systems in the floras
of oceanic islands. Evolution **20**, 433–455 (1966)

CARVALHO, A., MONACO, L. C.: The breeding of arabica coffee. In: Outlines of Perennial
Crop Breeding in the Tropics. FERWERDA, F. P. (ed.) Landbouwhoogesch. Wageningen.
Misc. Papers **4**, 200–202 (1969)

CATARINO, F.: Some effects of kinetin on sex expression in *Bryophyllum crenatum*. Port.
Acta Biol. Ser. A **8**, 267–284 (1964)

CHANDRARATHNA, M. F.: Genetics and Breeding of Rice. London: Longmans, Green and
Co. 1964, pp. 389

CHAUHAN, S. V. S., SINGH, S. P.: Pollen abortion in male sterile hexaploid wheat (Norin)
having *Aegilops ovata* L. cytoplasm. Crop Sci. **6**, 532–535 (1966)

CHAUHAN, S. V. S., SINGH, S. P.: Studies on pollen abortion in *Cucumis melo* L. Agra
Univ. J. Res. Sci. **17**, 11–22 (1968)

CHASE, S. S.: Monoploids in maize. In: Heterosis. GOWEN, J. W. (ed.). Ames: Iowa State
College Press 1956

CHEN, K., KUNG, S. D., GRAY, J. C., WILDMAN, S. G.: Polypeptide composition of fraction
I protein from *Nicotiana glauca* and from cultivars of *Nicotiana tabacum*, including
a male sterile line. Biochem. Genet. **13**, 771–778 (1975)

CHILDERS, W. R.: Male sterility in *Medicago sativa*. Sci. Agr. **32**, 351–364 (1952)

CHOWDHURY, J. B., VARGHESE, T. M.: Pollen sterility in crop plants—a review. Palynol.
Bull. **4**, 71–86 (1968)

CLAASEN, E., HOFFMAN, A.: The inheritance of the pistillate character in castors and
its possible utilization in the production of hybrid seed. Agron. J. **42**, 79–82 (1950)

CLARK, R. K., KENNEY, D. S.: Comparison of staminate flower production on gynoecious
strains of cucumber *(Cucumis sativus)* by pure gibberellins A_3, A_4, A_7 and A_{13} and
mixtures. J. Am. Soc. Hort. Sci. **94**, 131–132 (1969)

CLAUSEN, J.: Male sterility in *Viola orphanides*. Hereditas **14**, 53–72 (1930)

CLAYBERG, C. D.: Further data on aw−ms_{15} linkage. Tomato Genet. Coop. **16**, 7 (1966)

CLAYBERG, C. D., BUTLER, L., KERR, E. A., RICK, C. M., ROBINSON, R. W.: Third list
of known genes in the tomato. J. Heredity **57**, 189–196 (1966)

CLEIGH, G.: Influencing of cytoplasmic male sterility and fertility in beets. Euphytica **16**,
23–28 (1967)

CLEMENT, W. M.: Plasmon mutations in cytoplasmic male-sterile pearl millet, *Pennisetum
typhoides*. Genetics **79**, 583–588 (1975)

COLE, K.: Inheritance of male sterility in green sprouting broccoli. Can. J. Genet. Cytol.
1, 203–207 (1959)

COLLINS, G. B., LEGG, P. D., KASPERBAUER, M. J.: Chromosome numbers in anther-derived
haploids of two *Nicotiana* species: *N. tabacum* and *N. otophora*. J. Heredity **63**, 113–118
(1972)

COLLINS, G. B., LEGG, P. D., KASPERBAUER, M. J.: Use of anther-derived haploids in
Nicotiana. I. Isolation of reeding lines differing in total alkaloid content. Crop Sci.
14, 77–80 (1974)

COMSTOCK, V. E.: Possibilities of hybrid flax production. 35th Ann. Flax Inst. U. S.,
pp. 24–25 (1965)

COPE, F. W.: The mechanism of pollen incompatibility in *Theobroma cacao*. Heredity **17**, 157–182 (1962)

CORBETT, M. K., EDWARDSON, J. R.: Inter-graft transmission of cytoplasmic male sterility. Nature (London) **201**, 847–848 (1964)

CORRENS, C.: Experimentelle Untersuchungen über die Gynodiözie. Ber. Deut. Botan. Ges. **22**, 506–517 (1904)

CORRENS, C.: Die Vererbung der Geschlechtsformen bei den gynodiözischen Pflanzen. Ber. Deut. Botan. Ges. **24**, 459–474 (1906)

CORRENS, C.: Untersuchungen über Geschlechtsbestimmung der Distelarten. S. Ber. Preuss. Akad. Wiss. **20**, 448–477 (1916)

CORRENS, C.: Bestimmung, Vererbung und Verteilung des Geschlechtes bei den höheren Pflanzen. In: Handbuch der Vererbungswissenschaft. Berlin: Gebr. Borntraeger 1928, Vol. II, Part C, pp. 1–138

CORRENS, C.: Über nichtmendelnde Vererbung. Z. Induktive Abstamm. Vererb. Suppl. Vol. **I**, 131–168 (1928)

COTTRELL-DORMER, W.: An electric pollinator for tomatoes. Queensland J. Agr. Sci. **2**, 157–169 (1945)

COYAUD, Y.: Le riz – étude botanique, génétique, physiologique, agrologique et technologique appliquée à l'Indochine. Arch. l'Office Indochinois du Riz, **30**, pp. 312 (1950)

CRAIG, J. L.: Haploid plants from *in vitro* anther culture of *Triticum aestivum*. Can. J. Genet. Cytol. **16**, 697–700 (1974)

CRANE, M. B., THOMAS, P. T.: Reproductive versatility in *Rubus*. III. Raspberry-blackberry hybrids. Heredity **3**, 99–107 (1949)

CRILL, J. P., VILLALON, B., STROBEL, J. W.: An improved technique for crossing tomatoes in the field and greenhouse. Tomato Genetic Coop. Rept. **20**, 14–15 (1970)

CRISWELL, J. G., HUME, D. J., TANNER, J. W.: Effect of cytoplasmic male sterility on accumulation and translocation of 14C-labelled assimilates in corn. Crop Sci. **14**, 252–259 (1974)

CRONQUIST, A.: The Evolution and Classification of Flowering Plants. London: Thomas Nelson and Sons 1968

CROWE, L. K.: The evolution of outbreeding in plants. I. The angiosperms. Heredity **19**, 435–457 (1964)

CROWLEY, N. C.: Studies on the seed transmission of plant virus diseases. Australian J. Biol. Sci. **10**, 449–464 (1957)

CURRENCE, T. M.: Nodal sequence of flower type in the cucumber. Proc. Am. Soc. Hort. Sci. **29**, 477–479 (1932)

CURTIS, B. C., CROY, L. I.: The approach method of making crosses in small grains. Agron. J. **50**, 49–51 (1958)

CURTIS, G. J.: Graft transmission of male sterility in sugar beet (*Beta vulgaris* L.). Euphytica **16**, 419–424 (1967)

DARLINGTON, C. D.: Recent Advances in Cytology. London: Churchill 1932

DARLINGTON, C. D.: The Evolution of Genetic Systems. Cambridge: Univ. Press 1939

DARLINGTON, C. D.: Evolution of Genetic Systems. London: Oliver and Boyd 1958

DARWIN, C.: On the Origin of Species by Means of Natural Selection, or Preservation of Favoured Races in the Struggle for Life. London: John Murray 1859 1st ed. Chap. IV

DARWIN, C.: The Different Forms of Flowers on Plants of the Same Species. London: John Murray 1877

DARWIN, C.: The Effects of Cross- and Self-Fertilisation in the Vegetable Kingdom. London: John Murray 1878 2nd ed.

DASKALOFF, S.: Three new male sterile mutants in pepper (*Capsicum annuum* L.). Comp. Rend. Acad. Agric. Bulg. **6**, 39–41 (1973)

DAVEY, A. J., GIBSON, C. M.: Note on the distribution of the sexes in *Myrca gale*. New Phytologist **16**, 147–151 (1947)

DAVIS, E. W.: The distribution of male sterility genes in onion. Proc. Am. Soc. Hort. Sci. **70**, 316–318 (1957)

DAVIS, E. W.: An improved method of producing hybrid onion seed. J. Heredity **57**, 55–57 (1966)

DEAKIN, J. R., BOHN, G. W., WHITAKER, T. W.: Interspecific hybridization in *Cucumis*. Econ. Botan. **25**, 195–211 (1971)

DENNA, D. W.: The potential use of self-incompatibility for breeding F_1 hybrids of naturally self-pollinated vegetable crops. Euphytica **20**, 542–548 (1971)

DESJARDINS, P. R., LATTERELL, P. L., MITCHELL, J. E.: Seed transmission of tobacco ringspot virus in Lincoln variety of soybean. Phytophathology **44**, 86 (1954)

DEVREUX, M., LANERI, U., MAGNIEN, E., CELESTRE, M. R.: Biological screening method for mutated pollen at the *S*-locus by *in vitro* culture of pollinated pistil. Incompatibility Newslett. **5**, 17–18 (1975)

DICKINSON, H. G., LAWSON, J.: Pollen tube growth in the stigma of *Oenothera organensis* following compatible and incompatible intraspecific pollinations. Proc. Roy. Soc., Ser. B **188**, 327–344 (1975)

DILLMAN, A. C.: Natural crossing in flax. J. Am. Soc. Agron. **30**, 279–286 (1938)

DIXON, G. E. (ed.): Brassica Meeting of Eucarpia: Wellesbourne 1968: N. V. R. S., p. 65

DOAK, C. C.: A new technique in cotton hybridization. J. Heredity **25**, 201–204 (1939)

DOGGETT, H.: Sorghum. London: Longmans, Green and Co. 1970, 403 pp.

DOGGETT, H.: Recurrent selection in sorghum populations. Heredity **28**, 9–29 (1972)

DOIG, R. I., DONE, A. A., ROGERS, D. F.: Preharvest sprouting in bread wheat *(Triticum aestivum)* as influenced by cytoplasmic male-sterility derived from *T. timopheevi*. Euphytica **24**, 229–232 (1975)

DOROSSIEV, L.: A new technique for pollination of the flowers of tomatoes, eggplants and other crops in hybrid seed production (Bulgarian, English summary). Izv. Nauk. – izsled. Inst. Rasteniev Sofia **14**, 203–212 (1962)

DOULL, K. M.: An analysis of bee behaviour as it relates to pollination. In: The Indispensable Pollinators. Report 9th Pollin. Conf., Hot Springs, Arkansas 1970, pp. 5–18

DRESSLER, O.: Zytogenetische Untersuchungen an diploiden und polyploiden Spinat *(Spinacia oleracea)* unter besonderer Berücksichtigung der Geschlechtsvererbung als Grundlage einer Inzucht-Heterosis-Züchtung. Z. Pflanzenzücht. **40**, 385–424 (1958)

DRESSLER, O.: Erfahrungen bei der Vermehrung und Züchtung monözicher Spinatsorten *(Spinacia oleracea)*. Z. Pflanzenzücht. **70**, 108–128 (1973)

DRISCOLL, C. J.: X Y Z system of producing hybrid wheat. Crop Sci. **12**, 516–517 (1972)

DUBEY, D. K., SINGH, S. P.: Use of cytoplasmic male sterility for the production of hybrid seed in flax *(Linum usitatissimum)*. Crop Sci. **6**, 125–126 (1966)

DULBERGER, R.: Flower dimorphism and self-incompatibility in *Narcissus tazetta*. Evolution **18**, 361–363 (1964)

DULBERGER, R.: Floral dimorphism in *Anchusa hybrida*. Israel J. Botany **19**, 37–41 (1970a)

DULBERGER, R.: Tristyly in *Lythrum junceum*. New Phytologist **69**, 751–759 (1970b)

DULBERGER, R.: Intermorph structural differences between stigmatic papillae and pollen grains in relation to incompatibility in Plumbaginaceae. Proc. Roy. Soc., Ser. B **188**, 257–274 (1975)

DUPUIS, F.: Evolution of the plastidial system during the microsporogenesis in *Impatiens balsamina* L. In: Fertilization in Higher Plants. LINSKENS, H. F. (ed.). Amsterdam: North Holland Publ. Co. 1974, pp. 65–71

DURAND, B.: L'expression de sexe chez les mercuriales annuelles. Bull. Soc. Fr. Physiol. Veg. **13**, 195–202 (1967)

DUVICK, D. N.: The use of cytoplasmic male sterility in hybrid seed production. Econ. Botan. **13**, 167–195 (1959)

DUVICK, D. N.: Cytoplasmic pollen sterility in corn. Advan. Genet. **13**, 1–56 (1965)

DUVICK, D. N.: Influence of morphology and sterility on breeding methodology. In: Plant Breeding, FREY, K. J. (ed.). Iowa State Univ. Press, Ames, Iowa 1966, pp. 85–138

EAST, E. M.: The distribution of self-sterility in the flowering plants. Proc. Am. Phil. Soc. **82**, 449–518 (1940)

EAST, E. M., MANGELSDORF, A.: A new interpretation of the hereditary behaviour of self-sterile plants. Proc. Nat. Acad. Sci. **11**, 166–171 (1925)

EATON, F. M.: Selective gametocide opens way to hybrid cotton. Science **126**, 1174–1175 (1957)

ECHLIN, P.: The role of the tapetum during microsporogenesis of angiosperms. In: Pollen: Development and Physiology. HESLOP-HARRISON, J. (ed.) London: Butterworths 1971, pp. 41–61

ECOCHARD, R., RAMANNA, M. S., DE NETTANCOURT, D.: Detection and cytological analysis of tomato haploids. Genetica **40**, 181–190 (1969)

EDWARDSON, J. R.: Cytoplasmic male sterility. Botan. Rev. **22**, 696–738 (1956)

EDWARDSON, J. R.: Cytoplasmic male sterility and fertility restoration in *Crotolaria mucronata*. J. Heredity **58**, 266–268 (1967)

EDWARDSON, J. R.: Cytoplasmic male sterility. Botan. Rev. **36**, 341–420 (1970)

EDWARDSON, J. R., CORBETT, M. K.: Asexual transmission of cytoplasmic male sterility. Proc. Nat. Acad. Sci. **47**, 390–396 (1961)

EDWARDSON, J. R., WARMKE, H. E.: Fertility restoration in cytoplasmic male sterile petunia. J. Heredity **58**, 195–196 (1967)

EHRLICH, P. R., RAVEN, P. H.: Differentiation of populations. Science **165**, 1228–1232 (1969)

EKBERG, I.: Different types of sterility induced in barley by ionizing radiations and chemical mutagens. Hereditas **63**, 257–278 (1969)

ELLISON, W.: Meiosis and fertility in certain British varieties of the cultivated potato. Genetica **18**, 217–254 (1936)

EMERSON, R. A.: The present status of maize genetics. Proc. 6th Intern. Congr. Genet. **I**, 141–152 (1932)

EMERSON, R. A., BEADLE, G. W., FRASER, A. C.: A summary of linkage studies in maize. Mem. Cornell Univ. Agr. Exp. Stn. **180**, 1–83 (1935)

ENGLAND, F. J. W.: Isolation chambers for controlled pollination in grasses. Euphytica **21**, 523–526 (1972)

ENGLAND, F. J. W.: The use of incompatibility for the production of F_1 hybrids in forage grasses. Heredity **32**, 183–188 (1974)

ENGVILD, K. C.: Plantlet ploidy and flower bud size in tobacco anther culture. Hereditas **76**, 320–322 (1974)

ERICHSEN, A. W., ROSS, J. G.: Inheritance of colchicine induced male sterility in sorghum. Crop Sci. **3**, 335–388 (1963)

ERICHSEN, A. W., ROSS, J. G.: Irregularities at microsporogenesis in colchicine induced male sterile mutants of *Sorghum vulgare*. Crop Sci. **3**, 481–483 (1963)

ERICKSON, J. R.: Approach crossing of rice. Crop Sci. **10**, 610–611 (1970)

ESCHERICH, W.: Untersuchungen über den Ab- und Aufbau der Callose. Z. Botan. **49**, 153–218 (1961)

FABERGÉ, A. C.: The cytology of the male sterile *Lathyrus odoratus*. Genetica **19**, 423–430 (1937)

FAEGRI, K.: Reflections on the development of pollination systems in African Proteaceae. J. S. Afr. Botan. **31**, 133–136 (1965)

FAEGRI, K., PIJL, L. VAN DER: The Principles of Pollination Ecology (2nd ed.). Oxford: Pergamon Press 1971

FAHN, A., KLARMAN-KISLEV, N., ZIV, D.: The abnormal flower and fruit of May flowering dwarf Cavendish bananas. Bot. Gaz. **123**, 116–125 (1961)

FILION, W. G., CHRISTIE, B. R.: The mechanism of male sterility in a clone of orchard grass (*Dactylis glomerata* L.). Crop Sci. **6**, 345–347 (1966)

FINCH, R. A., BENNETT, M. D.: The duration of meiosis in diploid and autotetraploid barley. Can. J. Genet. Cytol. **14**, 507–515 (1972)

FISHER, J. E., SYMKO, S.: Teratological stamens in the flowers of *Triticale*. Can. J. Plant Sci. **53**, 61–64 (1973)

FISHER, R. A., MATHER, K.: The inheritance of style length in *Lythrum salicaria*. Ann. Eugen. **12**, 1–23 (1943)

FLEMING, A. A.: Effects of male cytoplasm on inheritance in hybrid maize. Crop Sci. **13**, 570–573 (1975)

FORD, L. E.: A genetic study involving male sterility in *Coleus*. Genetics **35**, 664 (1950)

FOSTER, A. S., GIFFORD, E. M., Jr.: Comparative Morphology of Vascular Plants. San Francisco: W. H. Freeman and Co. 1959

FOSTER, R. E.: F_1 hybrid muskmelons. J. Heredity **59**, 205–207 (1968)

FOSTER, R. E., BOND, W. T.: Abrachiate, an androecious mutant muskmelon. J. Heredity **58**, 13–14 (1967)

FRANCIS, R. R., BERNIS, W. P.: A cytomorphological study of sterility in a mutant of *Cucurbita maxima* Duch. Econ. Botan. **24**, 325–352 (1970)

FRANKEL, O. H.: Studies in *Hebe*. II. The significance of male sterility in the genetic system. J. Genet. **40**, 171–184 (1940)

FRANKEL, R.: Graft induced transmission to progeny of cytoplasmic male sterility in *Petunia*. Science **124**, 684–685 (1956)

FRANKEL, R.: Further evidence on graft-induced transmission to progeny of cytoplasmic male sterility in *Petunia*. Genetics **47**, 641–646 (1962)

FRANKEL, R.: Genetical evidence on alternative maternal and Mendelian hereditary elements in *Petunia hybrida*. Heredity **26**, 107–119 (1971)

FRANKEL, R.: The use of male sterility in hybrid seed production. In: Agricultural Genetics. MOAV, R. (ed.) New York: John Wiley and Sons 1973, pp. 85–94

FRANKEL, R., IZHAR, S., NITSAN, J.: Timing of callase activity and cytoplasmic male sterility in *Petunia*. Biochem. Genet. **3**, 451–455 (1969)

FRANKEN, A. A.: Sex characteristics and inheritance of sex in asparagus (*Asparagus officinalis*). Euphytica **19**, 277–287 (1970)

FREE, J. B.: The effect of distance from pollinizer varieties on the fruit set on trees in plum and apple orchards. J. Hort. Sci. **37**, 262–271 (1962)

FREE, J. B.: The effect on pollen collecting of feeding honeybee colonies with sugar syrup. J. Agr. Sci. Camb. **64**, 167–168 (1965)

FREE, J. B.: Factors determining the collection of pollen by honeybee foragers. Anim. Behav. **15**, 134–144 (1967)

FREE, J. B.: Insect Pollination of Crops. New York: Academic Press 1970

FREY, K. J.: Mass selection for seed width in oat populations. Euphytica **16**, 341–349 (1967)

FRISCH, K. VON: Duftgelenkte Bienen im Dienste der Landwirtschaft und Imkerei. Wien: Springer 1947, pp. 189

FRISCH, K. VON: The Dance Language and Orientation of Bees. London: Oxford University Press 1967

FRYXELL, P. A.: Effect of varietal mass on percentage of outcrossing in *Gossypium hirsutum* in New Mexico. J. Heredity **47**, 299–301 (1956)

FRYXELL, P. A.: Mode of reproduction of higher plants. Botan. Rev. **23**, 135–233 (1957)

FUJIEDA, K.: Cucumber breeding. III. Studies on the method of establishing the inbred gynoecious strains in cucumber. Bull. Hort. Res. Stn. Jap. Ser. D **1**, 101–116 (1963)

FUKASAWA, H.: Studies on restoration and substitution of nucleus in *Aegilotricum* I. Appearance of male sterile durum in substitution crosses. Cytologia **18**, 167–175 (1953)

FUKASAWA, H.: On the free amino acids in anthers of male sterile wheat and maize. Jap. J. Genet. **29**, 135–137 (1954)

FUKASAWA, H.: Biochemical mechanism of pollen abortion and other alterations in cytoplasmic male sterile wheat. Seiken Zihô **13**, 107–111 (1962)

FUKASAWA, H.: Constancy of cytoplasmic property during successive backcrosses. Am. Naturalist **101**, 41–46 (1967)

GABELMAN, W. H.: Male sterility in vegetable breeding. Brookhaven Symposia in Biology **9**, 113–122 (1956)

GABELMAN, W. H.: F_1 hybrids in vegetable production. Proc. 19th Intern. Hort. Congr., Warsaw **3**, 419–428 (1974)

GAIGNARD, L.: Sur les antherozoides et la double copulation sexuelle chez les végétaux angiosperms. C. R. Acad. Sci. Paris **128**, 864–871 (1899)

GAJEWSKI, W. A.: A contribution to the knowledge of the cytoplasmic influence on the effect of nuclear factors in *Linum*. Acta Soc. Botan. Polon. **14**, 205–214 (1937)

GALIANT, W. C.: Use of male sterile 1 gene to eliminate detasselling in production of hybrid seed of bicolor sweet corn. J. Heredity **66**, 387–388 (1975)

GALUN, E.: Effect of seed treatment on sex expression in cucumber. Experientia **12**, 218 (1956)

GALUN, E.: The role of auxins in sex expression of the cucumber. Physiol. Plantarum **12**, 48–61 (1959a)

GALUN, E.: Effects of gibberellic acid and naphthaleneacetic acid on sex expression and some morphological characters in the cucumber plant. Phyton **13**, 1–8 (1959b)

GALUN, E.: Study of the inheritance of sex expression in the cucumber. The interaction of major genes with modifying genetic and non-genetic factors. Genetica **32**, 134–163 (1961)

GALUN, E.: The use of genetic sex types for hybrid seed production in *Cucumis*. In: MOAV, R. (ed.) Agricultural Genetics—Selected Topics. New York–Toronto: John Wiley and Sons 1973, pp. 23–56

GALUN, E., ATSMON, D.: The leaf-floral bud relationship of genetic sex types in the cucumber plant. Bull. Res. Coun. Israel **9D**, 43–50 (1960)

GALUN, E., JUNG, Y., LANG, A.: Culture and sex modification of male cucumber buds *in vitro*. Nature (London) **194**, 595–598 (1962)

GALUN, E., JUNG, Y., LANG, A.: Morphogenesis of floral buds of cucumber cultured *in vitro*. Devel. Biol. **6**, 370–387 (1963)

GALUN, E., PORATH, D.: Morphogenesis of floral buds cultured *in vitro*: Autoradiographic studies. In: BERNIER, G. (ed.) Cellular and Molecular Aspects of Floral Induction. London: Longmans, Green and Co. 1970, pp. 431–445

GARDNER, V. R.: Principle of Horticultural Production. Michigan State Univ. Press 1966, 583 pp

GEIGER, H. H., MORGENSTERN, K.: Angewandt-genetische Studien zur cytoplasmatischen Pollensterilität bei Winterroggen. Theoret. Appl. Genet. **46**, 269–276 (1975)

GEORGE, W. L. Jr.: Dioecism in cucumbers, *Cucumis sativus* L. Genetics **64**, 23–28 (1970)

GERSTEL, D. U.: Self-incompatibility studies in guayule. II. Inheritance. Genetics **35**, 482–506 (1950)

GIANORDOLI, M.: A cytological investigation on gametes and fecundation among *Cephalstaxus drupacea*. In: Fertilization in Higher Plants. LINSKENS, H. F. (ed.) Amsterdam: North-Holland Publ. Co. 1974, pp. 221–232

GILMORE, E. C.: Suggested method of using reciprocal recurrent selection in some naturally self-pollinated species. Crop Sci. **4**, 323–325 (1964)

GOLDSMITH, G. A.: Current developments in the breeding of F_1 hybrid annuals. Hort Sci. **3**, 269–271 (1968)

GOODSELL, S. F.: Male sterility in corn by androgenesis. Crop Sci. **1**, 227–228 (1961)

GOODSPEED, K. H.: Tabak II. Biology and seed production. In: Breeding of Special Cultivated Plants. KAPPERT, H., RUDORF, W. (eds.) Berlin and Hamburg: Paul Parey 1961, pp. 131–135

GORSKA-BRYLASS, A.: Transitory callose envelope surrounding the generative cell in pollen grains. Acta Soc. Bot. Pol. **36**, 419–422 (1967)

GOSH, M. S., BOSE, T. K.: Sex modifications in cucurbitaceous plants by using CCC. Phyton **27**, 113–127 (1970)

GOSS, J. A.: Development, physiology and biochemistry of corn and wheat pollen. Botan. Rev. **34**, 333–358 (1968)

GOTTSBERGER, G.: The structure and function of the primitive Angiosperm flower—a discussion. Acta Bot. Neerl. **23**, 461–471 (1974)

GOTTSCHALK, W., JAHN, A.: Cytogenetische Untersuchungen an desynoptischen und männlich-sterilen Mutanten von Pisum. Z. Vererbungsl. **95**, 150–167 (1964)

GOTTSCHALK, W., KAUL, M. L. H.: The genetic control of microsporogenesis in higher plants. Nucleus (Calcutta) **17**, 133–166 (1974)

GOWERS, S.: Methods of producing F_1 hybrid swedes *(Brassica napus* ssp. *rapifera)*. Euphytica **24**, 537–541 (1975)

GRACEN, P. E., GROGEN, C. O.: Diversity and suitability for hybrid production of different sources of cytoplasmic male sterility in maize. Agron. J. **66**, 654–657 (1974)

GRANT, K. A., GRANT, A.: Hummingbirds and their Flowers. New York: Columbia Univ. Press 1968

GRANT, V.: Pollination systems as isolating mechanisms in angiosperms. Evolution **3**, 82–97 (1949)

GRANT, V.: The influence of breeding habit on the outcome of natural hybridisation in plants. Am. Naturalist **90**, 319–322 (1956)

GRANT, V.: The regulation of recombination in plants. Cold Spring Harbor Symp. Quant. Biol. **23**, 337–363 (1958)

GRANT, V.: The Origin of Adaptations. New York: Columbia Univ. Press 1963

GREEN, J. M., JONES, M. D.: Isolation of cotton for seed increase. Agron. J. **45**, 366–368 (1953)

GREGORY, F. G., PURVIS, O. N.: Abnormal flower development in barley involving sex reversal. Nature (London) **160**, 221–222 (1947)

GREGORY, P. H.: The Microbiology of the Atmosphere. New York: Interscience Publishers 1961

GRIFFITHS, D. J.: Standards of spacial isolation for seed production of herbage crops. Herb. Abstr. **26**, 205–212 (1956)

GRIGGS, W. H., IWAKIRI, B. T.: Orchard tests of bee hive pollen dispensers for cross pollination of almonds, sweet cherries and apples. Proc. Am. Soc. Hort. Sci. **75**, 114–128 (1960)

GROGAN, C. O., SARVELLA, P.: Morphological variations in normal, cytoplasmic male sterile, and restored counterparts in maize, *Zea mays* L. Crop Sci. **4**, 567–570 (1964)

GUHA, S., MAHESHWARI, S. C.: *In vitro* production of embryos from anthers of *Datura*. Nature (London) **204**, 497 (1964)

GUHA, S., MAHESHWARI, S. C.: Cell division and differentiation of embryos in pollen grains of *Datura in vitro*. Nature (London) **212**, 97–98 (1966)

GUHA, S., MAHESHWARI, S. C.: Development of embryoids from pollen grains of *Datura in vitro*. Phytomorphology **17**, 454–461 (1967)

GUIGNARD, L.: Sur les anthérozoides et la double copulation sexuelle chez les végétaux angiosperms. Comp. Rend. Acad. Sci. Paris **128**, 864–871 (1899)

GUSTAFSSON, A.: Apomixis in higher plants. I. The mechanism of apomixis. Lunds Univ. Arsskr. **42**, 1–67 (1946)

GUSTAFSSON, A.: Apomixis in higher plants. II. The causal aspect of apomixis. Lunds Univ. Arsskr. **43**, 69–179 (1947a)

GUSTAFSSON, A.: Apomixis in higher plants. III. Biotype and species formation. Lunds Univ. Arsskr. **43**, 183–370 (1947b)

HAFEN, L., STEVENSON, E. C.: New male sterile and stamenless mutants. Tomato Genet. Coop. Rep. **5**, 17 (1955)

HAFEN, L., STEVENSON, E. C.: Preliminary studies of five stamenless mutants. Tomato Genet. Coop. Rep. **8**, 17–18 (1958)

HAGEMANN, R.: Plasmatische Vererbung. Jena: Gustav Fischer 1964

HAGERUP, O.: Rain-pollination. D. Kgl. danske Vidensk. Selsk. Biol. Medd. **18**, 5 (1950)

HALEVY, A. H., RUDICH, Y.: Modification of sex expression in muskmelon by treatment with the growth retardant B-995. Physiol. Plantarum **20**, 1052–1058 (1967)

HANNA, W. W., SCHERTZ, K. F., BASHAW, E. C.: Apospory in *Sorghum bicolor* (L.) Moensch. Science **170**, 338–339 (1970)

HANSCHE, P. E., GABELMAN, W. H.: Phenotypic stability of pollen sterile carrots. Proc. Am. Soc. Hort. Sci. **82**, 341–350 (1963)

HANSON, A. A., CARNAHAN, H. L.: Breeding perennial forage grasses. U. S. D. A. Tech. Bull. **1145** (1956) 116 pp.

HANSON, C. H.: Cleistogamy and the development of the embryo sac in *Lespedeza stipulacea*. J. Agr. Res. **67**, 265–272 (1943)

HANSON, C. H.: *Lespedeza stipulacea*: Stamen, morphology, meiosis, microgametogenesis and fertilisation. Agron. J. **45**, 200–203 (1953)

HANSON, W. D., PROBST, A. H., CALDWELL, B. E.: Evaluation of a population of soybean genotypes with implications for improving for improving self-pollinated crops. Crop Sci. **7**, 99–103 (1967)

HARDING, J., ALLARD, R. W., SMELTZER, D. G.: Population studies in predominantly self pollinating species. IX. Frequency-dependent selection in *Phaseolus lunatus*. Proc. Nat. Acad. Sci. **56**, 99–104 (1967)

HARLAN, H. V., MARTINI, M. L.: The effect of natural selection on a mixture of barley varieties. J. Agr. Res. **57**, 189–199 (1938)

HARLAN, J. R.: Cleistogamy and chasmogamy in *Bromus carinatus* Hook & Am. Am. J. Botany **32**, 66–72 (1945)

HARLAND, S. C.: The use of haploids in cotton breeding. Indian J. Genet. **15**, 15–17 (1955)

HARN, C., KIM, M. J.: Studies on the anther culture of *Nicotiana tabacum* Korean J. Botany **14**, 33–35 (1971)

HARRIES, H. C.: Isolation bags filled with screw-caps to simplify cross pollination. Euphytica **21**, 117–120 (1972)

HARRIS, H. B.: A new instrument for emasculating sorghum. Agron. J. **47**, 236–237 (1955)

HARRISON, D. P., BROWN, D. T., BODE, V. C.: The Lambda head-tail joining reaction: purification, properties and structure of biologically active heads and tails. J. Mol. Biol. **79**, 437–499 (1973)

HARTMANN, H. T., KESTER, D. E.: Plant Propagation, Principles and Practices. (2nd ed.) Englewood Cliffs, N. J.: Prentice Hall Inc. 1968

HARUTA, T.: Studies on the genetics of self- and cross incompatibility in cruciferous vegetables. Minneapolis, Minn.: Northrop, King and Co. 1966, p. VII + 67 (+135 Tables, 29 Figures)

HASHIZUME, T., IIZUKA, M.: Induction of female organs in male flowers of *Vitis* species by zeatin and dihydrozeatin. Phytochemistry **10**, 2653–2655 (1971)

HAUNOLD, A.: Self fertilization in a normally dioecious species *Humulus japonicus*. J. Heredity **63**, 238–286 (1972)

HAYES, J. D.: Varietal resistance to spray damage in barley. Nature (London) **183**, 551–552 (1959)

HAYMAN, D. L.: The genetical control of incompatibility in *Phalaris coerulescence*. Australian J. Biol. Sci. **9**, 321–331 (1956)

HAYWARD, M. D., MANTHRIRATNA, M. A. P. P.: Pollen development and variation in the genus *Lolium*. I. Pollen size and tapetal relationships in male-fertiles and male-steriles. Z. Pflanzenzücht. **67**, 131–144 (1972)

HAYWARD, M. D., WRIGHT, A. J.: The genetic control of incompatibility in *Lolium perenne*. Genetica **42**, 414–421 (1971)

HECTOR, J. M.: Introduction to the Botany of Field Crops. Centr. News Agency (Johannes-burg) (1936)

HEIDE, O. M.: Environmental control of sex expression in *Begonia*. Z. Pflanzenphysiol. **61**, 279–285 (1969)

HERMENSEN, J. G. TH.: The incompatibility genotype of an autotetraploid cultivar of *Solanum tuberosum* producing both self-compatible and self-incompatible dihaploids. Incompatibility Newslett. **2**, 24–27 (1973)

HERTZSCH, W.: Futtererbsen. In: Breeding of Forage Plants. KAPPERT, H., RUDORF, W. (eds.) 1959, pp. 96–102

HERVÁS, J. P.: Male sterile emmer wheat: development of meiosis with *Aegilops caudata* cytoplasm. Biol. Zbl. **93**, 649–654 (1974)

HESLOP-HARRISON, J.: Auxin and sexuality in *Cannabis sativa*. Physiol. Plantarum **9**, 588–597 (1956)

HESLOP-HARRISON, J.: The experimental modification of sex expression in flowering plants. Biol. Rev. **32**, 38–90 (1957)

HESLOP-HARRISON, J.: The unisexual flower—a reply to criticism. Phytomorphology **8**, 177–184 (1958)

HESLOP-HARRISON, J.: Growth substances and flower morphogenesis. J. Linn. Soc. London **56**, 269–281 (1959)

HESLOP-HARRISON, J.: The experimental control of sexuality and inflorescence structure in *Zea mays*. Proc. Linn. Soc. London **172**, 108–123 (1961)

HESLOP-HARRISON, J.: Sex expression in flowering plants. Brookhaven Symp. Biol. **16**, 109–122 (1963)

HESLOP-HARRISON, J.: Synchronous pollen mitosis and formation of the generative cell in massulate orchids. J. Cell. Sci. **3**, 457–466 (1968)

HESLOP-HARRISON, J.: Cell walls, cell membranes and protoplasmic connections during meiosis and pollen development. In: Pollen Physiology and Fertilization. LINSKENS, H. F. (ed.) Amsterdam: North Holland Publ. Co. 1969, pp. 39–47

HESLOP-HARRISON, J.: Sexuality in angiosperms. In: Plant Physiology—A Treatise. STEWARD, F. C. (ed.) **VI**(c), 133–289. New York: Academic Press 1972

HESLOP-HARRISON, J.: Incompatibility and the pollen-stigma interaction. A. Rev. Plant Physiol. **26**, 403–425 (1975)

HESLOP-HARRISON, J., HESLOP-HARRISON, Y.: The effect of carbon monoxide on sexuality in *Mercurialis ambigua*. New Phytologist **56**, 352–555 (1957)

HESLOP-HARRISON, J., HESLOP-HARRISON, Y.: Long-day and auxin-induced male sterility in *Silene pendula*. Portugaliae Acta Biol. Ser. A **5**, 79–94 (1958a)

HESLOP-HARRISON, J., HESLOP-HARRISON, Y.: Studies on flowering plant growth and organogenesis. III. Leaf shape changes associated with flowering and sex differentiation in *Cannabis sativa*. Proc. Roy Irish Acad. Ser. B **59**, 257–283 (1958b)

HESLOP-HARRISON, J., HESLOP-HARRISON, Y.: Studies on flowering plant growth and organogenesis. IV. Effects of gibberellic acid on flowering and secondary sexual differentiation in stature in *Cannabis sativa*. Proc. Roy Irish Acad. Ser. B **61**, 219–232 (1961)

HESLOP-HARRISON, J., HESLOP-HARRISON, Y.: Enzymatic removal of the proteinuceous pellicle of the stigma papilla prevents pollen tube entry in the Caryophyllaceae. Ann. Botan. **39**, 163–165 (1975)

HESLOP-HARRISON, J., HESLOP-HARRISON, Y., BARBER, J.: The stigma surface in incompatibility responses. Proc. Roy. Soc. Ser. B **188**, 287–297 (1975)

HICKMAN, J. C.: Pollination by ants: A low energy system. Science **184**, 1290–1292 (1974)

HIROSE, T., FUJIME, Y.: A new male sterility in pepper. HortScience **10**, 314 (1975)

HO, T., ROSS, M. D.: Maintenance of male sterility in plant populations. II. Heterotic models. Heredity **31**, 282–286 (1973)

HOCKETT, E. A., ESLICK, R. F.: Genetic male sterile genes, useful in hybrid barley production. In: Barley Genetics II. NILAN, R. A. (ed.) Washington State Univ. Press 1971, pp. 298–307

HOFF, B. J., CHANDRAPANYA, D.: Inheritance of two male sterile characters in rice. Agron. Abstr. 65th Ann. Meet. Am. Soc. Agron (1973), p. 7

HOFFMANN, W.: Gleichzeitig reifender Hanf. Züchter **13**, 277–283 (1941)

HOFFMANN, W.: Die Vererbung der Geschlechtsformen des Hanfes *(Cannabis sativa)*. II. Hanf. Züchter **22**, 147–158 (1952)

HOFMEYR, J. D. J.: Genetical studies of *Carica papaya*. I, II. Un. S. Afr. Dept. Agr. For. Sci. Bull. **187**, 5–64 (1938)

HOFMEYR, J. D. J.: Sex reversal as a means of solving breeding problems of *Carica papaya*. S. Afr. J. Sci. **49**, 228–238 (1953)

HOFMEYR, J. D. J.: Some genetic and breeding aspects of *Carica papaya*. Agron. Trop. (Maracay) **17**, 345–351 (1967)

HOGENBOOM, N. G.: Breaking breeding barriers in *Lycopersicon*. 5. The inheritance of the unilateral incompatibility between *L. peruvianum* and *L. esculentum* and the genetics of its breakdown. Euphytica **21**, 405–414 (1972)

HOGENBOOM, N. G.: Incompatibility and incongruity: two mechanisms for the non-functioning of intimate partner relationships. Proc. Roy. Soc. Ser. B **188**, 361–375 (1975)

HONDELMANN, W., WILBERG, B.: Breeding male varieties of asparagus by utilization of anther and tissue culture. Z. Pflanzenzücht. **69**, 19–24 (1973)

HORN, W.: Neue Wege in der Züchtung winterblühender Begonien. Gartenwelt **71**, 101–102 (1971)

HORN, W.: Induktion und züchterische Nutzung der Parthenogenese. Z. Pflanzenzücht. **67**, 39–44 (1972)

HORN, W.: F_1 hybrids in floriculture. Proc. 19th Intern. Hort. Congr. **4**, 267–277 (1974)

HORNER, J. R., ROGERS, M. A.: A comparative light and electron microscopic study of microsporogenesis in male-fertile and cytoplasmic male-sterile pepper *(Capsicum annuum)*. Can. J. Botany **52**, 435–441 (1974)

HOSFIELD, G. L., VERNSMAN, E. A.: Effect of an alien cytoplasm and fertility restoring factor on growth, agronomic characters, and chemical constituents in a male-sterile variety of flue-cured tobacco. Crop Sci. **14**, 575–577 (1974)

HOWLETT, B. J., KNOX, R. B., HESLOP-HARRISON, J.: Pollen-wall proteins: release of the allergen Antigen E from intine and exine sites in pollen grains of ragweed and *Cosmos*. J. Cell. Sci. **13**, 603–619 (1973)

HOWLETT, B. J., KNOX, R. B., PAXTON, J. D., HESLOP-HARRISON, J.: Pollen wall proteins, physiochemical characterization and role in self-incompatibility in *Cosmos bipinnatus*. Proc. Roy. Soc. Ser. B **188**, 167–182 (1975)

HOWLETT, F. S.: The modification of flower structure by environment in varieties of *Lycopersicum esculentum*. J. Agr. Res. **58**, 79–117 (1939)

HUGHES, K. W., BELL, S. L., CAPONETTI, J. D.: Anther-derived haploids of African Violet. Can. J. Botany **53**, 1442–1444 (1975)

HUGHES, M. B., BABCOCK, E. B.: Self-incompatibility in *Crepis foetida* subsp. *rhaeadifolia*. Genetics **35**, 570–588 (1950)

HUGHES, W. G., BENNETT, M. D., BODDEN, J. J., GALUNOPOULOU, S.: Effects of time of application of ethrel on male sterility and ear emergence in wheat, *Triticum eastivum*. Ann. Appl. Biol. **76**, 243–252 (1974)

HUTCHINS, A. E.: A male and female variant in squash, *Cucurbita maxima* Duch. Proc. Am. Soc. Hort. Sci. **44**, 494–496 (1944)

IMAM, A. G., ALLARD, R. W.: Population studies in predominantly self-pollinated species. VI. Genetic variability between and within natural populations of wild oats, *Avena fatua* L. from different habitats in California. Genetics **51**, 49–62 (1965)

ITO, H., SAITO, T.: Factors responsible for the sex expression of the cucumber plant. XII. Physiological factors associated with the sex expression of flowers. Tohoku J. Agr. Res. **11**, 287–308 (1960)

IWAHORI, S., LYONS, J. M., SIMS, W. L.: Induced femalenesss in cucumber by 2-chloroethane-phosphonic acid. Nature (London) **222**, 171–172 (1969)

IWAHORI, S., LYONS, J. M., SMITH, O. E.: Sex expression in cucumber as affected by chloroethylphosphonic acid, ethylene and growth regulators. Plant Physiol. **46**, 412–415 (1970)

IYER, R. D., RAINA, S. K.: The early ontogeny of embryoids and callus from pollen and subsequent organogenesis in anther culture of *Datura metel* and rice. Planta **104**, 146–156 (1972)

IZHAR, S.: Promotion of flowering in cytoplasmic male sterile petunia by gibberellin. HortScience **7**, 555 (1972)

IZHAR, S., FRANKEL, R.: Mechanism of male sterility in *Petunia:* The relationship between pH, callase activity in the anthers and the breakdown of the microsporogenis. Theoret. Appl. Genet. **44**, 104–108 (1971)

IZHAR, S., FRANKEL, R.: Duration of meiosis in petunia anthers *in vivo* and in floral bud culture. Acta Bot. Neerl. **22**, 14–22 (1973)

IZHAR, S., FRANKEL, R.: Mechanism of male sterility in *Petunia*. II. Free amino acids in male fertile and male sterile anthers during microsporogenesis. Theoret. Appl. Genet. **43**, 13–17 (1973)

IZHAR, S., FRANKEL, R., ARAQ, Y.: Pollen collector—An instrument for separation of particles from air stream. HortScience **10**, 426 (1975)

JACOBSEN, P.: The sex chromosomes in *Humulus*. Hereditas **43**, 357–370 (1957)

JAIN, S. K.: Male sterility in flowering plants. Bibli. Genet. **18**, 101–166 (1959)

JAIN, S. K.: On the possible adaptive significance of male sterility in predominantly inbreeding populations. Genetics **46**, 1237–1240 (1961)

JAIN, S. K.: Gynodioecy in *Origanum vulgare:* Computer simulation of a model. Nature (London) **217**, 764–765 (1968)

JAIN, S. K.: Increased recombination and selection in barley populations carrying a male sterility factor. II. Genotypic frequency at marker loci. Heredity **29**, 457–464 (1969)

JAIN, S. K., BRADSHAW, A. D.: Evolutionary divergence among adjacent plant populations. I. The evidence and its theoretical analysis. Heredity **22**, 407–441 (1966)

JAIN, S. K., MARSHALL, D. R.: Population studies in predominantly self-pollinated species. X. Variation in natural populations of *Avena fatua* and *A. barbata*. Am. Naturalist **101**, 19–33 (1967)

JAIN, S. K., SUNESON, C. A.: Increased recombination and selection in barley populations carrying a male sterility factor. I. Quantitative variability. Genetics **54**, 1215–1224 (1966)

JAKOB, K. M., ATSMON, D.: Sex inheritance in *Ricinus communis:* evidence for a genetic change during the ontogeny of female sex reversion. Genetica **36**, 253–259 (1965)

JAN, C. C., QUALSET, C. O., VOGT, H. E.: Chemical induction of sterility in wheat. Euphytica **23**, 78–85 (1974)

JANICK, J., IIZUKA, M.: Sex determination in spinach. 19th Intern. Hort. Congr. Brussels, 82–88 (1962)

JASMIN, J. J.: Male sterility in *Solanum melongena*: Preliminary report on a functional type of male sterility in eggplants. Proc. Am. Soc. Hort. Sci. **63**, 443 (1954)

JAYAKARAN, M.: Suppression of stamens in *Capsicum annuum* by a morphactin (EMD-IT 7839). Curr. Sci. **41**, 849–850 (1972)

JAYCOX, E. R., OWEN, F. W.: Honeybees and pollen insects can improve apple yields. Am. Bee J. **105**, 96–97 (1965)

JEFFREY, C.: Cucurbitaceae. In: Flora of Tropical East Africa. London: Crown Agents 1967

JENNINGS, P. R., BEACHELL, H. M., CHUAVIROJ, M.: An improved rice hybridisation technique. Crop Sci. **4**, 524–526 (1964)

JENNINGS, P. R., DE JESUS, J.: Studies on competition in rice. I. Competition in mixtures of varieties. Evolution **22**, 119–124 (1968)

JENSEN, N. F.: Multiline superiority in cereals. Crop Sci. **5**, 567–568 (1965)

JENSEN, N. F.: A diallel selective mating system for cereal breeding. Crop Sci. **10**, 629–630 (1970)

JENSEN, W. A.: Reproduction in flowering plants. In: Dynamic Aspects of Plant Ultrastructure. ROBARDS, A. W. (ed.). London: McGraw-Hill 1974, pp. 481–503

JOHANNSEN, W.: Elemente der exakten Erblichkeitslehre. 3rd ed. Jena: Fischer 1926

JOHN, B., LEWIS, K. R.: The Meiotic Mechanism. London: Oxford Univ. Press 1973, p. 32

JOHNSON, I. J.: Forage crop breeding. In: Forages. HUGHES, H. D., HEATH, M. E., METCALFE, D. S. (eds.) 2nd ed. Iowa State Univ. Press 1962, pp. 98–99

JOHNSON, V. A., SCHMIDT, J. W.: Hybrid Wheat. Advan. Agron. **20**, 199–233 (1968)

JOHRI, M., VASIL, I. K.: Physiology of pollen. Botan. Rev. **27**, 325–381 (1961)

JONES, D. F.: Unisexual maize plants and their bearing on sex differentiation in other plants and animals. Genetics **19**, 552–567 (1934)

JONES, D. F.: Sex intergrades in dioecious maize. Am. J. Botany **26**, 412–415 (1939)

JONES, D. F.: Genic and cytoplasmic control of pollen abortion in maize. In: Genetics in Plant Breeding. Brookhaven Symp. Biol. **9**, 101–112 (1956)

JONES, D. F., CLARKE, A. E.: Inheritance of male sterility in the onion and the production of hybrid seed. Proc. Am. Soc. Hort. Sci. **43**, 189–194 (1943)

JONES, D. F., EVERETT, H. L.: Hybrid field corn. Conn. Agr. Expt. Stn. Bull. **532**, 35–38 (1949)

JONES, H. A.: Pollination and life history studies in lettuce (*Lactuca sativa* L.). Hilgardia **2**, 425–478 (1927)

JONES, H. A., CLARKE, A. E.: Inheritance of male sterility in the onion and the production of hybrid seed. Proc. Am. Soc. Hort. Sci. **43**, 189–194 (1943)

JONES, H. A., DAVIS, G. N.: Inbreeding and heterosis and their relation to the development of new varieties of onion. Tech. Bull. U. S. Dep. Agr. 874 (1944) 28 pp.

JONES, H. A., EMSWELLER, S. L.: The use of flies of onion pollinators. Proc. Am. Soc. Hort. Sci. **31**, 160–164 (1934)

JONES, H. A., EMSWELLER, S. L.: A male sterile onion. Proc. Am, Soc. Hort. Sci. **34**, 582–585 (1937)

JONG, A. W. DE, BRUINSMA, J.: Pistil development in *Cleome* flowers. III. Effects of growth-regulating substances on flower buds of *Cleome iberidella* grown *in vitro*. Z. Pflanzenphysiol. **73**, 142–151 (1974a)

JONG, A. W. DE, BRUINSMA, J.: Pistil development in *Cleome* flowers. IV. Effects of growth-regulating substances on female abortion in *Cleome spinosa*. Z. Pflanzenphysiol. **73**, 152–159 (1974b)

JONG, A. W. DE, SMIT, A. L., BRUINSMA, J.: Pistil development in *Cleome* flowers. II. Effects of nutrients on flower buds of *Cleome iberidella* grown *in vitro*. Z. Pflanzenphysiol. **72**, 227–236 (1974)

JOPPA, L. R., McNEAL, F. H., BERG, M. A.: Pollen production and pollen shedding of hard red spring (*Triticum aestivum* L. em Thell.) and durum (*T. durum* Desf.) wheats. Crop Sci. **8**, 487–490 (1968)

JORDAN, H. D.: Hybridisation of rice. Trop. Agr. Trinidad **34**, 133–136 (1957)

JOSEPHSON, L. M.: Deletion of rf gamete in male sterile crosses in corn. 56th Ann. Meet. Am. Soc. Agron. Kansas City, Nov. 1964. Agron. Abstr. p. 70 (1964)

JUSTUS, N., LEINWEBER, C. L.: A heritable partially male sterile character in cotton. J. Heredity **51**, 191–192 (1960)

KAMPMEIJER, P.: Fluorescence pattern of the sex chromosomes of *Melandrium dioicum* stained with quinacrine mustard. Genetica **43**, 201–206 (1972)

KANNENBERG, L. W., ALLARD, R. W.: Population studies in predominantly self-pollinated species. VIII. Genetic variability in the *Festuca microstachys* complex. Evolution **21**, 227–240 (1967)

KAPPERT, H.: Untersuchungen über die Plasmonwirkung bei *Aquilegia*. Flora (Jena) **37**, 95–105 (1944)

KARPER, R. E., STEPHENS, J. C.: Floral abnormalities in *Sorghum*. J. Heredity **27**, 183–194 (1936)

KASHA, K. J., KAO, K. N.: High frequency haploid production in barley *(Hordeum vulgare)*. Nature (London) **225**, 874–876 (1970)

KATZNELSON, J.: Semi-natural interspecific hybridisation in plants. Euphytica **20**, 266–269 (1971)

KELLER, W.: Emasculation and pollination technics. Proc. Intern. Grassland Congr. **6**, 1613–1619 (1952)

KENDER, W. J., RAMAILY, G.: Regulation of sex expression and seed development in grapes with 2-chloroethylphosphonic acid. HortScience **5**, 491–492 (1970)

KERN, J. J., ATKINS, R. E.: Free amino-acid content of the anthers of male-sterile and fertile lines of grain, sorghum, *Sorghum bicolor* (L) Moench. Crop Sci. **12**, 835–839 (1972)

KHAN, M. N., HEYNE, E. G., ARP, A. L.: Pollen distribution and the seed set on *Triticum aestivum* L. Crop Sci. **13**, 223–226 (1973)

KHO, Y. O., DE BRUYN, J. W.: Gametocidal action of dichloroacetic acid. Euphytica **11**, 287–292 (1962)

KHOO, V., STINSON, H.: Free amino acid differences between cytoplasmic male sterile and normal fertile anthers. Proc. Nat. Acad. Sci. **43**, 603–607 (1957)

KIHARA, H., TSUNEWAKI, K.: Genetic principles applied to breeding of crop plants. In: BRINK, R. A. (ed.) Heritage from Mendel. Madison: Univ. Wisconsin Press 1967, pp. 403–418

KIMATA, M., SAKAMOTO, S.: Callus induction and organ differentiation of *Triticum, Aegilops* and *Agropyron* by anther culture. Jap. J. Palynol. **8**, 1–7 (1971)

KIMBER, G., RILEY, R.: Haploid angiosperms. Botan. Rev. **29**, 480–531 (1963)

KING, J. R.: The rapid collection of pollen. Proc. Am. Soc. Hort. Sci. **66**, 155–156 (1955)

KING, J. R.: The storage of pollen—particularly by the freeze drying method. Bull. Torrey Bot. Club **92**, 270–287 (1965)

KING, J. R., BROOKS, R. M.: The terminology of pollination. Science **105**, 379–380 (1947)

KINOSHITA, T., NAGAO, S.: Use of male sterility in triploid sugar beet. Proc. 12th Intern. Congr. Genet. Tokyo **2**, 232–233 (1968)

KIRK, J. T. O., TIENEY-BASSETT, R. A. E.: The Plastids. San Francisco: Freeman 1967

KIRK, L. E.: Abnormal seed development in sweet clover species crosses—A new technique for emasculating sweet clover flowers. Sci. Agr. **10**, 321–327 (1930)

KLEIN, H. D., MILUTINOVIC, M.: Genbedingte Störungen der Infloreszenz- und Blütenbildung. Theoret. Appl. Genet. **41**, 255–258 (1971)

KNIGHT, R.: A technique for controlled pollination in the production of grass seed. Euphytica **15**, 374–376 (1966)

KNIGHT, R., ROGERS, H. H.: Incompatibility in *Theobroma cacao*. Heredity **9**, 69–77 (1955)

KNIGHT, T. A.: Upon the effects of very high temperatures on some species of plants. (Read before the Roy. Hort. Soc., Dec. 7, 1819.) In: A Selection from the Physiological and Horticultural Papers by the Late T. A. Knight, pp. 238–241. London: Longman, Orme, Brown, Green and Longmans 1841

KNOWLES, P. F.: Improving an annual bromegrass, *Bromus mollis* L. for range purposes. J. Am. Soc. Agron. **35**, 584–594 (1943)

KNUTH, P.: Handbuch der Blütenbiologie. Leipzig: Wilhelm Englemann 1898–1905, Vols. I–III

KOBABE, G.: Entwicklungsgeschichtliche und genetische Untersuchungen an einem männlich sterilen Mutanten der Küchenzwiebel (*Allium cepa* L.). Z. Pflanzenzücht. **40**, 353–384 (1958)

KOBABE, G.: Einfache Hilfsmittel für die Bestäubungsregulierung bei der Züchtung von Fremdbefruchtern. Züchter **35**, 299–307 (1965)

KOHEL, R. J., RICHMOND, T. R.: Test for cytoplasmic genetic interaction involving a genetic male sterile stock of genotype ms_2ms_2. Crop Sci. **3**, 351–362 (1963)

KÖHLER, D.: Geschlechtsbestimmung bei Blütenpflanzen. Ergebn. Biol. **27**, 98–115 (1964a)

KÖHLER, D.: Veränderung des Geschlechts von *Cannabis sativa* durch Gibberellinsäure. Ber. deut. Botan. Gesell. **78**, 275–281 (1964b)

KÖLREUTER, D. J. G.: Vorläufige Nachricht von einigen das Geschlecht der Pflanzen betreffenden Versuchen und Beobachtungen. Fortsetzung 1. Ostwalds Klassiker der exakten Wissenschaften Nr. 41, Leipzig: Engelman 1763

KOOISTRA, E.: Femaleness in breeding glasshouse cucumbers. Euphytica **16**, 1–17 (1967)

KRAAI, A.: The use of honey-bees and bumble bees in breeding work. Euphytica **3**, 97–107 (1954)

KREMP, G. O. W.: Morphological Encyclopedia of Palynology. Tucson, Ariz.: Univ. Arizona Press 1965

KRISHNAMOORTHY, H. N.: Effect of GA_3, GA_{4+7}, G_5 and G_9 on the sex expression of *Luffa acutangula* var. H-2. Plant Cell Physiol. **13**, 381–382 (1972)

KUBICKI, B.: Investigations on sex determination in cucumbers (*Cucumis sativus*). I. The influence of 1-naphthaleneacetic acid and gibberellin on differentiation of flowers in monoecious cucumbers. Genet. Pol. **6**, 153–176 (1965a)

KUBICKI, B.: New possibilities of applying different sex types in cucumber breeding. Genet. Pol. **6**, 241–250 (1965b)

KUBICKI, B.: Investigations on sex determination in cucumber (*Cucumis sativus*). Genet. Pol. **10**, 3–143 (1969a)

KUBICKI, B.: Sex determination in muskmelon (*Cucumis melo*). Genet. Pol. **10**, 145–166 (1969b)

KUBICKI, B.: Cucumber hybrid seed production based on gynoecious lines multiplied with the aid of complementary hermaphroditic lines. Genet. Pol. **11**, 181–186 (1970)

KUBICKI, B., POTOCZEK, H.: Gibberellin-induced perfect flowers in a tomato female form and the possibility to obtain gynoecious lines for hybrid seed production. Genet. Pol. **13**, 67–74 (1972)

KUGLER, H.: Einführung in die Blütenökologie. (2nd ed.) Stuttgart: Fischer 1970

KUNG, S. D.: Tobacco fraction 1 protein: A unique genetic marker. Science **191**, 429–434 (1976)

KWASINIKOW, B. W., DOLGICH, S. T., STOZAROWA, I. A., TARASENKOW, L. L.: Induzierung von Sterilitätsmutationen bei Gemüsenpflanzen. In: La stérilité mâle chez les plantes horticoles. Eucarpia Meeting, Versailles 1970, pp. 115–122

LACADENA, J. R.: Hybrid wheat: VII. Test on the transmission of cytoplasmic male sterility in wheat by embryo-endosperm grafting. Euphytica **17**, 439–444 (1968)

LAIBACH, F., KRIBBEN, F. J.: Der Einfluss von Wuchsstoff auf die Bildung männlicher und weiblicher Blüten bei einer monözischen Pflanze (*Cucumis sativus*). Ber. deut. botan. Gesell. **62**, 53–55 (1949)

LAIBACH, F., KRIBBEN, F. J.: Die Bedeutung des Wuchsstoffs für die Bildung und Geschlechtsbestimmung der Blüten. Beitr. Biol. Pflanzen **28**, 131–144 (1951)

LANGE, A. H.: Factors affecting sex changes in the flowers of *Carica papaya*. Proc. Am. Soc. Hort. Sci. **77**, 252–264 (1961)

LAPUSHNER, D., FRANKEL, R.: Practical aspects, and the use of male sterility in the production of hybrid tomato seed. Euphytica **16**, 300–310 (1967)

LARSEN, K.: Four loci governing self-incompatibility in *Beta vulgaris*. Incompatibility Newslett. **4**, 20–22 (1974)

LARSON, D. A.: Cytoplasmic dimorphism within pollen grains. Nature (London) **200**, 911–912 (1963)

LARSON, D. A.: Fine structural changes in the cytoplasm of germinating pollen. Am. J. Botany **52**, 139–154 (1965)

LARSON, R. E., PAUR, SH.: The description and inheritance of functionally sterile flower mutant in tomato and its probable value in hybrid tomato seed production. Proc. Am. Soc. Hort. Sci. **52**, 355–364 (1948)

LASER, K. D., LERSTEN, N. R.: Anatomy and cytology of microsporogenesis in cytoplasmic male sterile angiosperms. Botan. Rev. **38**, 425–454 (1972)

LAUGHNAN, J. R., GABAY, S. J.: Reaction of germinating maize pollen to *Helminthosporium maydis* pathotoxins. Crop Sci. **13**, 681–684 (1973)

LAW, J., STOSKOPF, N. C.: Further observations on ethephon (Ethrel) as a tool for developing hybrid cereals. Can. J. Plant Sci. **53**, 765–766 (1973)

LAWRENCE, M. J.: The genetics of self-incompatibility in *Papaver rhoeas*. Proc. Roy. Soc. Ser. B **188**, 275–285 (1975)

LECHNER, L.: Wicken (Vicia-) Arten. In: Breeding of Forage Plants. KAPPERT, H., RUDORF, W. (eds.) Berlin and Hamburg: Paul Parey 1959, pp. 52–95.

LEONARD, K. J.: Factors effecting rates of stem rust increase in mixed plantings of resistant and subceptible oat varieties. Phytopathology **59**, 1845–1850 (1969)

LEPPIK, E. E.: Searching for gene centers of the genus *Cucumis* through host-parasite relationship. Euphytica **15**, 323–328 (1966)

LEVIN, D. A.: Plant density, cleistogamy, and self fertilisation in natural populations of *Lithospermum caroliniense*. Am. J. Botan. **59**, 71–77 (1972a)

LEVIN, D. A.: Competition for pollinator service: A stimulus for the evolution of autogamy. Evolution **26**, 668–669 (1972b)

LEVINGS, C. S., PRING, D. R.: Restriction endonuclease analysis of mitochondrial DNA from Normal and Texas cytoplasmic male-sterile maize. Science **193**, 158–160 (1976)

LEWIS, D.: Male sterility in natural populations of hermaphrodite plants. The equilibrium between females and hermaphrodites to be expected with different types of inheritance. New Phytologist **40**, 53–63 (1941)

LEWIS, D.: The evolution of sex in flowering plants. Biol. Rev. **17**, 46–67 (1942)

LEWIS, D.: Physiology of incompatibility. III. Autopolyploids. J. Genet. **45**, 171–185 (1943)

LEWIS, D.: Incompatibility in flowering plants. Biol. Rev. **24**, 472–492 (1949a)

LEWIS, D.: Structure of the incompatibility gene. II. Induced mutation rate. Heredity **3**, 339–355 (1949b)

LEWIS, D.: Comparative incompatibility in angiosperms and fungi. Advan. Genet. **6**, 235–285 (1954)

LEWIS, D.: Incompatibility and plant breeding. In: Genetics and Plant Breeding. Brookhaven Symp. Biol. **9**, 89–100 (1956)

LEWIS, D.: Genetic control of specificity and activity of the S-antigen in plants. Proc. Roy. Soc. Ser. B **151**, 468–477 (1960)

LEWIS, D.: A protein dimer hypothesis on incompatibility. Proc. 11th Intern. Contr. Genet. **3**, 657–663 (1965)

LEWIS, D.: Heteromorphic incompatibility system under disruptive selection. Proc. Roy. Soc. Ser. B **188**, 247–256 (1975)

LEWIS, D., CROWE, L. K.: Male sterility as an outbreeding mechanism in *Origanum vulgare*. Heredity Abstr. **6**, 136 (1952)

LEWIS, D., CROWE, L. K.: The genetics and evolution of gynodioecy. Evolution **10**, 115–125 (1956)

LEWIS, K. R., JOHN, B.: The chromosomal basis of sex determination. Intern. Rev. Cytol. BOURNE, G. H., DANIELLI, J. F. (eds.) **23**, 277–379. New York, London: Academic Press 1968

LIM, S. M., HOOKER, A. L., KINSEY, J. G., SMITH, D. R.: Comparative grain yields of corn hybrids in normal and in Texas male sterile cytoplasm (cms-T) infected with *Helminthosporium maydis* race T and disease components of cms-T corn hybrids. Crop Sci. **14**, 190–195 (1974)

LINDAUER, M.: Ein Beitrag zur Frage der Arbeitsteilung im Bienenstaat. Z. vergl. Physiol. **34**, 299–345 (1952)

LINDEN, A. J. TER: Investigations into cyclone dust collectors. Inst. Mech. Engin. Proc. **160**, 233–251 (1949)

LINDQUIST, K.: Inheritance studies in lettuce. Hereditas **46,** 387–470 (1960)

LINNERT, G.: Kerngesteuerte Gynodiözie bei *Salvia nemorosa.* Z. Vererb. Lehre **89,** 36–51 (1958)

LINSKENS, H. F.: Zur Frage der Entstehung der Abwehr-Körper bei der Inkompatibilitäts-reaktion von *Petunia.* III: Serologische Tests mit Leitgewebs- und Pollen-Extrakten. Z. Botan. **48,** 126–135 (1960)

LINSKENS, H. F.: Pollen Physiology. A. Rev. Plant Physiol. **15,** 255–270 (1964)

LINSKENS, H. F.: The physiological basis of incompatibility in angiosperms. Biol. J. Linn. Soc. London **6,** suppl. 1 (1974)

LINSKENS, H. F., KROH, M.: Inkompatibilität der Phanerogamen. In: RUHLAND, W. (ed.) Encyclopedia of Plant Physiology **18,** 506–530 (1967)

LITTLE, T. M., JONES, H. A., CLARKE, A. E.: The distribution of the male sterility gene in varieties of onion. Herbertia **11,** 310–312 (1944)

LLOYD, D. G.: The maintenance of gynodioecy and androdioecy in angiosperms. Genetica **45,** 325–339 (1975)

LOEHWING, W. F.: Physiological aspects of sex in angiosperms. Botan. Rev. **4,** 581–625 (1938)

LOMBARDO, G., GEROLA, F. M.: Cytoplasmic inheritance and ultrastructure of the male generative cell of higher plants. Planta **82,** 105–110 (1968)

LÖTTER, J. DE V.: Recent developments in pollination techniques of deciduous fruit trees. Decid. Fruit Grow. **10,** 182–190; 212–224; 304–311 (1960)

LÖVE, A., LÖVE, D.: Experiments on the effects of animal sex hormones on dioecious plants. Ark. Botan. **32,** 1–60 (1945)

LÖVE, A., SARKAR, N.: Cytotaxonomy and sex determination of *Rumex paucifolius.* Can. J. Botany **34,** 261–268 (1956)

LÖVE, D.: Cytogenetic studies on dioecious *Melandrium.* Bot. Notiser 125–213 (1944)

LUNDQVIST, A.: Studies on self-sterility in rye, *Secale cereale.* Hereditas **40,** 278–294 (1954)

LUNDQVIST, A.: Auto-incompatibility and breeding. 5th Congr. Europ. Ass. Plant Breeding. Eucarpia, 365–380 (1968)

LUNDQVIST, A.: Complex self-incompatibility systems in angiosperms. Proc. Roy. Soc. Ser. B **188,** 235–245 (1975)

LUNDQVIST, A., ØSTERBYE, U., LARSEN, K., LINDE-LAURSEN, J.: Complex self-incompatibi-lity systems in *Ranunculus acris* and *Beta vulgaris.* Hereditas **74,** 161–168 (1973)

LUNSFORD, J. N., FUTRELL, M. C., SCOTT, G. E.: Maternal influence on response of corn to *Fusarium moniliforme.* Phytopathology **65,** 223–225 (1975)

MAAN, S. S., LUCKEN, K. A.: Interacting male sterility–male fertility restoration systems for hybrid wheat research. Crop Sci. **12,** 360–364 (1972)

MACDONALD, I. M., GRANT, W. F.: Anthers culture of pollen containing ethrel induced micronuclei. Z. Pflanzenzücht. **73,** 292–297 (1974)

MACIOR, L. M.: The pollination ecology of *Pedicularis* in Colorado. Am. J. Botany **57,** 716–728 (1970)

MADJOLELO, S. D. P., GROGAN, C. O., SARVELLA, P. A.: Morphological expression of genetic male sterility in maize (*Zea mays* L.). Crop Sci. **6,** 379–380 (1966)

MADGE, M. A. P.: Spermatogenesis and fertilisation in the cleistogamous flower of *Viola oderata* var. *praecox* Gregory. Ann. Botan. **43,** 545–577 (1929)

MAEKAWA, T.: On the phenomena of sex transition in *Arisaema japonica.* J. Coll. Agr. Hokkaido Imp. Univ. **13,** 217–305 (1924)

MAESEN, L. J. G. VAN DER: *Cicer* L., A Monograph of the genus, with special reference to the chickpea (*Cicer arietinum* L.) its ecology and cultivation. Landbouwhogesch. Wageningen, 72–10, 243–245 (1972)

MAHESHWARI, J. K.: Cleistogamy in angiosperms. In: Proc. Summer School Botany, Darjeel-ing, June 1960. MAHESHWARI, P., JOHRI, B. M., VASIL, I. K. (eds.) New Delhi (1962), pp. 145–155

MAHESHWARI, P.: An Introduction to the Embryology of Angiosperms. New York: McGraw-Hill 1950

MALINOVSKY, B. N., ZOZ, N. N., KITAEV, A. A.: Induction of cytoplasmic male sterility (CMS) in sorghum by chemical mutagens (in Russian). Genetika **9,** 19–27 (1973)

MANN, J., JONES, G. L., MATZINGER, D. F.: The use of cytoplasmic male sterility in flue cured tobacco hybrids. Crop Sci. **2**, 407–410 (1962)

MARCHESI, G. E.: Mechanical collector for tomato pollen. Tomato Genet. Coop. Rept. **20**, 25 (1970)

MARKS, G. E.: Selected asparagus plants as sources of haploids. Euphytica **22**, 310–316 (1973)

MARREWIJK, G. A. M. VAN: Cytoplasmic male sterility in petunia. I. Restoration of fertility with special reference to the influence of environment. Euphytica **18**, 1–20 (1969)

MARREWIJK, G. A. M. VAN: Cytoplasmic male sterility in petunia. II. A discussion on male sterility transmission by means of grafting. Euphytica **19**, 25–32 (1970)

MARSHALL, D. R., ALLARD, R. W.: Performance and stability of mixtures of grain sorghum. 1. Relationship between level of genetic diversity and performance. Theoret. Appl. Genet. **44**, 145–152 (1974)

MARTIN, F. W.: Some improvements in pollen collection and storage techniques. Tomato Genet. Coop. Rept. **10**, 23 (1960)

MARTIN, F. W.: Sex ratio and sex determination in *Dioscorea*. J. Heredity **57**, 95–99 (1966)

MARTIN, F. W., GREGORY, L. E.: Mode of pollination and factors affecting fruit set of *Piper nigrum* L. in Puerto Rico. Crop Sci. **2**, 295–299 (1962)

MASCARENHAS, J. P.: The biochemistry of angiosperm pollen development. Botan. Rev. **41**, 259–314 (1975)

MATHER, K.: Nucleus and cytoplasm in differentiation. Symp. Soc. Exp. Biol. **2**, 196–216 (1948)

MATHER, K.: Polymorphism as an outcome of disruptive selection. Evolution **9**, 52–61 (1955)

MATSUO, E., UEMOTO, S., FAKUSHIMA, E.: Studies on the photoperiodic sex differentiation in cucumber, *Cucumis sativus:* Ageing effect on the photoperiodic dependency of sex differentiation. J. Fac. Agr. Kyushu Univ. **15**, 287–303 (1969)

MATZINGER, D. F., WERNSMAN, E. A.: Four cycles of mass selection in a synthetic variety of an autogamous species *Nicotiana tabacum* L. Crop Sci. **8**, 239–243 (1968)

MCCOLLUM, G. D.: Hybrid origin of top onion, *Allium cepa* var. *viviparum*. Z. Pflanzenzücht. **71**, 222–232 (1974)

MCDANIEL, M. E., KIM, H. B., HATHCOCK, B. R.: Approach crossing of oats (*Avena* spp.) Crop Sci. **7**, 538–540 (1967)

MCDONALD, J. E.: Collection and washout of airborne pollen and spores by raindrops. Science **135**, 435–437 (1962)

MCKEE, H. S.: Nitrogen Metabolism in Plants. London: Oxford Univ. Press 1962

MCKINNEY, H. H.: Two strains of tobacco mosaic virus, one of which is seed-transmitted in an etch immune pungent pepper. Plant Disease Reptr. **36**, 184–187 (1952)

MCKINNEY, H. H., GREELEY, L. W.: Biological characteristics of barley stripe mosaic virus strains and their evolution. Tech. Bull. U. S. Dep. Agr. **1324** (1965)

MCLEAN, R. C., IVIMEY-COOK, W. R.: Textbook of Theoretical Botany. London, New York, Toronto: Longmans, Green & Co. 1958

MCMURRAY, A. L., MILLER, C. H.: Cucumber sex expression modified by 2-chloroethane-phosphonic acid. Science **162**, 1396–1397 (1968)

MEER, Q. P. VAN DER: Frequencies of heritable factors determining male sterility in onion (*Allium cepa* L.) and their importance for the breeding of hybrids. In: La stérilité mâle chez les plantes horticoles. Eucarpia Meeting, Versailles 1970, pp. 27–36

MEER, Q. P. VAN DER, VAN BENNEKOM, J. L.: Failure of graft transmission of male sterility substance in onion. Euphytica **19**, 430–432 (1970)

MEER, Q. P. VAN DER, VAN BENNEKOM, J. L.: Gibberellic acid as a gametocide for the common onion (*Allium cepa* L.). Euphytica **22**, 239–243 (1973)

MEINDERS, H. C., JONES, M. D.: Pollen shedding and dispersal in the castor plant *Ricinus communis* L. Agron. J. **42**, 206–207 (1950)

MELCHERS, G.: Haploid higher plants for plant breeding. Z. Pflanzenzücht. **67**, 19–32 (1972)

MELCHERS, G., LABIB, G.: Die Bedeutung haploider höherer Pflanzen für Pflanzenphysiologie und Pflanzenzüchtung. Die durch Anthernkultur erzeugten Haploiden, ein neuer Durchbruch für die Pflanzenzüchtung. Ber. deut. Botan. Gesell. **83,** 129–150 (1970)

MELETTI, P.: Induzione spermintale della maschiosterilita in *Triticum.* Nov. Gior. Bot. Ital. (N. S.) **68,** 299–307 (1961)

MEPHAM, R. H., LANE, G. R.: Formation and development of the tapetal periplasmodium in *Tradescantia bracteata.* Protoplasma **68,** 175–192 (1969)

MEPHAM, R. H., LANE, G. R.: Observation on the fine structure developing microspores of *Tradescantia bracteata.* Protoplasma **70,** 1–20 (1970)

MEREDITH, W. R., BRIDGE, R. R.: Natural crossing in cotton (*Gossypium hirsutum* L.) in the delta of Mississippi. Crop Sci. **13,** 551–552 (1973)

METZGE, E. B.: Inflorescence patterns and sexual expression in *Begonia semperflorens.* Am. J. Botany **25,** 465–478 (1938)

MEYER, V. G.: Flower abnormalities. Botan. Rev. **32,** 165–218 (1966)

MEYER, V. G.: Some effects of genes, cytoplasm and environment on male sterility in cotton *(Gossypium).* Crop Sci. **9,** 237–242 (1969)

MEYER, V. G.: Male sterility from *Gossypium harknessii.* J. Heredity **66,** 23–27 (1975)

MEYER, V. G., BUFFETT, M.: Cytoplasmic effects on external ovule production in cotton. J. Heredity **53,** 251–253 (1962)

MEYER, V. G., MEYER, J. R.: Cytoplasmically controlled male sterility in cotton. Crop Sci. **5,** 444–448 (1965)

MIAN, H. R., KUSPIRA, J., WALKER, G. W. R., MUNTGEWERFF, N.: Histological and cytochemical studies on five genetic male sterile lines of barley *(Hordeum vulgare).* Can. J. Genet. Cytol. **16,** 355–379 (1974)

MICHAELIS, P.: Über reziprok verschiedene Sippenbastarde bei *Epilobium hirsutum.* II. Über die Konstanz des Plasmons der Sippe Jena. Z. indukt. Abstamm. Vererb. Lehre **78,** 223–237 (1940)

MICHAELIS, P., MICHAELIS, G.: Über die Konstanz des Zytoplasmons bei *Epilobium.* Planta **35,** 467–512 (1948)

MILLER, J. F., SCHMIDT, J. W., JONSON, V. A.: Inheritance of genes controlling male sterility restoration in the wheat cultivar primépi. Crop Sci. **14,** 437–440 (1974)

MILLER, J. F., ROGERS, K. J., LUCKEN, K. A.: Male sterile wheat production in North Dakota. Crop Sci. **14,** 702–705 (1974)

MILLER, R. J., KOEPPE, D. E.: Southern corn leaf blight: susceptible and resistant mitochondria. Science **173,** 67–69 (1971)

MININA, E. G.: On the phenotypical modification of sexual characters in higher plants under the influence of the condition of nutrition and other external factors. Comp. Rend. Acad. Sci. USSR (Doklady) **21,** 298–301 (1938)

MININA, E. G., TYLKINA, L. G.: Physiological study of the effect of gasses upon sex differentiation in plants. Comp. Rend. Acad. Sci. USSR (Doklady) **55,** 165–168 (1947)

MIRAVELLE, R. J.: A new bulked-pollen method for cotton cross pollination. J. Heredity **55,** 276–280 (1964)

MITCHELL, W. D., WITTWER, S. H.: Chemical regulation of sex expression and vegetative growth in *Cucumis sativus.* Science **136,** 880–881 (1962)

MITTWOCH, U.: Do genes determine sex? Nature **221,** 446–448 (1969)

MITTWOCH, U.: Genetics of Sex Differentiation. New York: Academic Press 1973

MODJELELO, S. D. P., GROGAN, C. O., SARVELLA, P. A.: Morphological expression of genetic male sterility in maize (*Zea mays* L.). Crop Sci. **6,** 379–380 (1966)

MOFFETT, J. O., STITH, L. S.: Pollination by honey bees of male sterile cotton in cages. Crop Sci. **12,** 476–478 (1972)

MOHAM RAM, H. Y., JAISWAL, V. S.: Induction of female flowers on male plants of *Cannabis sativa* by 2-chloroethane phosphonic acid. Experientia **26,** 214–216 (1970)

MOHAM RAM, H. Y., JAISWAL, V. S.: Feminization of male flowers of *Cannabis sativa* by a morphactin. Naturwissenschaften **58,** 149–150 (1971)

MOHAM RAM, H. Y., JAISWAL, V. S.: Induction of male flowers on female plants of *Cannabis sativa* and its inhibition by abscisic acid. Planta **105,** 263–266 (1972a)

MOHAM RAM, H. Y., JAISWAL, V. S.: Sex reversal in male plants of *Cannabis sativa* by ethyl-hydrogen-1-propylphosphonate. Z. Pflanzenphysiol. **68,** 181–183 (1972b)

MOLLIARD, M.: Sur la détermination du sex chez le chanvre. Comp. Rend. Acad. Sci. Paris **125**, 792–794 (1897)

MOORE, J. F.: Male sterility induced in tomato by sodium 2,3-dichloroisobutyrate. Science **129**, 1738–1739 (1959)

MOORE, J. F.: Male sterility induced in field grown tomatoes with sodium α,β-dichloroisobutyrate. Proc. Am. Soc. Hort. Sci. **84**, 474–479 (1964)

MURASHIGE, T.: Plant propagation through tissue cultures. Ann. Rev. Plant Physiol. **25**, 135–166 (1974)

MURTHI, A. N., WEAVER, J. B.: Histological studies on five male-sterile strains of upland cotton. Crop Sci. **14**, 658–663 (1974)

MYERS, W. M.: Effects of cytoplasm and gene dosage on expression of male sterility in *Dactylis glomerata*. Genetics **31**, 225–226 (1946)

NAGAI, I.: Japonica Rice, its Breeding and Culture. Tokyo: Yokendo Ltd. 1959, pp. 843

NAGAICH, B. B., UPADHYA, M. D., PRAKASH, O., SINGH, S.: Cytoplasmically determined expression of symptoms of potato virus X in crosses between species of *Capsicum*. Nature (London) **220**, 1341–1342 (1968)

NAPP-ZINN, K.: Modifikative Geschlechtsbestimmung bei Spermatophyten. In: RUHLAND, W. (ed.) Encyclopedia of Plant Physiology **18**, 153–213 (1967)

NASRALLAH, M. E., BARBER, J. T., WALLACE, D. H.: Self-incompatibility proteins in plants: detection, genetics and possible mode of action. Heredity **25**, 23–27 (1970)

NAVASHIN, S. G.: Resultate einer Revision der Befruchtungsvorgänge bei *Lilium martagon* und *Fritillaria tenella*. Izo. Imp. Akad. Nauk **9**, 377–382 (1898)

NAYLOR, F. L.: Effect of length of induction period on floral development in *Xanthium pennsylvanicum*. Botan. Gaz. **103**, 146–154 (1941)

NEGI, S. S., OLMO, H. P.: Sex conversion in a male *Vitis vinifera* by a kinin. Science **152**, 1624–1625 (1966)

NEGI, S. A., OLMO, H. P.: Certain embryological and biochemical aspects of cytokinin SD 8339 in converting sex of male *Vitis vinifera* (sylvestris). Am. J. Botany **59**, 851–857 (1972)

NEI, M., SYAKUDO, K.: The estimation of outcrossing in natural populations. Jap. J. Genet. **33**, 46–51 (1958)

NELSON, P. M., ROSSMAN, E. C.: Chemical induction of male sterility in inbred maize by use of gibberellins. Science **127**, 1500–1501 (1958)

NETTANCOURT, D. DE: Radiation effects on the one locus-gametophytic system of self-incompatibility in higher plants. Theoret. Appl. Genet. **39**, 187–196 (1969)

NETTANCOURT, D. DE: Self-incompatibility in basic and applied researches with higher plants. Genetica Agraria (Pavia) **26**, 163–226 (1972)

NETTANCOURT, D. DE, DEVREUX, M., BOZZINI, A., CRESTI, M., PACINI, E., SARFATTI, G.: Ultrastructural aspects of the self-incompatibility mechanism in *Lycopersicum peruvianum*. J. Cell Sci. **12**, 403–419 (1973)

NETTANCOURT, D. DE, DEVREUX, M., CARLUCCIO, F., LANERI, U., CRESTI, M., PACINI, E., SARFATTI, G., GASTEL, A. J. G. VAN: Facts and hypotheses on the origin of S mutations and the function of the S gene in *Nicotiana alata* and *Lycopersicum peruvianum*. Proc. Roy Soc. Ser. B **188**, 345–360 (1975)

NICKELL, L. G., HEINZ, D. J.: Potential of cell and tissue culture techniques as aids in economic plant improvement. In: Genes, Enzymes and Populations. SRB, A. M. (ed.) New York: Plenum Press 1973, pp. 109–128

NICKERSON, N. H.: Sustained treatment with gibberellic acid of maize plants carrying one of the dominant genes teopod and corn-grass. Am. J. Botany **47**, 809–815 (1960a)

NICKERSON, N. H.: Studies involving sustained treatment of maize with gibberellic acid. II. Responses of plants carrying certain tassel-modifying genes. Ann. Mo. Botan. Gard. **47**, 243–261 (1960b)

NICKERSON, N. H., DALE, E. E.: Tassel modification in *Zea mays*. Ann. Mo. Botan. Gard. **42**, 195–212 (1955)

NIEDLE, E. K.: Nitrogen nutrition in relation to photoperiodism in *Xanthium pennsylvanicum*. Botan. Gaz. **100**, 607–618 (1938)

NIEUWHOF, M.: Single and double crosses of cole crops *(Brassica oleracea)*. In: Brassica Meeting of Eucarpia (1968), DIXON, G. E. (ed.). NVRS, Wellesbourne, Warwick 1968, pp. 2–4

NIEUWHOF, M.: Sex expression in onion *(Allium cepa* L.). In: La stérilité mâle chez les plantes horticoles. Eucarpia Meeting, Versailles 1970, pp. 47–61

NIGTEVECHT, G. VAN: Genetic studies in dioecious *Melandrium*. I. Sex-linked and sex-influenced inheritance in *Melandrium album* and *Melandrium dioicum*. Genetica 37, 281–306 (1966a)

NIGTEVECHT, G. VAN: Genetic studies in dioecious *Melandrium*. II. Sex determination in *Melandrium album* and *Melandrium dioicum*. Genetica 37, 307–344 (1966b)

NIIZEKI, H.: Induction of haploid plants from anther culture. Japan Agr. Res. Quart. 3, 41–45 (1968)

NIIZEKI, H., OONO, K.: Induction of haploid rice plants from anther culture. Proc. Jap. Acad. 44, 554–557 (1968)

NILSSON-TILLGREN, T., VON WETTSTEIN-KNOWLES, P.: When is the male plastone eliminated? Nature (London) 227, 1265–1266 (1970)

NITCH, C., NORRELL, B.: Factors favoring the formation of androgenetic embryos in anther culture. In: Genes, Enzymes and Populations. SRB, A. M. (ed.) New York: Plenum Press 1973, pp. 129–144

NITSCH, J., KURTZ, E. B., LIVERMANN, J. L., WENT, F. W.: The development of sex expression in *Cucurbit* flowers. Am. J. Botany 39, 32–43 (1952)

NOBLE, S. W., RUSSELL, W. A.: Effects of male sterile cytoplasm and pollen restorer genes on performance of hybrid corn. Crop Sci. 3, 92–96 (1963)

NORREEL, B.: Étude cytologique de l'androgénèse expérimentale chez *Nicotiana tabacum* et *Datura innoxia*. Bull. Soc. Botan. Fr. 117, 461–478 (1970)

NOTHMANN, J., KOLLER, D.: Morphogenetic effects of low temperature stress on flowers of eggplant, *Solanum melongena*. Israel J. Botany 22, 231–235 (1973a)

NOTHMANN, J., KOLLER, D.: Morphogenetic effects of growth regulators on flowers of eggplant *(Solanum melongena)*. Hort. Res. 13, 105–110 (1973b)

NOVAK, F. J.: Tapetal development in anthers of *Allium sativum* L. and *Allium longicuspis* Regel. Experientia 28, 1380–1381 (1972)

NOVAK, F. J., BETLACH, J., DUBROVSKY, J.: Cytoplasmic male sterility in sweet pepper. I. Phenotype and inheritance of male sterile character. Z. Pflanzenzücht. 65, 129–140 (1971)

NUNES, M. A.: Sex expression abnormalities in rice plants suffering from "Straighthead"—an attempt to hormonal interpretation. Portugaliae Acta Biol., Ser. A 8, 285–300 (1964)

NYGREN, A.: Apomixis in the angiosperm. Botan. Rev. 20, 577–649 (1954)

O'BRIEN, T. P., MCCULLY, M. E.: Plant Structure and Development. London: The Macmillan Comp. 1970

ODLAND, M. L., PORTER, A. W.: A study of natural crossing in peppers *(Capsicum frutescens)*. Proc. Am. Soc. Hort Sci. 38, 585–588 (1941)

OGAWA, J. M., ENGLISH, H.: The efficiency of a quantitative spore collector using the cyclone method. Phytopathology 45, 239–240 (1955)

OHTA, Y.: Grafting and cytoplasmic male sterility in *Capsicum*. Seiken Zihô 12, 35–39 (1961)

OHTA, Y.: Cytoplasmic male sterility and virus infection in *Capsicum annuum* L. Jap. J. Genet. 45, 277–283 (1970)

OHTA, Y.: Nature of a cytoplasmic entity causing male sterility in *Capsicum annuum* L. Ann. Fac. Sci. Univ. Torino 7, 229–238 (1971–72)

OHTA, Y.: Identification of cytoplasms of independent origin causing male sterility in red peppers *(Capsicum annuum* L.). Seiken Zihô 24, 105–106 (1973)

OHTA, Y.: Viruses, plasmids or episome-like elements and extrachromosomal inheritance in relation to plant breeding research. Seiken Zihô 25–26, 41–54 (1975)

OHTA, Y., CHUONG, P. V.: Hereditary changes in *Capsicum annuum* L. I. Induced by ordinary grafting. Euphytica 25, 355–368 (1975)

OLMO, H. P.: Pollination in Almeria grape. Proc. Am. Soc. Hort. Sci. 42, 401–406 (1943)

OLSON, R. J.: M. S. Thesis Washington State Univ. (cited by JOHNSON, V. A., SCHMIDT, J. W. 1968, Hybrid wheat. Advanc. Agron. 20, 222–223.) 1966

ONO, T.: Polyploidy and sex determination in *Melandrium*. I. Colchicine induced polyploids of *M. album*. Botan. Mag. (Tokyo) **53**, 549–556 (1939)

OSBORNE, D. J., WENT, F. W.: Climatic factors influencing parthenocarpy and normal fruit-set in tomatoes. Botan. Gaz. **114**, 312–322 (1953)

ØSTERBYE, U.: Self-incompatibility in *Ranunculus acris*. I. Genetic interpretation and evolutionary aspects. Hereditas **80**, 91–112 (1975)

OUDEJANS, J. H. M.: Date palm. In: Outline of Perennial Crop Breeding in the Tropics. FERWERDA, F. P. (ed.) Landbouwhoogesch. Wageningen, Misc. papers **4**, 243–257 (1969)

OVERLEY, F. L., BULLOCK, R. M.: Pollen diluents and application of pollen to tree fruits. Proc. Am. Soc. Hort. Sci. **49**, 163–169 (1947)

OVERMAN, M. A., WARMKE, H. E.: Cytoplasmic male sterility in *Sorghum*. II. Tapetal behavior in fertile and sterile anthers. J. Heredity **63**, 227–234 (1972)

OWEN, F. V.: Cytoplasmically inherited male sterility in sugarbeets. J. Agric. Res. **71**, 423–440 (1945)

PAKENDORF, K. W.: Male sterility in *Lupinus mutabilis* Sweet. Z. Pflanzenzücht. **63**, 227–236 (1970)

PALIWAL, R. L., HYDE, B. B.: The association of a single B-chromosome with male sterility in *Plantage coronopus*. Am. J. Botany **46**, 460–466 (1959)

PALMER, R.: Cytological studies of ameiotic and normal maize with reference to premeiotic pairing. Chromosoma (Berlin) **35**, 233–246 (1971)

PANDEY, K. K.: Genetics of self-incompatibility in *Physalis ixocarpa*. A new system. Am. J. Botany **44**, 879–887 (1957)

PANDEY, K. K.: Time and site of S-gene action, breeding systems and relationships in incompatibility. Euphytica **19**, 364–372 (1970a)

PANDEY, K. K.: Elements of the S-gene complex. VI. Mutations of the incompatibility gene, pseudo-compatibility and origin of new incompatibility alleles. Genetica **41**, 477–516 (1970b)

PANDEY, K. K.: Theory and practice of induced androgenesis. New Phytologist **72**, 1129–1140 (1973)

PANGALO, K. I.: On genes determining sex types in plants as illustrated by the *Cucurbitaceae*. Comp. Rend. Acad. Sci. URSS (Doklady) III (XII), pp. 83–85 (1936)

PAYNE, W. W.: The morphology of the inflorescence of rag weeds. Am. J. Botan. **50**, 872–880 (1963)

PEARSON, O. H.: Cytoplasmically inherited male sterility characters and flavor components from the species cross *Brassica nigra* (L) Koch × *B. oleracea* L. J. Am. Soc. Hort. Sci. **97**, 397–402 (1972)

PERCIVAL, M. S.: The presentation of pollen in certain angiosperms and its collection by *Apis mellifera*. New Phytologist **54**, 353–368 (1955)

PERETZ, A. T., CHANG, T. T., BEACHELL, B. S., VERGARA, B. S., MARCIANO, A. P.: Induction of male sterility in rice with ethrel and RH-531. Sabrao Newslett. **5**, 133–139 (1973)

PETERSON, C. E., ANHDER, L. D.: Induction of staminate flowers on gynoecious cucumbers with gibberellin A₃. Science **131**, 1673–1674 (1960)

PETERSON, C. E., DE ZEEW, D. J.: The hybrid pickling cucumber Spartan Dawn. Q. Bull. Mich. St. Univ. Agr. Exp. Stn. Bull. **46**, 267–273 (1963)

PETERSON, C. E., FOSKETT, R. L.: Occurrence of pollen sterility in seed fields of Scott County Globe onions. Proc. Am. Soc. Hort. Sci. **62**, 443–448 (1953)

PETERSON, P. A.: Cytoplasmically inherited male sterility in *Capsicum*. Am. Naturalist **92**, 111–119 (1958)

PETERSON, P. A., FLAVELL, R. B., BARRATT, D. H. P.: Altered mitochondrial membrane activities associated with cytoplasmically-inherited disease sensitivity in maize. Theoret. Appl. Genet. **45**, 309–314 (1975)

PETRU, E., HRABETOVA, E., TUPY, J.: The technique of obtaining germinating pollen without microbial contamination. Biol. Planta (Praha) **6**, 68–69 (1964)

PFEIFFER, N. E.: Effectiveness of certain apple pollen diluents in hand pollination test. Contrib. Boyce Thompson Inst. **15**, 119–125 (1948)

PHATAK, S. C., WITTWER, S. H., HONMA, S., BUKOVAC, M. J.: Gibberellin-induced anther and pollen development in a stamenless tomato mutant. Nature (London) **209**, 635–636 (1966)

PHILOUZE, J.: Gènes marqueurs liés aux gènes de stérilité mâle ms_{32} et ms_{35} chez la tomate. Ann. Amélior. Plantes **24**, 77–82 (1974)

PICARD, E., BUYSER, J. DE: Nouveaux résultats concernant la culture d'anthères de *Triticum aestivum* L. Conditions de régéneration de plantes haploide et production de lingées entièrement homozygotes. C. R. Acad. Sci. (Paris) Ser. D. **281**, 989–992 (1975)

PIENAAR, R. DE V.: Male sterility in South African onion varieties. Proc. 1st Congr. S. Afr. Genet. Soc. July 1958, pp. 87–90

PIJL, L. VAN DER: Ecological aspects of flower evolution. I. Phyletic evolution. Evolution **14**, 403–416 (1960)

PIJL, L. VAN DER: Ecological aspects of flower evolution. II. Zoophilous flower classes. Evolution **15**, 44–59 (1961)

PIKE, L. M., MULKEY, W. A.: Use of hermaphrodite cucumber lines in development of gynoecious hybrids. HortScience **6**, 339–340 (1971)

PIKE, L. M., PETERSON, C. E.: Gibberellin A_4/A_7 for induction of staminate flowers on the gynoecious cucumber. Euphytica **18**, 106–109 (1969)

POEHLMAN, J. M.: Breeding Field Crops. New York: Henry Holt and Co 1959

POGLAZOR, B. F.: Morphogenesis of T-even bacteriophages. Monographs in Developmental Biology **7**, p. 105. Basel: Krager 1973

POOLE, C. F., GRIMBALL, P. C.: Inheritance of new sex forms in *Cucumis melo*. J. Heredity **30**, 21–25 (1939)

POPE, D. A., SIMPSON, D. M., DUNCAN, E. N.: Effect of corn barriers on natural crossing in cotton. J. Agr. Res. **68**, 347–361 (1944)

POPE, M. N.: Some notes on technique in barley breeding. J. Heredity **35**, 99–111 (1944)

PORATH, D., GALUN, E.: *In vitro* culture of hermaphrodite floral buds of *Cucumis melo*: Microsporogenesis and ovary formation. Ann. Botan. **31**, 283–290 (1967)

PORTER, K. B., LAHR, K. A., ATKINS, M.: Cross pollination of male sterile winter wheat (*Triticum aestivum* L.) having *Aegilops caudata* L. and *Aegilops ovata* L. cytoplasm. Crop Sci. **5**, 161–163 (1965)

POVILAITIS, B.: Effects of cytoplasmic male sterility on certain characters of flue-cured tobacco, *Nicotiana tabacum*. Can. J. Genet. Cytol. **14**, 403–409 (1972)

PRITCHARD, A. J., HUTTON, E. M.: Anther and pollen development in male-sterile *Phaseolus atropurpureus*. J. Heredity **63**, 280–282 (1972)

PROCTER, M., YEO, P.: The Pollination of Flowers. London: Collins 1973

PUNNETT, R. C.: Further studies of linkage in the sweet pea. J. Genet. **26**, 97–112 (1932)

PURSEGLOVE, J. W.: Tropical Crops. London: Longmans, Green and Co. 1968, Vol. I

QUINBY, J. R., MARTIN, J. H.: Sorghum improvement. Advan. Agron. **6**, 305–359 (1954)

QUINBY, J. R., SCHERTZ, K. F.: Sorghum genetics breeding, and hybrid seed production. In: Sorghum Production and Utilization. WALL, J. S., ROSS, W. M. (eds.) Westport: Avi Publishing Co. 1970, Chap. 3, pp. 73–117

RAEBER, J. G., BOLTON, A.: A new form of male sterility in *Nicotiana tabacum* L. Nature (London) **176**, 314–315 (1955)

RAI, R. K., STOSKOPF, N. C.: Amino acid comparisons in male sterile wheat derived from *Triticum timopheevi* Zuk. Cytoplasm and its fertile counterpart. Theoret. Appl. Genet. **44**, 124–127 (1974)

RAMAER, H.: Cytology of Hevea. Genetica **17**, 193–236 (1935)

RAMAGE, R. T.: Balanced tertiary trisomics for use in hybrid seed production. Crop Sci. **5**, 177–178 (1965)

RAMANNA, M. S., HERMSEN, J. G. T.: Unilateral "eclipse sterility" in reciprocal crosses between *Solanum verrucosum* Schlechtd. and diploid *S. tuberosum* L. Euphytica **23**, 417–421 (1974)

RAMIREZ, W.: Fig wasps: mechanism of pollen transfer. Science **163**, 580–581 (1969)

RASHID, A., STREET, H. E.: Growth, embryogenetic potential and stability of a haploid cell culture of *Atropa belladonna*. Plant Sci. Lett. **2**, 89–94 (1974a)

RASHID, A., STREET, H. E.: Segmentation of microspores of *Nicotiana sylvestris* and *Nicotiana tabacum* which lead to embryoid formation in anther culture. Protoplasma **80**, 323–334 (1974b)

REDDEN, R. J., JENSEN, N. F.: Mass selection and mating systems in cereals. Crop Sci. **14**, 345–350 (1974)

REHM, S.: Male sterile plants by chemical treatment. Nature (London) **170**, 38–39 (1952)

REIMANN-PHILIPP, R.: Entwicklungsarbeiten zur Züchtung von F_1 Hybriden bei Blumen. Z. Pflanzenzücht. **51**, 249–314 (1964)

REIMANN-PHILIPP, R.: The application of incompatibility in plant breeding. Proc. 11th Intern. Congr. Genet. **3**, 649–656 (1965)

REIMANN-PHILIPP, R.: Die Ausnützung genetischer Mechanismen zur Verhinderung von Selbstbefruchtungen in der Züchtung von F_1-Hybriden bei Gartenpflanzen. Biologia (Bratislava) **29**, 487–496 (1974)

REIMANN-PHILIPP, R., FUCHS, G.: F_1-Hybriden bei *Ageratum houstonianum*. Gartenwelt **71**, 443–444 (1971)

RESEARCH GROUP 301: A sharp increase of the frequency of pollen plant induction of wheat with potato medium. Acta Genetica Sinica (Peking) **3**, 25–31 (1976)

RESENDE, F.: Auxina e sexo em *Hyoscyamus*. Boll. Soc. Portug. Ci. Nat., 2 Ser. **4**, 248–251 (1953)

RESENDE, F.: General principles of sexual and asexual reproduction and life cycles. In: RUHLAND, W. (ed.) Encyclopedia of Plant Physiology. **18**, 257–281 (1967)

RESENDE, F., VIANA, M. J.: Gibberellin and sex expression. Portugaliae Acta Biol., Ser. A **6**, 77–78 (1959)

RETIG, B.: Hybridization methods in chickpeas (*Cicer arietinum* L.). II. Crossing without emasculation. Israel J. Agr. Res. **21**, 113 (1971)

RHOADES, M. M.: The cytoplasmic inheritance of male sterility in *Zea mays*. J. Genet. **27**, 71–93 (1933)

RHOADES, M. M.: Gene induced mutation of heritable cytoplasmic factor producing male sterility in maize. Proc. Nat. Acad. Sci. **36**, 634–635 (1950)

RICHEY, F. D.: Corn breeding. Advan. Genet. **3**, 159–192 (1950)

RICHEY, F. D., SPRAGUE, G. F.: Some factors affecting sex expression at the tassel of maize. Am. Naturalist **66**, 433–443 (1932)

RICK, C. M.: Field identification of genetically male sterile tomato plants for use in producing F_1 hybrid seed. Proc. Am. Soc. Hort. Sci. **46**, 277–283 (1945a)

RICK, C. M.: A survey of cytogenetic causes of unfruitfulness in the tomato. Genetics **30**, 347–360 (1945b)

RICK, C. M.: Genetics and development of nine male sterile tomato mutants. Hilgardia **18**, 599–633 (1948)

RICK, C. M.: Pollination relations of *Lycopersicon esculentum* in native and foreign regions. Evolution **5**, 110–122 (1950)

RICK, C. M., BUTLER, L.: Cytogenetics of the tomato. Advan. Genet. **8**, 267–382 (1956)

RICK, C. M., BOYNTON, J. E.: A temperature sensitive male sterile mutant vms. Tomato Genet. Coop. Rep. **15**, 52–53 (1965)

RICK, C. M., DEMPSEY, W. H.: Position of the stigma in relation to fruit setting of the tomato. Botan. Gaz. **130**, 180–186 (1969)

RICK, C. M., HANNA, G. C.: Determination of sex in *Asparagus officinalis*. Am. J. Botany **30**, 711–714 (1943)

ROATH, W. W., HOCKETT, E. A.: Genetic male sterility in barley. III. Pollen and anther characteristics. Crop Sci. **11**, 200–203 (1971)

ROBERTS, H. F.: Plant Hybridization before Mendel. New Jersey: Princeton Univ. Press 1929

ROBINSON, R. W., SHANNON, S., GUARDIA, M. D. DE LA: Regulation of sex expression in the cucumber. BioScience **19**, 141–142 (1969)

ROBINSON, R. W., WHITAKER, TH. W.: *Cucumis*. In: KING, R. C. (ed.). Handbook of Genetics **2**, 145–150. New York: Plenum Press 1974

ROBINSON, R. W., WHITAKER, TH. W., BOHN, G. W.: Promotion of pistillate flowering in *Cucurbita* by 2-chloroethylphosphonic acid. Euphytica **19**, 180–183 (1970)

RODRIGUEZ, B. P., LAMBETH, V. N.: Synergism and antagonism of GA and growth inhibitors on growth and sex expression in cucumber. J. Am. Soc. Hort. Sci. **97**, 90–92 (1972)

ROGGEN, H. P. J. R.: Scanning electron/microscopal observations on compatible and incompatible pollen-stigma interactions in *Brassica*. Euphytica **21**, 1–10 (1972)

ROLLINS, R. C.: The evolutionary fate of inbreeders and non sexuals. Am. Naturalist **101**, 343–351 (1967)

RONALD, W. G., ASCHER, P. D.: Transfer of self-compatibility from garden to greenhouse strains of *Chrysanthemum morifolium*. J. Am. Soc. Hort. Sci. **100**, 351–353 (1975 a)

RONALD, W. G., ASCHER, P. D.: Effects of high temperature treatments on seed yield and self-incompatibility in *Chrysanthemum*. Euphytica **24**, 317–322 (1975 b)

ROSA, J. T.: Sex expression in spinach. Hilgardia **1**, 259–275 (1925)

ROSA, J. T.: The inheritance of flower types in *Cucumis* and *Citrullus*. Hilgardia **3**, 233–250 (1928)

ROSENQUIST, C. E.: An improved method of producing F_1 hybrid seeds of wheat and barley. J. Am. Soc. Agron. **19**, 968–971 (1927)

ROSS, M. D.: Two genetic mechanisms governing outbreeding in *Plantago lanceolata*. Genetics Abstr. **56**, 584 (1967)

ROSS, M. D.: Evolution of dioecy from gynodioecy. Evolution **24**, 827–828 (1970)

ROSS, M. D., SHAW, R. F.: Maintenance of male sterility in plant populations. Heredity **26**, 1–8 (1971)

ROWELL, P. L., MILLER, D. G.: Induction of male sterility in wheat with 2-chloroethylphosphonic acid (Ethrel). Crop Sci. **11**, 629–631 (1971)

ROWLANDS, D. G.: The nature of the breeding system in the field bean (*Vicia faba* L.) and its relationship to breeding for yield. Heredity **12**, 113–126 (1958)

ROWLANDS, D. G.: Self-incompatibility in sexually propagated cultivated plants. Euphytica **13**, 157–162 (1964)

RUDICH, J., HALEVY, A. H., KEDAR, N.: Increase in femaleness of three cucurbits by treatment with Ethrel, an ethylene-releasing compound. Planta **86**, 69–76 (1969)

RUDICH, J., HALEVY, A. H., KEDAR, N.: Interaction of gibberellin and SADH on growth and sex expression of muskmelon. J. Am. Soc. Hort. Sci. **97**, 369–372 (1972)

RUDICH, J., HALEVY, A. H.: Involvement of abscisic acid in the regulation of sex expression in the cucumber. Plant Cell Physiol. **15**, 635–642 (1974)

RUDICH, J., KEDAR, N., HALEVY, A. H.: Changes in sex expression and possibilities for F_1 hybrid seed production in some cucurbits by application of Ethrel and Alar (B-995). Euphytica **19**, 47–53 (1970)

RUDORF, W., JOB, M.: Estudio sobre ecologia floral del trigo. Rev. Museo de la Plata **34**, 195–253 (1934)

RUHLAND, H. (ed.): Handbuch der Pflanzenphysiologie. 18: Sexualität, Fortpflanzung, Generationswechsel. Berlin, Heidelberg, New York: Springer 1967

RUSINOVA-KONDAREVA, I.: (Results of the investigation of pepper male sterile form.) (Bulgarian). In: Proc. Conf. Use of Male Sterility in Hybrid Seed Production. Bulgarian Acad. Sci., Sofia (1965), pp. 147–152

RUTISHAUSER, A.: Fortpflanzungsmodus und Meiose apomiktischer Blütenpflanzen. Protoplasmatologia. Wien: Springer 1967, Vol. VI, F, 3

RYDER, E. J.: An epistatistically controlled pollen sterile in lettuce (*Lactuca sativa* L.). Proc. Am. Soc. Hort. Sci. **83**, 585–589 (1963)

RYDER, E. J.: A recessive male sterility gene in lettuce (*Lactuca sativa* L.). Proc. Am. Soc. Hort. Sci. **91**, 366–368 (1967)

RYDER, E. J., JONSON, A. S.: Mist depollination of lettuce flowers. Hort Science **9**, 589 (1974)

RYLSKI, I.: Effect of night temperature on shape and size of sweet pepper (*Capsicum annuum*). J. Am. Soc. Hort. Sci. **98**, 149–152 (1973)

SAGE, G. C. M.: The inheritance of fertility restoration in male sterile wheat carrying cytoplasms from *Triticum timopheevi*. Theoret. Appl. Genet. **42**, 233–243 (1972)

SAGE, G. C. M., ISTURIZ, M. J. DE: The inheritance of anther extrusions in two spring wheat varieties. Theoret. Appl. Genet. **45**, 126–133 (1974)

SAITO, T., ITO, H.: Factors responsible for the sex expression of the cucumber plant. XIII Physiological factors associated with the sex expression of flowers. (2) Role of gibberellin. J. Jap. Soc. Hort. Sci. **32**, 278–290 (1963)

SAITO, T., ITO, H.: Factors responsible for the sex expression of the cucumber plant. XIV Auxin and gibberellin content in the stem apex and the sex pattern of flowers. Tohoku J. Agr. Res. **14**, 227–239 (1964)

SALAMAN, R. N.: Male sterility in potatoes: a dominant Mendelian character. J. Linn. Soc. (London) **39**, 301–312 (1910)

SAMPSON, D. R.: Intergeneric pollen stigma incompatibility in the Cruciferae. Can. J. Genet. Cytol. **4**, 38–49 (1962)

SAMPSON, D. R.: Close linkage of genes for male sterility and anthocyanin synthesis in *Brassica oleracea* promising for F_1 hybrid seed production: multivalents at meiosis not involved in linkage. Can. J. Genet. Cytol. **12**, 677–684 (1970)

SAND, S. A.: Autonomy of cytoplasmic male sterility on grafted scions of tobacco. Science **131**, 665 (1960)

SAND, S. A., CHRISTOFF, G. T.: Cytoplasmic-chromosomal interactions and altered differentiation in tobacco. J. Heredity **64**, 24–30 (1973)

SANGER, J. M., JACKSON, W. T.: Fine structure study of pollen development in *Haemanthus katherinae*. I. Formation of vegetative and generative cells. J. Cell Sci. **8**, 289–301 (1971)

SANTHANAM, V., SCRINIVASAN, K., RAJASEKHARAN, S.: Performance of hybrid cotton involving male sterile line. Curr. Sci. India **41**, 423–424 (1972)

SARVELLA, P., GROGAN, C. O.: Morphological variations at different stages of growth in normal, cytoplasmic male-sterile and restored versions of *Zea mays* L. Crop Sci. **5**, 235–238 (1965)

SASSEN, M. M. A.: Fine structure of petunia pollen grain and pollen tube. Acta Bot. Neerl. **13**, 175–181 (1964)

SAWHNEY, V. K.: Morphogenesis of the stamenless-2 mutant in tomato. III. Relative levels of gibberellins in the normal and mutant plants. J. Exp. Botany **25**, 1004–1009 (1974)

SAWHNEY, V. K., GREYSON, R. J.: Morphogenesis of the stamenless-2 mutant in tomato. I. Comparative description of the flowers and ontogeny of stamens in the normal and mutant plants. Am. J. Botany **60**, 514–523 (1973)

SCHAFFNER, J. H.: Observations and experiments on sex in plants. Bull. Torrey Botan. Club **62**, 387–401 (1935)

SCHERTZ, K. F., CLARK, L. E.: Controlling dehiscence with plastic bags for hand crosses in sorghum. Crop Sci. **7**, 540–542 (1967)

SCHOOLER, A. B.: A form of male sterility on barley hybrids. J. Heredity **58**, 206–211 (1967)

SCHÖTZ, F.: Beobachtungen zur Plastidenkonkurrenz bei Oenothera und Beiträge zum Problem der Plastidenvererbung. Planta **51**, 173–185 (1958)

SCHÜRHOFF, P. N.: Die Geschlechtsbegrenzte Vererbung der Kleistogamie by *Plantago* Sect. Novorbis. Ber. deut. botan. Gesell. **42**, 311–321 (1924)

SCHWARTZ, D.: The interaction of nuclear and cytoplasmic factors in the inheritance of male sterility in maize. Genetics **36**, 676–696 (1951)

SCHWEISGUTH, B.: Étude d'un nouveau type de stérilité mâle chez l'oignon *Allium cepa* L. Ann. Amel. Pl. **23**, 221–233 (1973)

SCHWEPPENHAUSER, M. A., MANN, T. J.: Restoration of staminal fertility in *Nicotiana* by introgression. Can. J. Gent. Cytol. **10**, 401–411 (1968)

SCOTT, D. H., RINER, M. E.: Inheritance of male sterility in winter squash. Proc. Am. Soc. Hort. Sci. **47**, 375–377 (1946)

SCOTT, G. E., FUTRELL, M. C.: Reaction of diallel crosses of maize in T and N cytoplasms to *Bipolaris* race T. Crop Sci. **15**, 779–782 (1975)

SCOTT, J. W., BAKER, L. R.: Inheritance of sex expression from the crosses of dioecious cucumber *(Cucumis sativus)*. J. Am. Soc. Hort. Sci. **100**, 457–461 (1975)

SEN, N. K., BHOWAL, J. G.: A male sterile mutant cowpea. J. Heredity **53**, 44–46 (1962)

SHIFRISS, C.: Additional spontaneous male-sterile mutant in *Capsicum annuum* L. Euphytica **22**, 527–529 (1973)

SHIFRISS, O.: Male sterilities and albino seedlings in cucurbits. J. Heredity **36**, 47–52 (1945)

SHIFRISS, O.: Sex instability in *Ricinus*. Genetics **41**, 265–280 (1956)

SHIFRISS, O.: Conventional and unconventional systems controlling sex variations in *Ricinus*. J. Genet. **57**, 361–388 (1960)

SHIFRISS, O.: Gibberellin as sex regulator in *Ricinus communis*. Science **133**, 2061–2062 (1961a)

SHIFRISS, O.: Sex control in cucumbers. J. Heredity **52**, 5–12 (1961b)

SHIFRISS, O., GALUN, E.: Sex expression in the cucumber. Proc. Am. Soc. Hort. Sci. **67**, 479–486 (1956)

SHIFRISS, O., GEORGE, W. L. Jr.: Sensitivity of female inbreds of *Cucumis sativus* to sex reversion by gibberellin. Science **143**, 1452–1453 (1964)

SHIFRISS, O., GEORGE, W. L. Jr., QUINONES, J. A.: Gynodioecism in cucumbers. Genetics **49**, 285–291 (1964)

SHINJYO, C.: Distribution of male sterility-inducing cytoplasms and fertility-restoring genes in rice. 1. Commercial lowland-rice cultivated in Japan. Jap. J. Genet. **47**, 237–243 (1972a)

SHINJYO, C.: Distributions of male sterility-inducing cytoplasms and fertility-restoring genes in rice. II. Varieties introduced from sixteen countries. Jap. J. Breed. **22**, 334–339 (1972b)

SHULL, G. H.: Inheritance of sex in *Lychnis*. Botan. Gaz. **52**, 110–125 (1910)

SHUMWAY, L. K., BAUMAN, L. F.: The effect of hot water treatment, X-ray irradiation and mesocotyl grafting on cytoplasmic male sterility of maize. Crop Sci. **6**, 341–342 (1966)

SILVOLAP, JN. M.: Amino acid composition of reproductive organs of plants with different types of male sterility (Russian). Sel. Skokhozioaistvennaia Biologiia **3**, 434–437 (1968)

SIMMONDS, N. W.: Variability in crop plants, its use and conservation. Biol. Rev. **37**, 422–465 (1962)

SIMPSON, D. M.: Natural cross pollination in cotton. Tech. Bull. U. S. Dept. Agr. 1094 (1954)

SIMPSON, D. M., DUNCAN, E. N.: Varietal response to natural crossing in cotton. Agron. J. **48**, 74–75 (1956)

SINGH, A., LAUGHNAN, J. R.: Instability of S male sterile cytoplasm in maize. Genetics **71**, 607–620 (1972)

SINGH, I. S.: Induced pollen sterility in *Petunia*. Mode of inheritance and tapetal behavior. Ann. Amel. Pl. **25**, 303–319 (1975)

SINGH, R. B., SMITH, B. W.: The mechanism of sex determination in *Rumex acetosella*. Theoret. Appl. Genet. **4**, 360–364 (1971)

SINGH, S. P., RHODES, A. M.: A morphological and cytological study of male sterility in *Cucurbita maxima*. Proc. Am. Soc. Hort. Sci. **78**, 375–378 (1961)

SINGH, T. P., MALHOTRA: Crossing technique in mung bean (*Phaseolus aureus* Roxb.). Curr. Sci. **44**, 64–65 (1975)

SINGLETON, W. R., JONES, D. F.: Heritable characters in maize. XXXV Male sterile. J. Heredity **21**, 266–268 (1930)

SINHA, S. K., KHANNA, R.: Physiological, biochemical and genetic basis of heterosis. Advan. Agron. **27**, 123–179 (1975)

SINK, K. C.: The inheritance of the apetalous mutant in the snapdragon. Proc. 17th Intern. Hort. Congr. I: 213 (1966)

SINK, K. C.: The inheritance of apetalous flower type in *Petunia hybrida* Vilm. and linkage tests with the genes for flower doubleness and grandiflora characters and use in hybrid seed production. Euphytica **22**, 520–526 (1973)

SLADKY, Z.: Role of growth regulators in differentiation processes in maize (*Zea mays*) organs. Biol. Plant. Praha **11**, 208–215 (1969)

SMITH, B. W.: Cytogeography and cytotaxonomic relationships of *Rumex paucifolium*. Am. J. Botany **55**, 673–683 (1968)

SMITH, D. C.: Pollination and seed formation in Grasses. J. Agr. Res. **68**, 79–95 (1944)

SMITH, H. H.: Recent cytogenetic studies in the genus *Nicotiana*. Advan. Genet. **14**, 1–54 (1968)

SMITH, H. H.: Model systems for somatic plant genetics. BioScience **24**, 269–276 (1974)

SMITH, O.: Relation of temperature to anthesis and blossom drop of the tomato together with a histological study of the pistil. J. Agr. Res. **44**, 183–190 (1932)

SNAYDON, R. W.: Rapid population differentiation in a mosaic environment. I. The response of *Anthoxanthum odoratum* populations to soils. Evolution **24**, 257–269 (1970)

SOPER, M. H. R.: A study of the principal factors affecting the establishment and development of the field bean (*Vicia faba*). J. Agr. Sci. **42**, 335–346 (1952)

SOPORY, S. K., MAHESHWARI, S. C.: Production of haploid embryos by anther culture technique in *Datura innoxia*—a further study. Phytomorphology **22**, 87–90 (1972)

SORRESSI, G. P.: Brown seed: a recessive genetic marker useful in tomato hybrid seed production. Rept. Meeting Eucarpia Tomato Working Group. Wageningen 24–26 (1968)

SPARKS, W. C., BLINKLEY, A. M.: Natural crossing in Sweet Spanish onions as related to distance and direction. Proc. Am. Soc. Hort. Sci. **47**, 320–322 (1946)

SPECTOR, W. S. (ed.): Handbook of Biological Data. Philadelphia: W. B. Saunders, Co. 1956, pp. 134–135

SPLITTSTOESSER, W. E.: Effects of 2-chloroethylphosphonic acid and gibberellic acid on sex expression and growth of pumpkins. Physiol. Plantarum **23**, 762–768 (1970)

SPRAGUE, G. F.: Heritable characters in maize: Vestigal glume. J. Heredity **30**, 143–145 (1939)

STANLEY, R. G., LINSKENS, H. F.: Pollen: Biology, Biochemistry and Management. Berlin: Springer 1974

STEBBINS, G. L.: Apomixis in the angiosperm. Botan. Rev. **7**, 507–542 (1941)

STEBBINS, G. L.: Self fertilization and population variability in the higher plants. Am. Naturalist **91**, 337–354 (1957)

STEBBINS, G. L.: Adaptive radiation of reproductive characteristics in angiosperms. I. Pollination mechanisms. A. Rev. Ecol. System. **1**, 307–326 (1970)

STEIN, H.: A gene for unfruitfulness in the castor bean plant and its utilization in hybrid seed production. Crop Sci. **5**, 90–93 (1965)

STERN, H., HOTTA, Y.: Biochemical control of meiosis. A. Rev. Genet. **7**, 37–66 (1974)

STEVENSON, T. M., KIRK, L. E.: Studies in interspecific crossing with *Melilotus* and in intergeneric crossing with *Melilotus medicago* and *Trigonella*. Sci. Agric. **15**, 580–589 (1935)

STIEGLITZ, H., STERN, H.: Regulation of β-1,3-glucanase activity in developing anthers of *Lilium*. Devel. Biol. **34**, 169–173 (1973)

STINSON, H. T.: Extranuclear barriers to interspecific hybridisation between *Oenothera hookeri* and *Oenothera argillicola*. Genetics **45**, 819–839 (1960)

STOKES, G. W.: Development of complete homozygotes of tobacco. Science **141**, 1185–1186 (1963)

STOREY, W. B.: Genetics of the papaya. J. Heredity **44**, 70–78 (1953)

STOREY, W. B.: Theory of the deviations of the unisexual flowers of Caricaceae. Agron. Trop. (Maracay) **17**, 273–321 (1967)

STOSKOPF, N. C., RAI, R. K.: Cross pollination in male sterile wheat in Ontario. Can. J. Plant Sci. **52**, 387–394 (1972)

STOUT, A. B.: Fertility in *Cichorium intybus*. The sporadic appearance of self-fertile plants among the progeny of self-sterile plants. Am. J. Botany **4**, 375–395 (1917)

STOUT, A. B.: Dichogamy in flowering plants. Bull. Torrey Bot. Club **55**, 141–153 (1933)

STRASBURGER, E.: Neue Untersuchungen über den Befruchtungsvorgang bei den Phanerogamen. Jena: Fischer 1884

STRASBURGER, E.: Versuche mit diöizischen Pflanzen in Rücksicht auf Geschlechtsverteilung. Biol. Zbl. **20**, 657–875 (1900)

STUBBE, W.: Genetische Analyse der Zusammenwirkung von Genom und Plastom bei *Oenothera*. Z. Vererb. Lehre **90**, 288–298 (1959)

STUBBE, W.: Untersuchungen zur genetischen Analyse des Plastoms von *Oenothera*. Z. Botan. **48**, 191–218 (1960)

STUBBE, W.: The role of the plastome in evolution of the genus *Oenothera*. Genetica **35**, 28–33 (1964)

SUNDERLAND, N.: Pollen and anther culture. In: Plant Tissue and Cell Culture. STREET, H. (ed.). Oxford: Blackwell Sci. Publ. 1973, pp. 205–234

SUNDERLAND, N.: Anther culture as a means of haploid induction. In: Haploids in Higher Plants. Advances and Potential. KASHU, K. J. (ed.), Univ. Guelph. 1974, pp. 91–122

SUNDERLAND, N., DUNWELL, J. M.: Pathways in pollen embryogenesis. In: Tissue Culture and Plant Science. STREET, H. (ed.) London, New York: Academic Press 1974, pp. 141–167

SUNESON, C. A.: Emasculation of wheat by chilling. J. Am. Soc. Agron. **29**, 247–249 (1937)

SUNESON, C. A.: Survival of four barley varieties in a mixture. Agron. J. **41**, 459–461 (1949)

SUNESON, C. A.: Male sterile facilitated synthetic hybrid barley. Agron. J. **43**, 234–236 (1951)

SUNESON, C. A.: An evolutionary plant breeding method. Agron. J. **48**, 188–191 (1956)

SUNESON, C. A.: Genetic diversity—a protection against plant disease and insects. Agron. J. **52**, 317–319 (1960)

SUNESON, C. A., WIEBE, G. A.: Survival of barley and wheat varieties in mixtures. J. Am. Soc. Agron. **34**, 1052–1056 (1942)

SYNGE, A. D.: Pollen collection by honeybees (*Apis mellifera*). J. Anim. Ecol. **16**, 122–138 (1947)

TAKEBE, I., LABIB, G., MELCHERS, G.: Regeneration of whole plants from isolated mesophyll protoplasts of tobacco. Naturwissenschaften **58**, 318–320 (1971)

TALIAFERRO, C. M., BASHAW, C.: Inheritance and control of obligate apomixis in breeding buffelgrass, *Pennisetum ciliare*. Crop Sci. **6**, 473–476 (1966)

TAMMES, P. M. L., WHITEHEAD, R. A.: Coconut (*Cocus nucifera* L.). In: Outlines of perennial crop breeding in the tropics. FERWERDA, F. P. (ed.) Landbouwhoogesch. Wageningen. Misc. papers **4**, 175–188 (1969)

TATUM, L. A.: The southern corn leaf blight epidemic. Science **171**, 1113–1116 (1971)

TAUBER, H.: Differential pollen dispersion and filtration. In: Quarternary Paleoecology. CUSHING, E. J., WRIGHT, H. E. (eds.). New Haven, Conn.: Yale Univ. Press 1967, pp. 131–141

THEURER, J. C., HECKER, R. J., OTTLEY, E. H.: Attempted graft transmission of cytoplasmic male sterility in sugar beets (*Beta vulgaris* L.) J. Am. Soc. Sugar Beet Tech. **14**, 695–703 (1968)

THI DIEN, N., TRAN THANH VAN, M.: Differentiation *in vitro* et *de novo* d'organes floraux directement à partir des couches mines de cellules type épidermique de *Nicotiana tabacum*. Étude au niveau cellulaire. Can. J. Botany **52**, 2319–2322 (1974)

THOMAS, E., HOFFMANN, F., WENZEL, G.: Haploid plantlets from microspores of rye. Z. Pflanzenzücht. **75**, 106–113 (1975)

THOMAS, E., WENZEL, G.: Embryogenesis from microspores of rye. Naturwissenschaften **62**, 40–41 (1975 a)

THOMAS, E., WENZEL, G.: Embryogenesis from microspores of *Brassica napus*. Z. Pflanzenzücht. **74**, 77–81 (1975 b)

THOMPSON, A. E.: Methods of producing first generation hybrid seed in spinach. Mem. Cornell Univ. Agr. Exp. Stn. **336**, 1–48 (1955)

THOMPSON, K. F.: Breeding better kales. Agriculture (London) **65**, 487–491 (1959)

THOMPSON, K. F.: Triple cross hybrid kale. Euphytica **13**, 173–177 (1964)

THOMPSON, K. F.: Classified S-alleles for *Brassica* breeders. Brassica Meeting of Eucarpia (1968) DIXON, G. E. (ed.). N. V. R. S. Wellesbourne, Warwick 1968, pp. 25–28

THOMPSON, K. F.: Cytoplasmic male-sterility in oil-seed rape. Heredity **29**, 253–257 (1972)

THOMPSON, K. F., HOWARD, H. W.: Self-incompatibility in marrow-stem kale, *Brassica oleracea* var. *acephala*. II. Recognition of plants homozygous for S alleles. J. Genet. **56**, 325–340 (1959)

THOMPSON, T. E., AXTELL, J. D.: An investigation of the graft transfer-ability of cytoplasmic male sterility in tetraploid alfalfa. Agron. Abstr., 65th Ann. Meet. Am. Soc. Agron., Las Vegas 1973, p. 15

TKACHENKO, N. N.: Preliminary results of a genetic investigation of the cucumber, *Cucumis sativus*. Bull. Appl. Biol. Plant Breed. (USSR) **9**, 311–356 (1935)

TOKUMASU, S.: Expression of male sterility in *Pelargonium crispum* l'Her ex Ait. Euphytica **23**, 209–217 (1974)

TOURNOIS, J.: Anomales florales du houblon japonais et du chauvre déterminées par des semis hâtifs. Comp. Rend. Acad. Sci. Paris **153**, 1017–1020 (1911)

TOWNSEND, G. F., RIDDELL, R. T., SMITH, M. V.: The use of pollen insects for tree fruit pollination. Can. J. Plant Sci. **38**, 39–44 (1958)

TSUNEWAKI, K., ENDO, R. T., MUKAI, Y.: Further discovery of alien cytoplasms inducing haploids and twins in common wheat. Theoret. Appl. Genet. **45**, 104–109 (1974)

TYSDAL, H. M., GARL, J. R.: A new method for alfalfa emasculation. J. Am. Soc. Agron. **32**, 405–407 (1940)

ULLSTRUP, A. J.: The impacts of the southern corn leaf blight epidemics of 1970–1971. Ann. Rev. Phytopathol. **10**, 37–50 (1972)

UPHOF, J. C. TH.: Cleistogamous flowers. Botan. Rev. **4**, 21–49 (1938)

UPHOF, J. C. TH.: Dictionary of Economic Plants. Lehre: J. Cramer 1968

VALDEYRON, G.: On the development of incompatibility and sex systems in higher plants and their evolutive meaning. Symp. Biol. Hung. **12**, 83–91 (1972)

VALDEYRON, G., ASSONAD, W., DOMMEE, B.: Coexistence des déterminismes géniques et cytoplasmiques de la stérilité mâle. Recherche d'une hypothèse explicative. In: La stérilité mâle chez les plantes horticoles. Eucarpia Meeting Versailles, 1970, pp. 175–186

VALLEAU, W. W.: Seed transmission and sterility studies of two strains of tobacco ringspot. Bull. Kentucky Agr. Exp. Stn. **327**, 43–80 (1932)

VASEK, F. C.: Outcrossing in natural populations: A comparison of outcrossing estimation methods. In: Evolution and Environment. DRAKE, E. T. (ed.) New Haven, Conn.: Yale Univ. Press 1968, pp. 369–385.

VASIL, I. K.: Physiology and cytology of anther development. Biol. Rev. **4**, 327–373 (1967)

VASIL, I. K., JOHRI, M. M.: The style, stigma and pollen tube. Phytomorphology **14**, 352–369 (1964)

VIRMANI, S. S., ATHWAL, D. S.: Genetic variability in floral characteristics influencing outcrossing in *Oryza sativa* L. Crop Sci. **13**, 66–67 (1973)

VIRMANI, S. S., ATHWAL, D. S.: Inheritance of floral characteristics influencing outcrossing in rice. Crop Sci. **14**, 350–353 (1974)

VIS, E. G., PERKINS, R. M., BROWN, G. K.: Mechanical pollination experiments with the Deglet Noor date palm in 1970. Date Growers Inst. Rep. **48**, 19–22 (1971)

VISSER, T.: Germination and storage of pollen. Landbouwhoogesch. Wageningen **55**, 1–68 (1955)

VRIES, A. PH. DE: Flowering biology of wheat, particularly in view of hybrid seed production—a review. Euphytica **20**, 152–170 (1971)

VRIES, A. PH. DE: Some aspects of cross-pollination in wheat (*Triticum aestivum* L.). 1. Pollen concentration in the field as influenced by variety, diurnal pattern, weather condition and level as compared to the height of the pollen donor. Euphytica **21**, 185–203 (1972)

VRIES, A. PH. DE: Some aspects of cross pollination in wheat (*Triticum aestivum* L.). 2. Anther extrusion and ear and flowering pattern and duration. Euphytica **22**, 445–456 (1973)

VRIES, A. PH. DE: Some aspects of cross-pollination in wheat (*Triticum aestivum* L.). 4. Seed set on male sterile plants as influenced by distance from the pollen source, pollinator: male sterile ratio and width of the male sterile strip. Euphytica **23**, 601–622 (1974)

VRIES, A. PH. DE, IE, T. S.: Electron microscopy of anther tissue and pollen of male sterile and fertile wheat (*Triticum aestivum* L.). Euphytica **19**, 103–120 (1970)

VUILLEUMIER, B. S.: The origin and evolutionary development of heterostyly in Angiosperms. Evolution **21**, 210–226 (1967)

WALLES, B.: Plastid inheritance and mutations. In: Structure and Function of Chloroplasts. GIBBS, M. (ed.) Berlin: Springer 1971, pp. 51–88

WARDLOW, C. W.: The floral meristem as a reaction system. Proc. Roy. Soc. Ser. B. Edinb. **66**, 394–408 (1957)

WARMKE, H. E.: Seed determination and sex balance in *Melandrium*. Am. J. Botan. **33**, 640–660 (1946)

WARMKE, H. E., BLAKESLEE, A. F.: Sex mechanisms in polyploid *Melandrium*. Science **89**, 391–392 (1939)

WARMKE, H. E., BLAKESLEE, A. F.: The establishment of a 4n dioecious race in *Melandrium*. Am. J. Botany **27**, 751–762 (1940)

WARMKE, H. E., OVERMAN, M. A.: Cytoplasmic male sterility in *Sorghum*. I. Callose behavior in fertile and sterile anthers. J. Heredity **63**, 103–108 (1972)

WASHINGTON, Q. J., MAAN, S. S.: Disease reaction of wheat with alien cytoplasms. Crop Sci. **14**, 903–905 (1974)

WASTRUD, L. D., HOOKER, A. L., KOEPPE, D. E.: The effects of nuclear restorer genes of Texas male-sterile cytoplasm on host response to *Helminthosporium maydis* Race T. Phytopathology **65**, 178–182 (1975)

WATERKEYN, L.: Callose microsporocytaire et callose pollinique. In: Pollen Physiology and Fertilization. LINSKENS, H. F. (ed.) Amsterdam: North Holland Publ. Co. 1964, pp. 52–58

WATTS, V. M.: A marked male sterile mutant in watermelon. Proc. Am. Soc. Hort. Sci. **81**, 498–505 (1962)

WEATHERWAX, P.: Cleistogamy in *Poa chapmaniana*. Torreya **29**, 123–124 (1929)

WEAVER, J. R.: Analysis of a genetic double recessive completely male sterile cotton. Crop Sci. **8**, 597–600 (1968)

WEBER, H.: Vegetative Fortpflanzung bei Spermatophyten. In: Handbuch der Pflanzenphysiologie. RUHLAND, W. (ed.). Berlin: Springer 1967, Vol. XVIII, pp. 787–804

WEIER, T. E., STOCKING, C. R., BARBOUR, M. G.: Botany: An Introduction to Plant Biology. London: John Wiley and Sons 1974

WEISS, E. A.: Castor, Sesame and Safflower. London: Leonard Hill 1971

WELLS, D. G., CAFFEY, H. R.: Scissor emasculation of wheat and barley. Agron. J. **48**, 496–499 (1956)

WELZEL, G.: Entwicklungsgeschichtlich-genetische Untersuchungen an pollensterilen Mutanten von Petunien. Z. Ind. Abst. Vererb. **86**, 35–53 (1954)

WENT, J. L. VAN: The ultrastructure of *Impatiens* pollen. In: Fertilization in Higher Plants. LINSKENS, H. F. (ed.). Amsterdam: North Holland Publ. Co. 1974, pp. 81–88

WESTERGAARD, M.: Studies on the cytology and sex determination in polyploid forms of *Melandrium album*. Dansk. Bot. Arkiv **10**, 1–131 (1940)

WESTERGAARD, M.: Aberrant Y-chromosomes and sex expression in *Melandrium album*. Hereditas **32**, 419–443 (1946)

WESTERGAARD, M.: The mechanism of sex determination in dioecious flowering plants. Advan. Genet. **9**, 217–281 (1958)

WESTON, E. W.: Changes in sex in hop caused by plant growth substances. Nature (London) **188**, 81–82 (1960)

WHELAN, E. D. P.: A cytogenetic study of a radiation induced male sterile mutant of cucumber. J. Am. Soc. Hort. Sci. **97**, 506–508 (1972)

WHELAN, E. D. P.: Linkage of male sterility, glabrate seedling and determinate habit in cucumber. Hort Science **9**, 576–577 (1974)

WHITAKER, T. W.: Sex ratio and sex expression in the cultivated cucurbits. Am. J. Botany **18**, 359–366 (1931)

WHITAKER, T. W., DAVIS, G. N.: Cucurbits, Botany, Cultivation and Utilization. London: Leonard Hill 1962, 249 pp.

WHITEHEAD, D. R.: Wind pollination in the angiosperms: Evolutionary and environmental considerations. Evolution **23**, 28–35 (1969)

WHITEHOUSE, H. L. K.: Multiple-allelomorph incompatibility of pollen and style in the evolution of the angiosperms. Ann. Botan. **14**, 198–216 (1950)

WIEBE, G. A.: A proposal for hybrid barley. Agron. J. **52**, 181–182 (1960)

WIEBE, G. A.: Use of genic male sterility in hybrid barley breeding. Proc. 12th Intern. Congr. Genet. Tokyo **2**, 234 (1968)

WIERING, D.: The use of insects for pollinating *Brassica* crops in small isolation cages. Euphytica **13**, 24–28 (1969)

WILLIAMS, I. H., FREE, J. B.: The pollination of onion (*Allium cepa* L.) to produce hybrid seed. J. Appl. Ecol. **11**, 409–417 (1974)

WINGE, Ø.: On sex chromosomes, sex determination and preponderance of females in some dioecious plants. Comp. Rend. Trav. Lab. Carlsberg **15**, 1–26 (1923)

WIT, F.: Chemically induced male sterility, a new tool in plant breeding? Euphytica **9**, 1–9 (1960)

WITSCH, H. VON: Die Entwicklung männlicher und weiblicher Inflorescenzen by *Xanthium* unter Einfluss von Tageslänge und Gibberellin. Planta **57**, 357–369 (1961)

WITTWER, S. H.: Maximum production capacity of food crops. BioScience **24**, 216–224 (1974)

WITTWER, S. H., BUKOVAC, M. J.: The effects of gibberellin on economic crops. Econ. Botan. **12**, 213–255 (1958)

WITTWER, S. H., BUKOVAC, M. J.: Staminate flower formation on gynoecious cucumber as influenced by the various gibberellins. Naturwissenschaften **49**, 305–306 (1962)

WOLFENBARGER, D. O.: Dispersion of small organisms. III. Pollen and seeds. In: Biology Data Book. ALTMANN, P. L., DITTMER, O. S. (eds.). Bethesda, Md.: Federation Am. Soc. Exp. Biol. 1973, Vol. II, pp. 885–888

WOO, S. C., SU, H. Y., NG, C. M., TANG, L. Y.: Seed formation on induced haploid plants and cytology of anther callus from hybrid rice. Botan. Bull. Acad. Sinica **14**, 61–64 (1973)

WOO, S. C., TANG, L. Y.: Induction of rice plants from hybrid anthers of *indica* and *japonica* cross. Botan. Bull. Acad. Sinica **13**, 67–69 (1972)

WORKMAN, P. L., ALLARD, R. W.: Population studies in predominantly self-pollinated species. V. Analysis of differential and random viabilities in mixtures of competing pure lines. Heredity **19**, 181–189 (1964)

WRICKE, G. V.: Untersuchung zur Vererbung des Geschlechts bei *Asparagus officinalis*. Z. Pflanzenzücht. **60**, 201–211 (1968)

WRICKE, G. V.: Untersuchung zur Vererbung des Geschlechts bei *Asparagus officinalis*. Z. Pflanzenzücht. **70**, 91–98 (1973)

WRIGHT, J. W.: Genetics of Forest Tree Improvement. F. A. O. For. Forest Prod. Stud. **16** (1962)

WRIGHT, S.: Isolation by distance under diverse systems of mating. Genetics **31**, 39–59 (1946)

YAMADA, I.: The sex chromosomes of *Cannabis sativa*. Rep. Kihara Inst. Biol. Res. **2**, 64–68 (1943)

YAMAMOTO, Y.: Karyogenetische Untersuchungen bei der Gattung *Rumex*. Mem. Coll. Agr. Kyoto Imp. Univ. **43** (1938)

YAMPOLSKY, C., YAMPOLSKY, H.: Distribution of sex forms in the phanerogamic flora. Biblioth. Genet. Leipzig **3**, 1–62 (1922)

ZENKTELER, M.: *In vitro* development of embryos and seedlings from grains of *Solanum dulcamara*. Z. Pflanzenphysiol. **69**, 189–192 (1973)

ZEVEN, A. C.: Transfer and inactivation of male sterility and source of restorer genes in wheat. Euphytica **16**, 183–189 (1967)

ZEVEN, A. C.: Inheritance of functional male sterility in *Streptocarpus* Constant Nymph and its mutants. Euphytica **21**, 265–270 (1972)

ZUK, J.: An investigation on polyploidy and sex determination within the genus *Rumex*. Acta Soc. Botan. Pol. **32**, 5–67 (1963)

Subject Index

The reader is referred to the detailed table of contents to locate pages for topical terms; these are not indexed.

Terms which occur throughout the book with high frequency, or are of an unspecific or general nature, have not been indexed. In certain cases reference is given only to pages where the term is defined or is of special importance in the context.

Abies spp. 27, 34, 103
Abscisic acid 150, 154
Abutilon spp. 187
Acer spp. 33
  negundo 17
Acetylene 150
Aegilops spp. 88, 202, 225
  caudata 202
  ovata 202, 208
  speltoides 202
Aegilotricum 216
Agamospermy 15, 16
Agave spp. 15, 16, 31, 36
Ageratum spp. 226
Agropyron spp. 53, 88
  cristatum 24
  desertorum 24
  elongatum 24
  intermedium 24
  repens 24
  smithii 24
  trachycaulum 18, 59
  trichophorum 24
Agrostemma githago 182
Agrostis alba 24
Aleurites montana 24
Alfalfa see *Medicago sativa*
Alkaloid 93
Allele, fertility restoring 200—201, 204, 206, 222
  male sterility 199—200
  mutaplasmic 201
  plasmon sensitive 199—203, 204, 214
Allergen 101
Allium spp. 14, 15, 98
  cepa 3, 16, 26, 31, 32, 33, 39, 40, 41, 69, 75, 201, 209, 216, 223, 226, 233
    var. *viviparum* 16
  fistulosum 16
  porrum 26, 39
  sativum 16
  schoenoprasum 26

Allogamy, abiotic 54
  biotic 54
Alloplasm 233
Almond see *Prunus* spp.
Amaryllidaceae 173, 174
Ambrosia spp. 34, 101
Amino acids, free 220
Anacardiaceae 22
Ananas comosus 3, 16, 22, 31
Anaphase 82, 214
Anchusa hybrida 174
Androdioecy see Dioecy, Andro
Androecious flower 11
  plant 11, 12
Androecium 7, 135, 136, 211
Androgenesis 15, 88, 89, 92, 93, 206
Androgynomonoecy 16
Andromonoecy 11, 12, 16
Andropogon furcatus 25
  scoparius 25
Anemophily 34—35, 52, 54
Annonaceae 22
Annona cherimola 22, 38
Anther culture 85—88
  dehiscence 35, 37, 38, 211, 217
  extrusion 59, 233
  less 213
  petalloid 211
Antheridal cell 104
Anthophily 31, 52
Antigen A 101
  pollen 184
  stigmatic 182
Antipodal cell 100
Antirrhinum spp. 68, 85, 167, 205, 210, 212, 226
Ants 33
Aphids 34
Apium graveolens 21, 39
Apomixis 5, 15
Apospory 15
Apple see *Pyrus* spp.

Approach crossing 75, 76
Apricot see *Prunus armeniaca*
Aquilegia vulgaris 199, 226
Arabidopsis 88
    thaliana 59, 61
Araceae 127
Arachis hypogaea 19, 59
Archegonium 104
Archesporal cell 83, 97
    tissue 97
Archesporium 79
Arisaema japonica 12, 127
Armeria maritima 198
Arrhenatherum avenaceum 25
Artichoke, Globe see *Cynara scolymus*
    Jerusalem see
        Helianthus tuberosus
Artocarpus communis 52
    heterophyllus 52
Ash see *Fraxinus* spp.
Asparagine 220
Asparagus see *Asparagus officinalis*
Asparagus 88, 89
    officinalis 17, 122
Aster, China see *Callistephus chinensis*
Asynapsis 214
Atriplex spp. 17
Atropa spp. 88, 89
    belladonna 28
Autogamy 13
    facultative 54
    functional 59—61
    habitual 54
Autoplasmy 201—203, 233
Avena spp. 5, 18, 39, 55, 59
Avocado see *Persea* spp.

Bacula 84
Bamboo see *Bambusa* spp.
Bambusa spp. 36
Banana see *Musa* spp.
Barley see *Hordeum vulgare*
    foxtail see *Hordeum jubatum*
Bean, broad see *Vicia faba*
    common see *Phaseolus vulgaris*
    lima see *Phaseolus lunatus*
    mung see *Phaseolus aureus*
    urd see *Phaseolus mungo*
Bees 33
Beet, fodder see *Beta vulgaris*
    red see *Beta vulgaris*
    sugar see *Beta vulgaris*
Beetles 74
Begonia spp. 129
    cheimantha 129
    semperflorens 127, 128, 210, 226, 234

Belladonna see *Atropa belladonna*
Bellis perennis 18
Benzyladenine 139
6-(benzylamino)-4-(2-tetra-hydropyranyl)-
    9H purine (SD 8339) 140
6-benzylamino-9-β-D-ribofuranosylpurine
    (6-BAR) 140
Bermuda grass see *Cynodon dactylon*
Beta spp. 180
    vulgaris 22, 34, 39, 40, 68, 172, 206, 216,
        223, 226, 233, 234
Betula spp. 34
    odorata 83
Birch see *Betula* spp.
Bisexuality 6
Bittersweet, American see
    Celastrus scandens
Blackberry see *Rubus* spp.
Black medic see *Medicago lupulina*
Blastophaga 30, 33
"Blindness" 225
Blue grass see *Poa* spp. Texas see *Poa arachnifera*
Blue stem see *Andropogon* spp.
"Bolting" 3
Boraginaceae 174
Border rows 48, 49
Bougainvillea spp. 184
Bouteloua curtipendula 25
    gracilis 25
Brassica spp. 3, 16, 23, 33, 37, 38, 39, 75,
    88, 89, 95, 101, 177, 186, 187, 188, 189,
    191, 195, 229
    campestris 62
    nigra 214
    oleracea 176, 188, 192, 214
        var. *gemmifera* 102, 190, 192
        acephala 190, 193
    rapa 192
Breadfruit tree see *Artocarpus communis*
Broccoli see *Brassica oleracea*
Brome grass see *Bromus* spp.
    soft see *Bromus mollis*
Bromeliaceae 22, 168
Bromus spp. 53
    carinatus 57, 59
    cartharticus 18, 57
    erectus 25
    inermis 25
    marginatus 18
    mollis 18
    pumpellianus 25
Brussels sprouts see *B. oleracea* var. *gemmifera*
Bryonia alba 105
    dioica 105, 141
Bryophyllum spp. 14, 132, 139, 140

Buchloe dactyloides   17
Buckthorn, sea see *Hippophae
    rhamnoides*
"Bud blasting"   225
Buffalo grass see *Buchloe dactyloides*

Cabbage see *Brassica oleracea*
    chinese see *Brassica oleracea*
Cacao see *Theobroma cacao*
Cactaceae   97
Cajanus cajan   59
    indicus   26
Caladium spp.   14
California brome see *Bromus carinatus*
Callase   220
Callistephus chinensis   18, 31, 59
Callose   82, 83, 180, 182, 214, 217, 218, 220
Callus   86, 87
Camellia sinensis   28
Canarium ovatum   17
Canary grass, red see *Phalaris arundinacea*
Cannabis indica   158
    sativa   8, 9, 17, 38, 52, 120, 123, 127,
    130, 132, 136, 139, 157—159, 210
Cantaloupe- see *Cucumis melo*
Cantharophily   31, 52
Capparis spp.   16
Capsicum spp.   88, 209, 216
    annuum   21, 33, 37, 61, 63, 74, 75, 76,
    199, 203, 205, 210, 216, 223, 224, 226, 232
    frutescens   21, 61, 134, 203
    pendulum   203
    peruvianum   203
Carbon monoxide (CO)   135, 189
Cardamine spp.   14
    argentatum   17
Carica papaya   17, 39, 41, 111, 125, 132,
    133, 136
Carnation see *Dianthus caryophyllus*
Carob see *Ceratonia siliqua*
Carolina see *Lithospermum caroliniense*
Carpel   96
Carpelloidy   211, 212, 213
Carpillate flower   11
Carrot see *Daucus carota*
Carthamus tinctoria   22, 32
Carya pecan   25, 34, 40
Caryophyllaceae   113, 182
Cassava see *Manihot esculenta*
Castanea spp.   24, 33
Castor bean see *Ricinus communis*
Catalpa spp.   33
Cauliflower see *Brassica* spp.
Ceiba pentandra   31, 33, 37, 39
Celastrus scandens   17
Celery see *Apium graveolens*
Cells, mixed   222
Centaurea cyanus   16

Ceratonia siligua   17, 197
Ceratostigma spp.   173
Chalaza   100
Chasmogamy   13, 56
Chenopodiaceae   22, 171
Chenopodium guinoa   22, 39
Cherimoya see *Annona cherimola*
Cherry see *Prunus* spp.
    avium
Chestnut see *Castanea* spp.
Chiasmata   110
Chichorium endivia   18, 59
    intybus   23
Chichory see *Cichorium intybus*
Chickpea see *Cicer arietinum*
Chimera   89
Chiroptera   31, 37, 39
Chive see *Allium schoenoprasum*
Chloris gayana   18
2-chloroethylphosphonic acid (Ethephon,
    Ethrel)   69, 135, 136, 137, 150, 151, 156,
    159
(2-chloroethyl)trimethylammonium chloride
    (CCC)   149
2-chlorophenyl(thio)proprionic acid   136
Christmas rose, see *Helleborus*
Chromosome, alien   230, 231
    hybridity of segments   51, 53
    numbers   51, 53
Chrysanthemum spp.   23
    morifolium   188
Cicer arietinum   16, 19, 59, 69
Cincinnus   128
Cirsium oleraceum   199
Citrullus vulgaris   23, 39, 40, 137, 141, 145,
    147, 151, 205, 216
Citrus spp.   3, 15, 16, 21, 41,
Cleistogamy   13, 233
    constitutional   57
    ecological   54, 56
    genetical   54
Cleome spp.   118, 128, 141
    iberidella   128
    spinosa   136
Clover see *Trifolium*
    bur see *Medicago hispida*
    sweet see *Melilotus* spp.
Cocoa, see *Theobroma cacao*
Coco palm see *Cocos nucifera*
Cocos nucifera   27, 38
Coffea arabica   28, 37
Coffee see *Coffea arabica*
Colchicine   86, 93, 114, 186
Colchicum   14
Coleoptera   31, 32
Coleus spp.   205
Collard see *Brassica* spp.
Columbine see *Aquilegia vulgaris*

Colza see *Brassica* spp.
Commelinaceae   168
Compatibility   22—28
  barriers   43
  pseudo   170, 187, 189
Competition between pure lines   55
  pollen   62
Compositae   18, 22, 23, 31, 98, 176, 180, 182,
  188, 191
Coniferopsida (Coniferales, Coniferae,
  conifers)   103, 104
Conifers see Coniferopsida
Convolvulaceae   23, 176
Corm (in Araceae)   127
Corn see *Zea mays*
Cornflower see *Centaurea cyanus*
Corrola, split   212
Corylaceae   23
Corylus spp.   23, 34, 35
Cosmos binnatus   85
Cotton see *Gossypium* spp.
Cotton, long staple see *Gossypium barbadense*
  Upland see *Gossypium hirsutum*
Crambe maritima   23
Cranberry see *Vaccinium* spp.
Crepis spp.   85
  foetida   175
Cress, garden see *Lepidium sativum*
  thale see *Arabidopsis thaliana*
Crosses, composite   204
  nuclear substitution   206, 207
  restoration   207
Crotolaria spp.   19, 223, 226
Cruciferae   33, 40, 88, 98, 176, 180, 182,
  190, 191, 192, 209, 216
Crucifers see Cruciferae
Cucumber see *Cucumis sativus*
Cucumis spp.   200, 216
  aculeatus   142
  africanus   142
  anguria   142, 145
  callosus   142
  dipsaceus   142
  ficifolius   142
  figarei   142
  flexuosis (= C. sativus)
  globosus   142
  hardwikii   142, 145
  heptadactylus   142
  hirsutus   142
  humifructus   142
  hystrix   142
  kalahariensis   142
  meeusei   142
  melo   23, 33, 124, 137, 138, 141—147,
  216
  metuliferus   142

  muriculatus   142
  myriocarpus   142
  prophetarum   142
  quintanilhae   142
  rigidus   142
  sacleuxii   142
  sagittatus   142
  sativus   3, 8, 9, 12, 23, 41, 105, 117, 118,
  120, 123, 129, 133, 135, 138, 139, 141—
  157, 210, 217
  sinensis (= C. sativus)
  zeyheri   142
Cucurbita spp.   24, 38, 39, 210, 215, 217
  maxima   156, 200
  mixta   156
  moschata   156
  pepo   141—157, 200
Cucurbitaceae   23, 24, 31, 32, 36, 38, 40,
  138, 140—157, 229
Cucurbits, see Cucurbitaceae
Culture, meristem   14
  tissue   14
Currant, black see *Ribes* spp.
Cutinase   180, 182
Cynara scolymus   16
Cynodon dactylon   25
Cystine   220
Cytokinesis, simultaneous   214, 218
  successive   214, 218
Cytokinin   140
Cytomix   82
Cytoplasm, paternal   206

Dactylis spp.   216
  glomerata   25, 199, 226
Dahlia, garden see *Dahlia rosea*
Dahlia spp.   14
  rosea   23
Daisy see *Bellis perennis*
Daucus carota   28, 31, 32, 33, 39, 75, 216,
  226, 233
Date palm see *Phoenix dactylifera*
Datura spp.   88, 89
  innoxia   86, 87, 90, 91, 93
Deoxyribonucleic acid (DNA)   6, 83
Desynapsis   214
Development, floral   7
Diakinesis   214
Diallel cross   171, 177
Dianthus spp.   31, 39
  caryophyllus   14
Dichasium   128
2,3-dichloroisobutyrate (DCIB)   68
Dichogamy   37, 38, 41, 54
  flower   13
  plant   13
Diclinous flower   9, 117, 135, 141, 143,
  145, 151, 152, 160

Dicliny 114
Dictyosome 84
Differentiation, floral 7
  sexual 1
  of species 54
Dihydrozeatin 140
N,N,-dimethylaminosuccinamic acid
  (SADH, B_{995}, B_9) 148, 149, 156
Dimorphism, sexual 6
Dioecy 3, 5, 11, 14, 16, 17, 38, 52, 105,
  108, 109
  Andro 11, 16
  Gyno 11, 16, 196—199, 203, 204, 208
Dioscorea spp. 17, 111
Diospyros spp. 17, 41
Diploidy 6
Diplospory 15
Diplotene 214
Diptera 31
Dispersal agents see vectors
  pollen 51
  seed 44, 51, 53, 54
Distyly 173
Domestication of plants 54
Dominance relations 1, 177
Drosophila 112, 115, 118, 120

Ear (maize) 109, 160, 161, 163
Efficiency, pollination 3
  reproductive 2
Eggplant see *Solanum melongena*
Eichhornia spp. 73
Elaeis guineensis 27
Elder, box see *Acer negundo*
Elodea spp. 35
Elymus spp. 18, 53
Emasculation, chemical 68—69, 226
  genetic 69
  mechanical 62—66, 226
  scissor 66
Embryo-ny, adventitious 15
  sac 86, 96, 100, 102, 113, 165, 183
  id 86, 87, 90—94
Endive see *Cichorium endivia*
Endonuclease, restriction 223
Endoplasmic reticulum 100, 101
Endosperm 100
Endothecium 211, 216
Entomophily 31—34, 52, 186, 233
Epigynous flower 143, 144
Epilobium spp. 223
Episome 224
Epistasis 169, 170, 187
Eragrostis trichoides 18
Ethyl alcohol 68
Ethylene 135, 150, 159
Eucalyptus spp. 27, 31, 33
Eugenia 16

Euphorbiaceae 24
Exine 84, 101, 165, 180, 182

Fagaceae 24
Feijoa sellowiana 27
Fertility, hybrid 53
  restoration 187, 233
Fertilization, double 4
  selective 43, 62
Fescue see *Festuca* spp.
Fescue, annual see *Festuca microstachys*
Festuca spp. 16, 53, 88
  elatior 25
  microstachys 57
  rubra 25
Ficus spp. 30
  carica 17, 31, 41, 52, 75
  purmila 35
Fig see *Ficus carica*
Filiform apparatus 100, 102
Fir see *Abies* spp.
  Douglas see *Pseudotzuga menziessii*
Flax see *Linum usitatissimum*
Flies, blow 33
Floral meristem 7
Flower, apetalous 57
  calender 33
  color 30
  ephemeral 37
  marker 78
  marking 232
  modifications 13, 18—28
  persistence 37
  structure 8, 9
Flowering, continous 36
  gregarious 36
  in cucumber 142—145
  pattern 12, 126, 128, 141, 145
  period 37
  seasonal 36
Formicophily 31
Fraction 1 protein 223
Fragaria spp. 14, 16, 27, 33, 39, 113
Fraxinus spp. 17, 27, 34
Freesia spp. 24, 165
Funiculus 97

Gamete 10, 163
  diploid 214
  plasmon sensitive 207
Gametocide, chemical 66—69
  temperature as 69
Gametophyte, female 98
Garlic see *Allium sativum*
Garousse see *Lathyrus cicera*
Gasteria spp. 165
Gene balance system 197

Generation, gametophyte   5
  sporophyte   5
  time, sexual   53
Generative cell   83, 90, 91, 217, 222
  nucleus   90, 91, 102, 217
Generiaceae   88
Genital ridges   10
Genotype, fertility restored   206, 209
  plasmon sensitive   209
Geophyte   165
Gerbera spp.   3
  viridifolia   62
gibberellic acid (GA₃)   138, 139, 140, 148,
  150, 151, 159, 163
Gibberellin (GA)   69, 137—139, 148, 151,
  153, 154, 157, 159, 162, 210, 225
Gladiolus spp. 14, 25
Glucanase   218
Glutamine   220
Glycine max   19, 55, 59, 68, 75, 88, 199,
  204, 226
Glycoprotein   101, 180
Gonad   112
  primordium   10
Gooseberry see Ribes spp.
Gossypium spp.   37, 40, 49, 63, 64, 65, 68,
  74, 76, 88, 199, 210, 212, 226, 229, 234
  anomalum   203
  arboreum   20, 203
  barbadense   20, 61, 62
  harknessii   203
  herbaceum   20
  hirsutum   20, 61, 203, 205
Gourd see Cucurbita spp.
Graft(ing)   14
Grama see Bouteloua spp.
Gramineae   18, 24, 25, 88, 98, 107, 169, 170,
  180, 191
Grape see Vitis vinifera
Grape, muscadine see Vitis rotundifolia
Grasshoppers   34
Grass peavine see Lathyrus sativus
Grevillea robusta   27
Growth retardants   68, 148, 157
Guard-rows   49
Guava see Psidium guajava
Guayule see Parthenium argentatum
Gynodioecy see Dioecy, Gyno
Gynoecious flower   11
  plant   11, 12
Gynoecium   7, 96, 136, 160
Gynogenesis   15
Gynomonoecy see Monoecy, Gyno

Haemanthus katherina   84
Haploids   86, 87, 93

Hardy-Weinberg equilibrium   44, 45
Hazelnut (Filbert) see Corylus spp.
Helianthus annus   22, 32, 39, 62, 226
  tuberosus   23
Helleborus spp.   8
Helminthosporium maydis   222, 224
Hemerocallis   11, 16
Hemicellulose   100
Hemizygous plant   92, 93, 94, 95
Hemp see Cannabis sativa
Hercogamy   13, 54
"Heredity, infectious"   223, 224
Hermaphrodite flower   11
  plant   11, 12
  species   11, 16
Heterogametic plants   110–120
Heteromorphic flower   13
  chromosomes   110—120
Heteromorphy   38, 53, 57
Heterosis   225
Heterostyly   69, 173—175
Heterozygosity, advantage of   55
  residual   53, 55
Hevea brasiliensis   24, 219
Hibiscus cannabinus   20
  sabdariffa   20
Hippophae rhamnoides   17
Holly, European see Ilex aquifolium
Homogamy   13
Homomorphy, flower   13
Homozygosity   1
Hop see Humulus lupulus
Hordeum spp.   53
  bulbosum   5, 86
  jubatum   19
  vulgare   5, 18, 39, 55, 57, 59, 63, 65, 86,
  88, 130, 131, 133, 137, 199, 216, 217, 219,
  226, 227, 229
Huang-hua tsai, see Taraxacum officinale
Humidity, atmospheric relative   36
Humulus lupulus   17, 52, 123, 130, 136
  japonicus   123
Hybrid, double cross   233
  seed production   3, 62, 73—78, 95,
  122—125, 137, 139, 141, 147, 153—157,
  164, 184—185, 187, 189, 190—196, 225—
  234
  single cross   234
  triploid   234
Hybridity, adjustment of   54
  optimum   200
Hydrophily   34—36
Hymenoptera   31, 32
Hyoscyamus spp.   136
  niger   135
Hypericum spp.   206
  perforatum   222
Hypogynous flower   143

Ilex aquifolium   17
Immunofluorescence   100
Impatiens spp.   222
Inbreeding, coefficient   45—58
   depression   4
Incompatibility   3, 13, 17, 52, 95, 164—
   177, 180—191, 195, 198
Incongruity   164, 186
β-indoleacetic acid (IAA)   135, 136, 157
Integuments   96, 97
Interphase   82
Intine   84, 85, 89, 90, 91, 180
Ipomoea batatas   14, 16, 23, 40, 184
Iridaceae   25
Iris spp.   14, 25
Isogenic line   194
Isolation, spatial   2, 6
   standards   48, 49
   temporal   2
   of flowers   70
   of plants   70
   of seed corps   48
2-isopropyl-4-dimethylamine-5-methyl-
   phenyl-1-piperidine methyl chloride
   (AMO 1618)   149, 150
Isozyme   185

Jack fruit see *Artocarpus heterophyllus*
Jojoba bean see *Simmondsia californica*
Juglandaceae   25
Juglans regia   26, 34

Kale see *Brassica* spp.
   Sea see *Crambe maritima*
Kapok see *Ceiba pentandra*
Kenaf see *Hibiscus cannabinus*
Kohlrabi see *Brassica* spp.

Labiaceae   26
Lactuca sativa   18, 37, 59, 61, 66, 75, 191,
   199, 205, 226
Larch see *Larix* spp.
Larix spp.   27, 34
Lathyrus cicera   19
   odoratus   199, 215
   sativus   19
   tingitanus   19
Lauraceae   26
Lecitin   101
Leek see *Allium porrum*
Leguminosae   19, 20, 26, 167, 186, 190
Lemma   59
Lens culinaris   19
Lentil see *Lens culinaris*
Lepidium sativum   59
Lepidoptera   31
Leptotene   214

Lespedeza spp.   19
   stipulacea   57, 58
Lettuce see *Lactuca sativa*
Liliaceae   26, 88, 98, 167, 168, 183
Lilium spp.   26
   longiflorum   183
Lily see *Lilium* spp.
Limonium spp.   173, 174
   meyeri   180, 182
   vulgare   198
Linaceae   20, 173
Linden see *Tilia* spp.
Linum spp.   173, 174
   floccosum   203
   usitatissimum   20, 37, 39, 59, 61, 63, 66,
   74, 226, 233
Lithospermum caroliniense   57
Loculus   82
Lodicule   57, 59
Lolium spp.   88
   italicum   25
   perenne   25
   temulentum   19, 59
Lotus corniculatus   26
Love grass, sand see *Eragrostis trichoides*
Luffa acutangula   139, 148
Lupine, see *Lupinus*
Lupinus spp.   216
   albus   20
   luteus   20
   perennis   20
Lychnis spp.   114
Lycopersicon spp.   53, 88, 89, 167, 184, 189
   esculentum   3, 21, 36, 59, 61, 63, 66, 68,
   69, 74, 75, 86, 93, 134, 137, 139, 167, 186,
   199, 205, 210, 213, 215, 216, 217, 219,
   226, 227, 229, 232
   peruvianum   167, 168, 183, 186, 195
   pimpinellifolium   86
Lythraceae   173, 174
Lythrum spp.   173
   salicaria   174, 175
   junceum   175

Macadamia ternifolia   27
Magnolia spp.   33
Maintainer line   226
Maintenance   227, 228
   cultivar   2, 3
Maize see *Zea mays*
Male sterility   13, 41
   functional   204, 205, 211, 219, 229
   pleitrotic effects of   205, 209, 215, 224,
   225, 229
   positional   205
Maleic hydrazide   68, 150
Malus see *Pyrus* spp.

Malvaceae   20, 98
Mangifera indica   16, 22, 31
Mango see *Mangifera indica*
Manihot esculenta   24
Manioc see *Manihot esculanta*
Maple see *Acer* spp.
Marigold see *Tagetes erecta*
Marjoran see *Origanum vulgare*
Marker genes, seedling   69, 75, 229
Marrow see *Cucurbita* spp.
Marrow-stem Kale, see *Brassica oleracea, var. acephale*
Mating systems, origin of   52—54
Matthiola incana   210
Medicago spp.   53
    hispida   5, 20
    lupulina   20
    sativa   5, 26, 31, 32, 39, 68, 215, 226
Megachile rotundata   33
Megaspore   99
    mother cell   97, 98
Megasporogenesis   99
Meiocytes, autonomy of   220
Meiosis   5, 82, 83, 214—216
Melandrium   10, 112, 113, 115, 116, 117, 118, 119
    album   114, 115
    dioecum   114
Melilotus dentata   20
    indica   20
Melittophily   31—33
Melon see *Cucumis melo*
Mentha piperita   26
Mercurialis   135, 140
Messenger ribonucleic acid (m RNA)   101
Metaphase   82, 214
Methionine   93
    sulfoximide   93
Microgamete, vacuolate   214
Micropyle   97, 100, 102
Microspore amitosis   217
    differentiation   83
    mitosis   217
    tetrad   214
    vacuolate   214
Microsporogenesis   211
    gametogenesis   209—211
Millet, pearl see *Pennisetum glaucum*
    proso see *Panicum miliaceum*
Mirabilis galapa   222
Mitochondrion   91, 100, 221—224
Mitosis   5
    pollen   216, 217
Mitotic spindle   83
Monoclinous flowers   141, 143, 151, 160
Monoecy   11, 12, 13, 16, 38, 52
    Gyno   11, 12, 16
    Tri   11, 12

"Morph" (in heteromorphic incompatibility)   173, 174, 180
Morphactin   150, 157, 210
Morus spp.   52
    alba   17
    nigra   17
Moths   34
Mountain brome grass see *Bromus marginatus*
Mulberry see *Morus alba, M. nigra*
Multiline cultivars   55
Multiplication   2
    rate, seed   195
Musa spp.   31, 121, 134, 136
Muskmelon see *Cucumis melo*
Mustard see *Brassica* spp.
Mutant, male sterile   52, 55
Mutation, permanent   168
    reversible   168
Myrca gale   127
Myrtaceae   27

α-naphthaleneacetic acid (NAA)   136, 137, 157, 161, 162
Narcissus spp.   14, 165, 173, 184
    tazetta   174
Nectar   40
    guide marks   33
    ies   33
Nectarine see *Prunus persica*
Needle grass see *Stipa* spp.
Nemesia   167
Neognathae   31, 32
Nexine   84, 85
"Nicking"   36, 41, 44, 48, 57, 73, 232, 233
Nicotiana spp.   31, 53, 55, 59, 63, 79, 85, 87, 88, 89, 93, 167, 186, 202, 205, 213
    alata   167, 168
    bigelovii   203, 212, 213
    debnyi   203, 212, 213
    glutinosa   212—214
    megalosiphon   203, 212, 213, 225
    plumbaginifolia   203
    rustica   21, 203
    sanderae   165
    suaveoleus   203, 212, 213, 223, 225
    tabacum   21, 61, 86, 87, 93, 167, 199, 203, 210, 212—214, 223, 226
    undulata   203, 210, 212
Nucellus   96
Nucleus, restitution   215

Oak see *Quercus* spp.
    silk *Grevillea robusta*
Oat see *Avena* spp.
    grass, tall see *Arrhenatherum avenaceum*
Oenothera spp.   167, 183, 186, 187, 188, 206

hookeri 222
organensis 182
Oil palm see *Elaeis guineensis*
Olea europaea 27, 34, 133
Oleaceae 27
Olive see *Olea europaea*
Onagraceae 98, 167
Onion see *Allium cepa*
 top see *Allium cepa* var. *viviparum*
 Welch see *Allium fistulosum*
Onobrychis vicifolia 20
Opuntia spp. 16
Orange see *Citrus* spp.
Orchard grass see *Dactylis glomerata*
Origanum vulgare 26, 197, 205, 210
Ornithine 220
Ornithophily 33
Oryza sativa 18, 55, 56, 57, 59, 65, 66, 69, 73, 75, 87, 88, 89, 92, 139, 199, 203, 209, 210, 219, 226
Outbreeding mechanism 196—199
Outcrossing, estimation of 45—58
 promotion of 48, 49
 restriction of 49, 50
Ovary 96, 97
Ovule, external 211, 213
Oxalidaceae 174
Oxalis acetosella 56, 58

Pachytene 214
Paeonia spp. 93
Palaenophily 31
Palea 59
Palmaceae 27
Panicum miliaceum 18
 virgatum 25
Papaver spp. 37, 38
 rhoeas 171
Papaya see *Carica papaya*
Papillae 98, 101, 102
Parietal layer 79, 211
Parsley see *Petroselium hortense*
Parsnip see *Pastinaca sativa*
Parthenium argentatum 16, 176
Parthenocarpy 3, 134, 144, 150, 151, 154
Parthenogenesis 86
 diploid 15
 haploid 15
Paspalum spp. 16, 63
Pastinaca sativa 21
Pathotoxin 222
Pea see *Pisum sativum*
Pea, sweet see *Lathyrus odoratus*
 cow see *Vigna sinensis*
Peach see *Prunus persica*
Peanut see *Arachis hypogaea*
Pear see *Pyrus* spp.
Pecan see *Carya pecan*

Pectin 100
Pectocellulosic material 34
Pedaliaceae 21
Pelargonium spp. 206
 zonale 222
Pellicle 180
Pemphis acidula 174
Pennisetum glaucum 25, 201, 226
Pepper see *Capsicum* spp.
 black see *Piper nigrum*
Peppermint see *Mentha piperita*
Perianth 35, 52
Perigynous flower 143, 144
Persea spp. 14, 26, 38, 41
Persimmon see *Diospyros* spp.
Personata type (of incompatibility) 165, 171
Petalloidy 211, 212
Petroselium hortense 28
Petunia spp. 88, 89, 167, 170, 183, 184, 186, 187
 hybrida 3, 28, 40, 76, 205, 206, 212, 216, 220, 222, 223, 225, 226
Phalaris spp. 169
 arundinacea 25, 169
Phaseolus spp. 53, 187
 aureus 19, 59
 coccineus 26
 lunatus 19, 49, 199
 mungo 20
 vulgaris 19, 59, 61, 66, 67
Phenotype, marker 205
 antherless 205, 210
 apetalous 205, 210
Phleum pratense 25
Phoenix dactylifera 5, 14, 17, 34, 38, 39, 40, 41, 75, 76
Photoperiod 5, 36, 54, 160, 161, 162
Photoperiodism 128—132
Phragmites spp. 14
Phyllostica spp. 223
Phylogeny of angiosperms 54
Physalis spp. 169
 ixocarpa 170
Phytocide 229
Picea spp. 27, 34, 103
Pigeon pea see *Cajanus indicus (cajan)*
Pili nut see *Canarium ovatum*
Pinaceae 27
Pine see *Pinus* spp.
Pineapple see *Ananas comosus*
"Pin" plant 174
Pinus spp. 27, 34, 39, 40, 103, 104
Piperaceae 27
Piper nigrum 17, 27, 35, 36
Pistacia vera 17, 35, 41
Pistacio see *Pistacia vera*
Pistil 97

Pistillate flower　11
Pistilloidy　211, 212
Pisum sativum　19, 59, 199, 205, 210, 215, 216, 217
Plantago spp.　57
　lanceolata　198
　ovata　199
Plantain see *Plantago* spp.
Plasmatype, male sterility inducing　199, 201—203
　variation　199, 222
Plasmodesmata　82
Plasmon-genome interaction　206
Plastid　91, 100, 221, 222
Pleiotropic gene　163
Plum see *Prunus* spp.
Plumbaginaceae　173, 180
Plumbago spp.　173
Poa spp.　15, 16, 53
　annua　57
　arachnifera　17
　bulbosa　16
　chapmaniana　57
Polar cell　100
Pollen, anemophilic　35, 39
　binucleate　39, 40, 217, 219
　collection　62, 70—73
　diluents　71, 76
　dispersion　34, 39, 41, 42, 43, 73
　entomophilic　39
　filtration　34, 35
　germination　43, 101
　insects　75
　interception　43, 75
　longevity　29, 37—40, 43
Pollen mother cell (PMC)　81, 186, 188
　parent (male parent)　74—78, 169, 177, 178, 225—234
　presentation　39
　release　39
　shedding　36, 38, 59, 232
　size　33, 35
　storage　40, 62, 70—73
　storage containers　77
　surface　33
　transfer tools　70, 77
　trinucleate　39, 217, 219
Pollination, abiotic　31, 34—36
　biotic　31—34, 43
　bud　56—58
　cages　75
　cleistantheric　58
　control of　48—50
　emergencey　56
　forced　76—78
　"illigitimate"　173
　intraclonal　43
　intraplant　43

natural cross　18—28, 43—47
　rate of　43
　syndromes　29—34
Polygonum spp.　99
Polymorphism　51
　sexual　6
　genetic　53
Polypeptides　6
Polyploidy　6
Polysaccharide　220
Polysome　91
Polyspory　214
Pontederiaceae　173, 174
Poplar see *Populus* spp.
Poppy see *Papaver* spp.
Population breeding size　51
Populus spp.　17, 34, 197
　alba　197
Portulaca spp.　37, 38
Potato, Irish see *Solanum tuberosum*
　sweet see *Ipomoea batatas*
Primexine　214, 220
Primula spp.　173, 174
Primulaceae　173
Progeny testing　45—48
Proline　220
Propagules, artificial　14
　asexual　5, 14, 16
　natural　14
Prophase, meiotic　82, 83
Protandry　4, 13, 38, 57, 59
　flower　28
　plant　28
Proteaceae　27
Prothalium　104
Protogyny　4, 13, 36, 38, 62
　flower　28
　plant　28
Protoplast　85
　haploid　93
Prunus spp.　28, 36, 38, 39, 41, 167, 186, 187
　armeniaca　20
　persica　20
　avium　167, 185
Pseudogamy　16
Pseudomonas tabaci　93
Pseudotsuga menziesii　27
Psidium guajava　27, 40
Psychophily　31
Pumkin see *Cucurbita* spp., *mixta, maxima, moschata*
Pyrethrum see *Chrysanthemum* spp.
Pyrus spp.　16, 28, 36, 37, 38, 39, 40, 41, 76, 167

Quercus spp.　24, 34
Quinoa see *Chenopodium quinoa*

Radish see *Raphanus sativus*
Ragweed see *Ambrosia* spp.
Ranunculaceae 171
Ranunculus spp. 35, 180
  acris 171
Rape see *Brassica* spp.
Raphanus spp. 182, 188
  sativus 23, 38, 39, 192, 226
Raspberry see *Rubus idaeus*
Rats 34
Receptacle 7, 8, 143
Recombination, adjustment of 51, 53, 54
Redtop see *Agrostis alba*
Reproduction, asexual 5, 6
  sexual 6, 103, 104
Reproductive fitness 200
Rescue grass see *Bromus catharticus*
Restorer genetics 208
Rhodes grass see *Chloris gayana*
Rhus vernicifera 17
Rib grass see *Plantago lanceolata*
Ribes spp. 31, 38, 39
Ribosome 100, 101
Rice see *Oryza sativa*
Ricinus communis 24, 34, 35, 37, 39, 121,
  122, 139, 213
Rogue 75, 226, 229
Rosa spp. 14, 28
Rosaceae 20, 27, 28, 98, 167, 186, 190, 191
Rose see *Rosa* spp.
Roselle see *Hibiscus sabdariffa*
Rosemary see *Rosmarinus officinalis*
Rosmarinus officinalis 26
Rubber tree see *Hevea brasiliensis*
Rubiaceae 28
Rubus spp. 28, 31
  idaeus 37, 38, 39, 62, 109, 110, 215
Rumex acetosa 118, 119, 158
  acetosella 118
  arifolia 120
  pausifolius 119
  thyrsifolia 120
"Running out" 2
Rutabaga see *Brassica* spp.
Rutaceae 21
Rye see *Secale cereale*
  wild see *Elymus* spp.
Rye grass see *Lolium* spp.
  annual see *Lolium temulentum*

Saccharum officinarum 3, 5, 16
Safflower see *Carthamus tinctoria*
Sage see *Salvia nemorosa*
Sainfoin (Esparcette) see *Onobrychis vicifo-
  lia*
Saintpaulia spp. 88
Salix spp. 17, 33
Saltbush see *Atriplex* spp.

Salvia nemorosa 197
Satureia hortensis 199
Savory see *Satureia hortensis*
Saxifraga spp. 14
Scape (in Araceae) 126
Scarlet runner see *Phaseolus coccineus*
Scent feeding 75
Scrophulariaceae 167
Sea lavender see *Limonium vulgare*
Secale cereale 24, 34, 88, 169, 202
Seed parent (female parent, ovule pa-
  rent) 73—78, 153, 164, 172, 177, 178,
  225—234
Selection, disruptive 51
  recurrent 204
Selenicereus grandiflorus 37
Separation of sexual organs 43
  physiological 43
  spatial 43, 52, 54
  temporal 13, 43, 52, 54
Sesame see *Sesamum indicum*
Sesamum indicum 21, 40, 59, 61
Setaria spp. 88
  italica 5
  sphacelata 5
Sex cords 10
  chromosomes 110—120
  determination 113
  dimorphism 110
  expression 141—163
  genetics of 8, 9, 10, 105, 109—120, 123,
  124
  modification 121, 124, 127—141, 145,
  147, 151
  reversion 132, 221
  tendency 121, 126, 135
  types 18—28
Sexine 84, 85
Silene spp. 114, 136
  pendula 137
Silk (in maize) 107, 163
Simmondsia californica 17
  sinensis 5, 16
Sinapsis alba 62
Sisal see *Agave* spp.
Snapdragon see *Antirrhinum* spp.
Sodium dichloroacetate 68
Sodium 1-(P-chlorophenyl)-1,2-dihydro-4,6-
  dimethyl-2-oxonicotinate 68
Solanaceae 21, 28, 88, 98, 167, 186, 188,
  190, 191
Solanum spp. 53, 88, 89, 167, 205, 216
  melongena 134, 139, 199, 219, 232
  tuberosum 3, 14, 16, 21, 37, 40, 186, 216
Solidago spp. 97
Sorghum see *Sorghum vulgare*
Sorghum papyrescens (membranaceum) 57
  vulgare 16, 18, 39, 40, 55, 56, 57, 63, 69,

71, 74, 199, 201, 204, 210, 216, 217, 220, 226, 233
  var. *sudanensis* 19
"Sorting out" 222
Soybean see *Glycine max*
Spadix (in Araceae) 126
Spike 107, 130, 131, 160
Spikelet 63, 65, 107, 109, 160, 161
Spinacea oleracea 5, 17, 39, 125
Spinach see *Spinacea oleracea*
Spindle, divergence of 214
Sporangium 82, 103, 211
  mega 99
  micro 79, 211—214
Sporophyll 11, 103
  macro 96
  micro 79, 103, 221
Sporophyte, diploid 10, 15
  haploid 15, 86
Sporopollenin 80, 83
Spruce see *Picea* spp.
Squash, winter see *Cucurbita* spp.
Squirrels 34
Staminate flower 11
Standards, isolation 18—28
  seed 22
Status albomaculatus 222
  paralbomaculatus 206, 222
Sterculiaceae 28
Sterility, female 41
Stigma 98
  "cob" 180, 181
  papillate 180, 181
  receptivity 36, 37, 38, 39, 40, 41, 56
Stigmoidy 211, 212
Stipa spp. 19
Strawberry see *Fragaria* spp.
Strelitzia spp. 31
Streptocarpus spp. 205
Strobilus 9, 104
  bisexual 54
Sudangrass see *Sorghum vulgare* var. *sudanensis*
Sunflower see *Helianthus annuus*
Sweet potato, see *Ipomoea batatas*
Sweet violet, see *Viola oderata*
Swisschard see *Beta vulgaris*
Switchgrass see *Panicum virgatum*
Sugar cane see *Saccharum officinarum*
Symphytophily 31
Syncytium 82
Synergid cell 100, 102, 103, 179, 183
Syngamy 5

Tagetes erecta 205, 226
Tangelo see *Citrus* spp.
Tapetum, invasive (amoeboid) 81

parietal (secretory) 80, 81
persistent 217, 218
Taraxacum officinale 16
Tassel 107, 108, 160, 161, 162, 163, 226
  seed 108
Tangier pea see *Lathyrus tingitanus*
TCA cycle 220
Tea see *Camellia sinensis*
Tectum 101
Telophase 82, 214
Testosterone 118
Theca 79
Theaceae 28
Theobroma cacao 28, 33, 38, 165, 179, 184
Thermoperiod 5, 54, 145
Thistle see *Cirsium oleraceum*
Thrips 34, 74
"Thrum" plant 174
Thyme see *Thymus vulgaris*
Thymus vulgaris 199, 212
Tilia spp. 33
Timothy see *Phleum pratense*
Tissue, sporogeneous 74, 80—82, 211, 213
  transmitting 97, 101, 183
Tobacco see *Nicotiana tabacum*
Tomato see *Lycopersicon esculentum*
Transformation, asexual 223
Trefoil see *Lotus corniculatus*
Tributyl-2,4-dichlorobenzylphosphonium chloride (Phosphon D) 149
Trifolium spp. 31, 32, 39, 53, 68, 186, 188, 192, 219
  alexandrinum 26
  fragiferum 5, 20
  glomeratum 20
  hybridum 26
  pratense 5, 26, 195
  procumbens 20
  repens 26, 167, 187
  subterraneum 20
Trigger gene 107, 109
  mechanism 109, 117, 133, 151, 152
Triiodoacetic acid (TIBA) 137, 140
Trimonoecy see Monoecy, Tri
Trioecy 16
Tripartite structure (of S locus) 168
"Tripping" 29
Trisomics, balanced tertiary 230, 231
Tristyly 173, 174
Triticale see Triticum spp.
Triticum spp. 5, 18, 39, 40, 49, 55, 59, 61, 65, 68, 69, 73, 74, 75, 88, 209, 217, 219, 220, 223, 226, 231, 233, 234
  aestivum 88, 134, 137
  araraticum 202
  boeoticum 202
  dicoccoides 202
  durum 208

timopheevi 202, 209, 225
  zhukovskyi 202
"Trueness to type" 2
Tulip see *Tulipa* spp.
Tulipa spp. 26
Tung tree see *Aleurites montana*
Turner's Syndrome 112
Turnip see *Brassica* spp.
  rapa

Umbelliferae 21, 28, 31
Unisexuality 6
Ustilago violaceae 118

Vaccinium spp. 38
Vallisneria spp. 35
Variability, genetic 53
Varnish tree see *Rhus vernicifera*
Vectors activity, manipulation of 48
  pollen 18—37, 42
  abiotic 34—36
  biotic 40
  involuntary 74
  voluntary 74
Vegetable marrow, see *Cucurbita pepo*
Vegetative cell 83
  nucleus 91, 102, 183, 217
Vetch see *Vicia* spp.
Verbascum spp. 164
Vespophily 31
Vicia angustifolia 58
  benghalensis 20
  faba 19, 37, 39, 61, 74, 75, 226
  pannonica 20
  sativa 5, 20, 58
  villosa 5
Vigna sinensis 215
Viola spp. 57, 215
  oderata 57, 58

Virus, barley stripe 224
  broad bean wilt 224
  coat protein 6
  potato X 223
  tobacco mosaic 224
  tobacco ring spot 224
Vitaceae 21
Vitis coignetiae 140
  rotundifolia 17, 41
  thumbergii 140
  vinifera 21, 40, 76, 136, 140, 219
Vivipary 15, 16

Walnut see *Juglans regia*
Watermelon see *Citrullus vulgaris*
Weed sorrel see *Oxalis acetosella*
Wheat see *Triticum* spp.
Wheatgrass see *Agropyron* spp.
  slender see *Agropyron trachycaulum*
Willow see *Salix* spp.

Xanthium spp. 129
  pennsylvanicum 129
  strumarium 139
X-ray radiation 168, 187, 188

Yam see *Dioscorea* spp.

Zea mays 12, 24, 38, 39, 48, 49, 69, 71, 85,
  88, 105, 108, 109, 120, 121, 124, 133, 137,
  139, 159—163, 200, 201, 205, 206, 207,
  209, 210, 215, 216, 217, 219, 220, 222,
  223, 224, 226, 230, 233, 234
Zeatin 140
Zinnia spp. 212, 226
Zygotene 214

Theoretical and Applied Genetics

Managing Editor: H. F. Linskens
Editorial Board: H. Abplanalp, Davis, Calif.; L. Alföldi, Szeged; R. W. Allard, Davis, Calif.; E. Andersson, Stockholm; S. Barbacki, Poznan; J. S. F. Barker, Sydney; D. K. Belyaev, Novosibirsk; A. Gustafsson, Lund; R. Hagemann, Halle/S.; R. C. Lewontin, Cambridge, Mass.; W. J. Libby, Berkeley, Calif.; H. F. Linskens, Nijmegen; J. MacKey, Uppsala; F. Mechelke, Stuttgart-Hohenheim; G. Melchers, Tübingen; B. R. Murty, New Delhi; O. E. Nelson, Madison; J. Rapacz, Madison; R. Riley, Cambridge, G. B.; A. Robertson, Edinburgh; W. Seyffert, Tübingen; H. Skjervold, Vollebekk; H. Stubbe, Gatersleben; L. D. Van Vleck, Ithaca, N. Y.; D. von Wettstein, Copenhagen DK.

Breeding genetics, with the aid of chemistry and mathematics, has become considerably more fundamental and general. This development has moved from the genetics of the individual to that of the group and, in turn, to the study of the evolution and origin of domesticated species. Improved mathematical models, which can be quantitatively solved or simulated by the computer, allow the new science of molecular genetics to study gene-enzyme interaction and the regulation of inherited characteristics. TAG fills a vital need for detailed research reports in this field and serves as an international vehicle for the exchange of scientific information.

Fields of Interest: Mathematical Genetics, Analysis of Genetic Models, Evolutionary Genetics, Genetics and Biochemistry of Development, Cytogenetics, Origin of Species, Population Genetics, Breeding Methods.

Subscription Information and sample copies available upon request.

Planta

Editorial Board: E. Bünning, Tübingen; H. Grisebach, Freiburg i. Br.; J. Heslop-Harrison, Kew; G. Jacobi, Göttingen; A. Lang, East Lansing; H. F. Linskens, Nijmegen; H. Mohr, Freiburg i. Br.; P. Sitte, Freiburg i. Br.; Y. Vaadia, Bet Dagan; M. B. Wilkins, Glasgow; H. Ziegler, München.

Planta publishes original articles in structural and functional botany, covering all aspects from biochemistry and ultrastructure to studies with tissues, organs and whole plants, but excluding evolutionary and population botany (taxonomy, floristics, ecology, etc.). Papers in cytology and genetics, and papers from applied fields such as phytopathology are accepted only if contributing to the understanding of specifically botanical problems.

Fields of Interest: Botany, Plant Physiology, Cytology, Genetics, Forestry, Agronomy.

Subscription Information and sample copies available upon request.

Springer-Verlag Berlin Heidelberg New York

J. Sybenga
Meiotic Configurations

A Source of Information for Estimating Genetic Parameters
65 figures, 64 tables. X, 251 pages. 1975
(Monographs on Theoretical and Applied Genetics, Vol. 1)

Contents: Introduction.—The analysis of crossing-over.—The analysis of chromosome pairing.—The analysis of distribution: centromere coorientation.
Meiotic configurations are viewed from a special angle in this book: as a source for the extraction of maximum quantitative information of genetic interest, primarily related to recombination. This involves the development of models and systems for estimating genetic parameters from relatively simple microscopic observations on normal as well as specially constructed material. There are four chapters. Since information on chiasma formation is required for the effective analysis of chromosome pairing, after an introductory Chapter 1, the analysis of crossing-over (Chapter 2) preceeds that on chromosome pairing (Chapter 3). Chromosome distribution (segregation) is treated in the final Chapter 4. Most of the material presented is based on published reports, some is new. Examples have been taken from various species, including plants and animals, and are not restricted to those with favorable chromosomes.

R. G. Stanley, H. F. Linskens
Pollen

Biology, Biochemistry, Management
64 figures, 66 tables. VIII, 307 pages. 1974

Contents: Biology: Development. Wall Formation. Dehiscence, Size and Distribution.—Management: Collection and Uses. Storage. Viability Tests. Nutritive Role.—Biochemistry: General Chemistry. Carbohydrates and Cell Walls. Organic Acids, Lipids and Sterols. Amino Acids and Proteins. Pollinosis. Nucleic Acids. Enzymes and Cofactors. Pollen Pigments. Growth Regulators.
This volume deals with the biology and chemistry of pollen, those cells in which the male genetic material is transported in the sexual reproduction of higher plants. The latest research results, as well as older data, are treated critically; practical questions on plant breeding are discussed, in addition to research projects in the fields of metabolic physiology of animals and plants and medicine. Directions are given for the handling of pollen.

Springer-Verlag Berlin Heidelberg New York